Communications
in Computer and Information Science    20

D.M. Akbar Hussain
Abdul Qadeer Khan Rajput
Bhawani Shankar Chowdhry
Quintin Gee (Eds.)

# Wireless Networks, Information Processing and Systems

International Multi Topic Conference, IMTIC 2008
Jamshoro, Pakistan, April 11-12, 2008
Revised Selected Papers

 Springer

Volume Editors

D.M. Akbar Hussain
Aalborg University
Department of Software Engineering & Media Technology
Niels Bohrs Vej 8, 6700 Esbjerg, Denmark
E-mail: akbar@aaue.dk

Abdul Qadeer Khan Rajput
Mehran University of Engineering & Technology
Jamshoro, Sindh 76062, Pakistan
E-mail: aqkrajput@hotmail.com

Bhawani Shankar Chowdhry
Mehran University of Engineering & Technology
Department of Electronics & Telecommunication Engineering
Jamshoro, Sindh 76062, Pakistan
E-mail: bsc_itman@yahoo.com

Quintin Gee
University of Southampton
School of Electronics and Computer Science
Learning Societies Lab
Southampton SO17 1BJ, UK
E-mail: qg2@ecs.soton.ac.uk

Library of Congress Control Number: 2008940286

CR Subject Classification (1998): C.2, H.2, H.5, K.6

ISSN       1865-0929
ISBN-10    3-540-89852-2 Springer Berlin Heidelberg New York
ISBN-13    978-3-540-89852-8 Springer Berlin Heidelberg New York

springer.com

© Springer-Verlag Berlin Heidelberg 2008
Printed in Germany

Typesetting: Camera-ready by author, data conversion by Scientific Publishing Services, Chennai, India
Printed on acid-free paper      SPIN: 12568673      06/3180      5 4 3 2 1 0

# Preface

The international multi-topic conference IMTIC 2008 was held in Pakistan during April 11–12, 2008. It was a joint venture between Mehran University, Jamshoro, Sindh and Aalborg University, Esbjerg, Denmark. Apart from the two-day main event, two workshops were also held: the Workshop on Creating Social Semantic Web 2.0 Information Spaces and the Workshop on Wireless Sensor Networks. Two hundred participants registered for the main conference from 24 countries and 43 papers were presented; the two workshops had overwhelming support and over 400 delegates registered.

IMTIC 2008 served as a platform for international scientists and the engineering community in general, and in particular for local scientists and the engineering community to share and cooperate in various fields of interest. The topics presented had a reasonable balance between theory and practice in multidisciplinary topics. The conference also had excellent topics covered by the keynote speeches keeping in view the local requirements, which served as a stimulus for students as well as experienced participants. The Program Committee and various other committees were experts in their areas and each paper went through a double-blind peer review process. The conference received 135 submissions of which only 46 papers were selected for presentation: an acceptance rate of 34%.

This CCIS book contains the revised manuscripts presented at IMTIC 2008. The topics include wireless sensor networks, satellite communication, tracking, grid computing, wireless communication and broadband networks, remote sensing, character recognition, FPGA, data mining, bioinformatics and telemedicine, switching networks, mobile IP, neural networks and neurofuzzy technology, automation and control, speech recognition, renewable energy, legal and ethical issues in IT.

This book presents a collection of excellent multi-topic manuscripts and it will serve as a reference work for researchers and students. This book is believed to be the first by Springer to be published in Pakistan.

July 2008

D.M. Akbar Hussain
A.Q.K. Rajput
B.S. Chowdhry
Q.H. Gee

# Organization

## Conference Chairs

| | |
|---|---|
| Abdul Qadeer Khan Rajput | MUET Jamshoro, Pakistan |
| D.M. Akbar Hussain | Aalborg University, Denmark |

## Steering Committee Chairs

| | |
|---|---|
| David L. Hicks | Aalborg University, Denmark (Convener) |
| Neil M. White | University of Southampton, UK |
| Mogens Pedersen | University of Miyazaki, Japan |
| Gul Agha | UIUC, USA |
| Mubark Shah | UCF, USA |
| M.Y. Sial | NTU, Singapore |
| Franz Wotawa | Graz University, Austria |

## Organizing Committee Chairs

| | |
|---|---|
| M.R. Abro | MUET, Pakistan (Convener) |
| R.A. Suhag | MUET, Pakistan |
| M.I. Panhwar | MUET, Pakistan |
| D.A. Khawaja | MUET, Pakistan |

## Organizing Committee Co-chairs

| | |
|---|---|
| A.K. Baloch | MUET, Pakistan (Convener) |
| B.S. Chowdhry | MUET, Pakistan |
| M. Aslam Uqaili | MUET, Pakistan |
| Engr. G.S. Kandhar | MUET, Pakistan |
| Muneer A. Shaikh | MUET, Pakistan |

## Program Committee Co-chairs

| | |
|---|---|
| D.M. Akbar Hussain | Aalborg University, Denmark |
| Bhawani Shankar Chowdhry | MUET Jamshoro, Pakistan |

## Local Arrangements Chairs

| | |
|---|---|
| M.A. Unar | MUET, Pakistan |
| Aftab Ahmed Memon | MUET, Pakistan |

## Local Arrangements Co-chairs

| | |
|---|---|
| M. Zahid Shaikh | MUET, Pakistan (Convener) |
| Qazi Abdul Razaq | MUET, Pakistan |
| Tahseen Hafiz | MUET, Pakistan |
| Abdul Waheed Umrani | MUET, Pakistan |

## Sponsor

Higher Education Commission of Pakistan (HEC)

# Table of Contents

# Keynote Speech: Role of PCSIR in Industrial Development of Pakistan

Javed Arshed Mirza

Former Chairman KRL & PCSIR Laboratories

PCSIR, since its inception and establishment in 1949, has always been very conscious and instrumental in contributing towards development of industrial sector in Pakistan. During its formative and evolving years, PCSIR Laboratories have served as nurseries of scientific research and a breeding base for innovation along-with characterization of natural / mineral deposits and raw Agro-Materials. Development of processes and technologies around local resources and capabilities has been the main focus of activities of PCSIR. Despite meager financial and human resources, the scientists / engineers of PCSIR have always come up to the rescue of the local industry in crucial / critical situations.

Due to changing world scenario after induction of ISO / WTO standards / requirements, PCSIR has realigned itself to cope with the new and emerging demands of quality and value addition to products, in order to help and serve the local industrial sector for competing in the world market in trade and export business.

## Contributions of PCSIR for Industrial Development

- Process / technology development for industrial applications
- Extension of test/analysis services for materials & products
- Extension of calibration services for test/measuring and monitoring equipment
- Undertaking monitoring of industries for ISO Certification
- Extension of Pilot Plant facilities to industrial sector
- Undertaking technical feasibility reports for industries
- Scientific/technical advisory /consultancy services to industrial sector
- Training of technical manpower from industry
- Feeding of technical trained technicians/associate engineers and bachelor engineers for the industrial sector
- Design and development of tools / equipment for industry
- Extension of repair /maintenance services for industrial equipment
- Development of specific industrial chemicals

D.M.A. Hussain et al. (Eds.): IMTIC 2008, CCIS 20, p. 1, 2008.
© Springer-Verlag Berlin Heidelberg 2008

# Keynote Speech:
# Computer Vision: Object Recognition and Scene Interpretation

Shaiq A. Haq

Air University, Islamabad Pakistan

This invited lecture is about Computer Vision. The lecture tries to answer questions like; what is computer vision, what are its applications, how objects recognition is done, and how computers can interpret a scene or action. Basic image processing techniques such as image enhancement, edge detection, segmentation, features extraction, objects recognition and tracking are described. The lecture also explains the techniques needed to develop real-time machine vision system, face recognition system, fingerprints identification system and action recognition system. Screenshots from actually developed software are shown in Power point slides.

D.M.A. Hussain et al. (Eds.): IMTIC 2008, CCIS 20, p. 2, 2008.
© Springer-Verlag Berlin Heidelberg 2008

# Electronic Automatic Gear Shifting System
# for a Motorcycle

Fahad Ahmed[1], Haider Ali Zaidi[1], Syed Waqar Hussain Rizvi[1],
Atiya Baqai[2], and Bhawani S. Chowdhry[2,3]

[1] Ex-Students Department of Electronics & Biomedical Engineering,
Mehran U.E.T. Jamshoro, Sindh, Pakistan
fwdc@hotmail.com
[2] Department of Electronics & Biomedical Engineering,
Mehran University of Engineering & Technology, Jamshoro, Pakistan
c.bhawani@ieee.org
[3] School of ECS, University of Southampton, UK
bsc06v@ecs.soton.ac.uk

**Abstract.** Motorcycle is widely used around the world and particularly in Pakistan. The gear shifting system of the motorcycle is conventionally manual. This paper covers development of an indigenous automatic gear shifting/changing system for the standard motorcycle. By this system the manual mechanical gear-shifting system will remain unchanged because an additional electro-mechanical system is placed on the vehicle to shift the gear and for automatic controlling the clutch. So the system has both the options manual as well as automatic. This system uses low-cost microcontrollers to make the accurate decision for shifting the gear up and down by observing the speed, and it controls the clutch transmission where necessary. The complete hardware and software has been tested and the functioning of the automatic gear shifting system is verified. This system is flexile and can be used with any motorcycle manufactured in Pakistan ranging from 50cc to 200cc.

## 1 Introduction

In our developed system, both the gear and the clutch are controlled electro-mechanically by a microcontroller-based computer system. This system does not require any modification to the engine. The equipment is mounted externally on the body of the motorcycle. The system shifts the gear up and down electro-mechanically like a human rider by sensing the speed of the vehicle, but the system shifts the gear at exactly the correct speed, which produces the smooth gear changing sequence. The engine runs smoothly and without any knocking, which increases the engine life. This system is different than the traditional automatic transmission used in cars, because it uses manipulators/electro-mechanical devices, which act similar to the human foot pressing the gear pedal up and down, and pressing the clutch lever like a human hand. Fig. 1 shows a block diagram of the system.

D.M.A. Hussain et al. (Eds.): IMTIC 2008, CCIS 20, pp. 3–10, 2008.

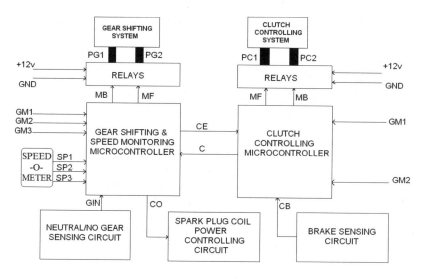

**Fig. 1.** Block Diagram of the system

## 2  Working Principle

### 2.1  Hardware Operation

We used a traditional mechanical speedometer, which has been modified to produce digital signals at specific speeds by using infrared LEDs and photodiodes [1]. The arrangement is shown in Figure 2.

**Fig. 2.** Showing Speed & Logic Outputs

The meter produces:

- three active high signals, when the speed is lower than 10 km/h
- one active low signal and two active high signals, when the speed is greater than 10 km/h
- two active low signals and one active high signal, when the speed is greater than 30 km/h
- three active low signals, when the speed is greater than 45 km/h

The signals from the meter are amplified by the transistor circuitry configured as common collector.

Table 1 shows the output from the photodiode and differed speed for this purpose and for digital logic signal which are named as the signal coming from the photodiode one placed at 10 km/h speed as "SP1", photodiode placed at 30 km/h speed as "SP2", and photodiode placed at 45 km/h speed as "SP3".

**Table 1.** Differed speed and digital logic signal

| SPEED | SP1 | SP2 | SP3 |
|-------|-----|-----|-----|
| 0 km/h to 10 km/h | 5 V, Logic 1 | 5 V, Logic 1 | 5 V, Logic 1 |
| 10 km/h less than 30 km/h | 0 V, Logic 0 | 5 V, Logic 1 | 5 V, Logic 1 |
| 30 km/h less than 45km/h | 0 V, Logic 0 | 0 V, Logic 0 | 5 V, Logic 1 |
| 45 km/h onwards | 0 V, Logic 0 | 0 V, Logic 0 | 0 V, Logic 0 |

For shifting of the gears, an electro-mechanical device is constructed, using a high power DC motor, to shift the gear up and down, which acts just like as human foot as shown in Fig. 3.

Three SPDT switches at specific ends to limit motion of the DC motor are used. Such a motor requires two high power relays to drive the devices; one relay for the forward movement, and one for the backward movement [2]. These high power relays (secondary relays) are activated at 220 volts AC, so two additional relays (primary relays) are used to drive the secondary relays. Since the controller does not provide enough power and voltage to activate the primary relays, a transistor circuit of emitter-follower (common collector configuration) as a switch (common emitter configuration) is used to drive the signals from the microcontroller to the primary relays for activating them.

For controlling the clutch, an electro-mechanical device using a high power DC motor is used which acts like a lever pressed by a human hand. Two SPDT switches at specific ends limit its motion. The configuration is the same as described above. The circuit of common emitter (as a Switch) and common collector as (emitter

**Fig. 3.** Picture showing Whole Gear Shifting System attached to motorcycle

**Fig. 4.** Clutch Controlling System installed on the motorcycle

follower) are used to drive the microcontroller signals to activate the primary relays. A low power relay is used to sense the neutral gear position. When the motorcycle is in neutral, this relay provides an active low signal to the microcontroller; and when the motorcycle is in any gear it provides an active high signal to the microcontroller.

In the same manner a brake-sensing relay is also used to sense brake activation and deactivation by producing an active high signal to the microcontroller when the brake is active and producing an active low signal when the brake is inactive. A relay is used to control the power of the ignition coil/spark plug coil. Fig. 3 shows the whole gear shifting system attached to motorcycle and Fig. 4 depicts the clutch controlling system.

A transistor circuit of common emitter and common collector is used to activate the relay when signalled from the microcontroller. The power of the ignition coil/spark plug coil is cut off in order to reduce the engine RPM.

Two AT89C51 microcontrollers are used [3], one (MCU1) being used to monitor the speed and shifting the gear up and down, controlling the power of the ignition coil, sensing the neutral position of the gear and producing the clutch enable signal for the second microcontroller. The second microcontroller (MCU2) is used to control the clutch system by receiving signals from the gear system microcontroller, and sensing the brake enable and disable condition for clutch pressing. Two voltage regulators, L7809 and L7805, produce 9 v for the microcontroller and 5 v for the relays respectively, and one voltage inverter converts the 12 v DC to 220 v AC. Fig. 5 gives a schematic diagram of the hardware.

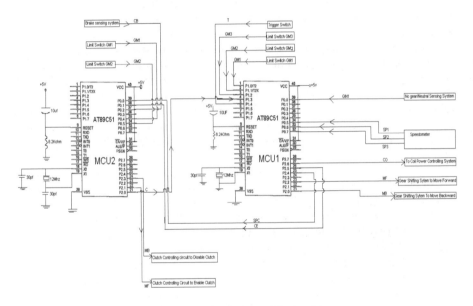

**Fig. 5.** Schematic diagram of the hardware

## 2.2  Program Logic for the Gear Shifting Microcontroller

When power is supplied to the system, MCU1 first brings the motorcycle gears into neutral by signalling the gear shifting system backward until the gears come to the neutral position. It will then wait for a triggering signal from the SPDT switch to execute its next instruction.

When an active low signal is received from the switch, MCU1 then waits for the active low signal from MCU2. MCU2 senses the brake operation by receiving an active high signal from the brake sensing system, and then sends a signal to the clutch controlling system to press the clutch, and produces an active low signal for MCU1.

The MCU1 sends a signal to the gear shifting system to shift the gear up and stores the value 1 recording that the motorcycle is in first gear. MCU1 constantly monitors the three signals coming from the speedometer.

When there are three active high signals coming from speedometer, this indicates that the vehicle's speed is below 10 km/h, and the motorcycle should be running in first gear. When running in first gear, the MCU1 provides an active high signal to MCU2, which causes MCU2 to release the half clutch, not the full clutch, because the motorcycle is starting from 0 km/h and its momentum is very low. So in order to run the engine smoothly, half clutch need to be released.

As the speed increases above 10 km/h, MCU1 recognises one active low signal and two active high signals. The MCU1 considers the motorcycle should be in second gear. MCU1 now signals the gear shifting system to shift the gear up and signals MCU2 to press the clutch. When the clutch is pressed, the active low signal is sent by MCU2 to MCU1, which then switches off the spark plug coil/ignition coil. As soon as the gear is shifted up, MCU1 switches on the spark plug coil, sends an active low signal to MCU2 to release the clutch, and stores the value 2, recording that the motorcycle is in second gear.

As the speed increases beyond 30 km/h, two active low signals and one active high signal are provided by the speedometer to MCU1. MCU1 now has to bring the motorcycle into third gear. It signals the gear shifting system to shift the gear up and signals MCU2 to press the clutch. When the clutch is pressed, the active low signal is sent by MCU2 to MCU1, which switches off the spark plug coil/ignition coil. When the gear is shifted up, MCU1 switches on the spark plug coil, sends an active low signal to MCU2 to release the clutch, and stores the value 3, recording that the motorcycle is in third gear.

As the speed increases above 45 km/h, the speedometer sends three active low signals to the MCU1. MCU1 consider the vehicle should be in fourth gear. MCU1 signals the gear shifting system to shift the gear up and signals MCU2 to press the clutch. When the clutch is pressed, the active low signal is sent by MCU2 to MCU1, which switches off the spark plug coil/ignition coil. When the gear is shifted up, MCU1 switches on the spark plug coil, sends an active low signal to MCU2 to release the clutch, and stores the value 4, recording that the motorcycle is in fourth gear.

Let us now suppose that the speed of the vehicle starts to decrease. As it becomes lower than 45 km/h, the speedometer signals two active low and one active high signal to the microcontroller which causes MCU1 to bring the motorcycle into third gear. MCU1 now signals the gear shifting system to shift the gear down and signals MCU2 to press the clutch. When the clutch is pressed, the active low signal is sent by MCU2 to MCU1. When the gear is shifted, MCU1 subtracts 1 from the stored value resulting in 3, and the motorcycle is in third gear.

Suppose the speed of the vehicle decreases below 10 km/h; then the speedometer sends three active high signals to MCU1. This signal causes MCU1 to bring back the engine into first gear. Now the MCU1 signals the gear shifting system to shift the gear down and it sends an active high signal to MCU2 to press the clutch. When the clutch is pressed, MCU2 produces an active low signal and sends it to MCU1. Now MCU1 subtracts 1 from the stored value.

If this value is 1, then MCU1 does nothing more than monitoring the speed of the vehicle. If the value is greater than 1, then MCU1 repeats the above process up to the subtraction of 1 from the stored value and checks whether it is now 1 or not. If it is still not 1, then MCU1 repeats the process until the value equals 1. It then stops performing the process and goes back to monitoring of speed of the vehicle.

If the speed of the vehicle is lower than the programmed speed limit for the current running gear, then MCU1 shifts the gear down by performing the operation defined above, for shifting the gear down until the specified gear is engaged. If the speed of the vehicle is greater than the programmed speed limit for the current gear, then MCU1 shifts the gear up in order to bring the motorcycle into the correct gear by performing the gear shifting up operation as defined previously.

### 2.3  Program Logic for the Clutch Controlling Microcontroller

MCU2 constantly monitors the active high signal from the brake sensing system and MCU1. If either of these systems sends an active high signal for MCU2, it presses the clutch. When there is no active high signal to MCU2, it releases the clutch and sends an active low signal to MCU1.When the clutch is not pressed it sends an active high signal to MCU1. When the motorcycle is in first gear, an additional active high signal is sent by MCU1 to MCU2. When MCU2 receives this active high signal after pressing the clutch, it releases the half clutch. This condition is only valid for first gear, and for all other gears, MCU2 will release the full clutch.

The main function of MCU2 is to monitor the active high signals sent by MCU1 and the brake sensing system, to press the clutch, and if there are no active high signals from any of the systems, MCU2 will release the clutch.

## 3  Future Developments

### 3.1  Future Improvement

In this system the automatic gear shifting is done by monitoring the vehicle speed. It could instead be done by monitoring the crankshaft RPM or the engine's acceleration. The electro-mechanical system has to be made more efficient, faster and more reliable. We have used two extra DC self-starter motors taken from 300 cc motorcycles, because they are powerful enough to shift the gear and control the clutch. A future improvement of the system could use a DC actuator, which is powerful enough to shift the gear and control the clutch, and consumes less power or only the power available in a traditional motorcycle battery.

The electro-mechanical system can be made more compact so it can easily be fitted in existing motorcycle chassis without disturbing the other components. The microcontrollers can be replaced by automotive microcontrollers. A more efficient gear indication sensor can be installed, which can show the gear of the motorcycle in which the engine is running.

### 3.2  Other Uses

The system is easy to implement on other vehicles such as cars, buses and trucks. A few modifications need to be made in the software and the electro-mechanical system to make the system compatible with those vehicles.

The manual electronic gear shifting system is still available in the system implemented, so a user can enjoy both manual and automatic gear transmission.

## 4 Conclusion

A complete indigenous automatic gear system for a motorcycle has been developed. This system is very useful for physically disabled persons, such as those who do not have a left leg and left hand. By using this system, only one hand is needed for controlling the motorcycle. Since this system changes the gear at the correct speed, motorcycle engine life is increased. Also, it provides a resistance against knocking, and it improves fuel consumption. This system is also very helpful for learner riders of the motorcycle.

This system is flexible and can be implemented on any motorcycle available in the Pakistani market without any major changes. The motorcycle manufacturing companies can also use this system in their vehicles because it can be easily fitted to the motorcycle and there is no need of internal modification of the gear system. By installing this low cost system in their motorcycle, companies may be able to increase their sales due to the availability of these new features. Some sport riders may also enjoy the benefits of the automatic transmission.

## Acknowledgment

Authors wish to thank the Department of Electronics & Biomedical Engineering, Mehran U.E.T. Jamshoro, Sindh, Pakistan for providing laboratory facilities for designing and implementing the project.

## References

1. Floyd, T.L.: Electronics Fundamentals: Circuits, Devices, and Applications, 6th edn. Prentice Hall, Englewood Cliffs (2003)
2. Theraja, B.L.: Basic Electronics Solid State. S. Chand & Company Ltd., NewDelhi (2002)
3. Mazidi, M.A.: The 8051 Microcontroller & Embedded System. Prentice Hall, Englewood Cliffs (1999)

# Segmentation of Arabic Text into Characters
# for Recognition

Noor Ahmed Shaikh[1], Zubair Ahmed Shaikh[2], and Ghulam Ali[1]

[1] Assistant Professor and PhD Scholar, Shah A Latif University, Khairpur, Pakistan
{noor_salu,ghulamali_salu}@yahoo.com
[2] Professor & Director, FAST-NU, Karachi
zubair.shaikh@nu.edu.pk

**Abstract.** One of the steps of character recognition systems is the segmentation of words/sub-words into characters. The segmentation of text written in any Arabic script is a most difficult task. Due to this difficulty, many systems consider sub-words instead of a character as the basic unit for recognition. We propose a method for the segmentation of printed Arabic words/sub-words into characters. In the proposed method, primary and secondary strokes of the sub-words are separated and then segmentation points are identified in the primary strokes. For this, we compute the vertical projection graph for each line, which is then processed to generate a string indicating relative variations in pixels. The string is scanned further to produce characters from the sub-words. In the proposed method we use Sindhi text for segmentation into characters as its character set is the super set of Arabic. This method can be used for any other Naskh-based Arabic script such as Persian, Pashto and Urdu.

**Keywords:** Arabic Text Segmentation, Sindhi OCR.

## 1 Introduction

The field of Optical Character Recognition (OCR) has been around since the inception of the digital computers. OCR systems for the languages like English, Chinese, Japanese and others in which characters are isolated from each other are commercially available. However, very few commercial OCR systems for the cursive languages like Arabic and Persian are available but are not much robust. It is pointed out by Abdullah [1] that the first published work for Arabic OCR carried in 1980 by Nouh *et al.* [2]. To the best of our knowledge, no commercial OCR system for Sindhi or Urdu language is available, but they are mostly in the research labs. It is due to the fact that the scripts either handwritten or printed used by these languages are cursive. The term *cursive* is used in a sense that the characters in the word/sub-word may or may not be connected with each other. It is like a joined up handwritten Latin/Roman scripts, where characters are usually connected with each other. It is very difficult to segment the connected characters from the sub-words written in Arabic style scripting languages because boundaries of the characters are not uniform. It is shown [3] that segmentation of Arabic text into characters and recognizing unconstrained off-line cursive writing is a very difficult task,

D.M.A. Hussain et al. (Eds.): IMTIC 2008, CCIS 20, pp. 11–18, 2008.

mainly due to the difficulty of character segmentation. So, many systems recognize isolated characters only like [4] or do not segment the text into characters but some other units or parts like sub-words or scripts etc. which are easier to segment. There are many approaches in the literature can be found that recognize sub-words [5] and the recent one is by S. Alma'adeed [6], he attempts to recognize the sub-words instead of characters using Neural Network. Such approaches can narrow down the sub-word candidates, because in the large-vocabulary several sub-words may have same features and there will be increase in the rate of rejection or misrecognition of the character. Elgammal *et al.* [7] segmented the words into small connected segments called '*scripts*'. A script does not necessarily correspond to a character; it may be a complete character, a part of a character or more than one character.

In this paper, a method for the segmentation of off-line printed text written in Arabic style scripting language into characters is proposed. Here, we consider printed text of Sindhi language for segmentation into characters because its characters' set is the superset of Arabic. The samples of Arabic and Sindhi cursive writing are presented in Fig. 1. The proposed technique for the segmentation of Sindhi text can be used for the segmentation of text written in Arabic style scripting languages.

سائینم سدائین ڪرين مٽي سنڌ سڪار
دو ست منادلدار عالم سپ آباد ڪرين

(a)

يولد جميع الناس أحراراً متساوين في الكرامة والحقوق. وقد وهبوا
عقلاً وضميراً وعليهم ان يعامل بعضهم بعضا بروح الإخاء.

(b)

**Fig. 1.** Samples of Cursive writing (a) Sindhi (b) Arabic

The rest of the paper is organized as follows: in section 2, introduction to Arabic text and the methods used to segment it into characters are presented; in section 3 our proposed method for the segmentation of Sindhi characters is presented. The conclusion is given in section 4 and section 5 is reserved for the presentation of future work.

## 2   Arabic Text and Methods for Its Segmentation into Characters

Arabic uses *Semitic Abjad* that represents consonants plus some vowels, and uses Naskh style of writing in which all text is written on an imaginary horizontal line called baseline. It is cursive in nature in which letters are connected with each other in sub-words on the baseline. Arabic characters that can be joined are always joined in both handwritten and printed text. Most letters change form depending on whether they appear at the beginning, middle or end of a word, or on their own.

The segmentation of Arabic sub-words into characters is the most delicate stage. The methods devised, so far, for the segmentation of such sub-word into characters are facing the problem of under-segmentation or over-segmentation.

Here, we present some methods for the segmentation of sub-words into characters. Most of them separate secondary strokes or diacritic marks i.e. *Nuqtas* and *Airabs* before performing character segmentation.

Some of the early work on off-line Arabic word segmentation was carried by Pakker *et al.* [8] who segment Arabic handwritten words by calculating the thickness of the stroke on each text line and this value is considered as the threshold for the determination of the segmentation points between characters on the baseline.

Najoua *et al.* [9], proposed the method of segmentation of printed and handwritten words into characters. In which the approximate limits of the characters in the Piece of Arabic Word (PAW) are estimated using vertical histogram modulated by the vertical width of writing. It is also determined that a distance equal to the one and a half of the vertical width of the writing is suitable.

Motawa *et al.* [10] uses morphological operations to segment the handwritten words into singularities. Regularities are found by subtracting the singularities from the original image. Regularities contain the information required for connecting a character to the next character. Hence, these regularities are the candidates for segmentation. Regularities close to the baseline are segmented using the rules for long and short regularities accordingly.

Elgammal *et al.* [7] performed the segmentation of Arabic text by using the topological relationship b/w the baseline and the text line. The baseline is identified by using the histogram; however the text line is represented by line adjacency graph (LAG). The LAG formed for each sub-word is then transformed into the compressed LAG or simply c-LAG which is homomorphic to the LAG and has minimum number of nodes and each node is either labelled as path or junction. The adjacent nodes are considered as the junction nodes as the path nodes are never adjacent to each other.

In [11] the value of the vertical profile of the middle zone is calculated and where it is less than two thirds of the baseline thickness, the area is considered a connection area between two characters. Then, any area that has a larger value is considered as the start of a new character, as long as the profile is greater than one third of the baseline.

Ymin *et al.* [12] proposed a method for the segmentation of printed multi-font Uygur text that uses Arabic style scripting in writing. In this method text line is divided into three zones. From the edge of character strokes of upper zone the algorithm searches for possible break points along the vertical projection, which may cause the under segmentation of the characters having loops in them. The quasi topological segmentation bases on the first segmentation, to segment a character on a combination of feature-extraction and character-width measurements.

In [13] a thinned sub-word is segmented into characters by following the baseline of the sub-word. The sub-word is segmented when pixels start to move above or below the baseline.

In [14] these connection points are characterized to be the locations where curvature of upper contour of sub-word changes from positive to negative.

In [15] the probable segmentation points are characterized as Valid Segmentation Points (VSP) if local minimum in the lower outer contour qualifies the acceptance rules. The acceptance and rejection rules are the morphological rules which have been extracted from the Arabic text characters.

Mehran *et al.* [16] investigates the Persian/Arabic scripts and found that the upper contour of the primary stroke of sub-words called PAWs (Piece of Arabic Word) has a high gradient at the junction points, and after most junction points, the vertical projection has a value larger than the mean. On the other hand, the pen tip is generally positioned near the baseline in the desired junction points. These features are considering together to identify the junction points.

# 3 Proposed Segmentation Method

In the proposed segmentation method, first of all, the connected components are extracted and each component is termed as it is either a primary stroke or a secondary stroke. Primary strokes are then segmented into their constituent root characters. Finally, the secondary strokes are linked with their associated root characters.

## 3.1 Connected Component Separation

The area Voronoi diagram [17] and the connected component region labelling [18] are the methods that are used for the extraction of connected components from the images. Here, in the images of Arabic text, the connected components are sub-words and/or the diacritic marks.

On the basis of area Voronoi diagram, Voronoi edges are selected which separate two neighboring connected components and words/sub-words are extracted by merging the connected components based on Voronoi edges between them.

A connected set of pixels in an image is defined as a Region and it may be a foreground or background region. A foreground pixel P is said to be eight-connected to another foreground pixel Q, if there exists a path of adjacent foreground pixels P and Q, traversing either horizontally, vertically or diagonally. It is also required to assign a unique label to each region and an arbitrary integer is selected as label for the connected foreground pixels.

### 3.1.1 Secondary Stroke Separation
The connected component region labelling method is used to find the isolated regions in the image and such black areas are assigned unique labels. The bounding box for each isolated region is determined by considering the height and width of the region. Each isolated region is characterized as either it is a primary or the secondary stroke of the sub-word. The following two criterions are set to consider the region as the secondary stroke:

1. **The Size:** A rectangular window is created considering the average size of the primary strokes. The isolated regions smaller than that window are characterized as secondary strokes, as usually the secondary strokes are smaller than the primary strokes.
2. **The Position:** If an isolated region lies either above or below the baseline, it will be considered as secondary stroke, as the dots and other diacritic marks are usually be positioned there. The baseline is the line with highest intensity of black pixels in the horizontal projection histogram of a line of text in the image.

The regions which fall below the fixed size and having the above position are characterize as the secondary strokes and separated from the image; the separated strokes will be processed in the final stage of segmentation where they will be linked with their corresponding primary strokes. Fig. 2 shows the bounding box of the various isolated regions in an image.

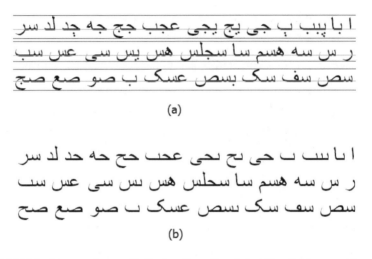

**Fig. 2.** The bounding box for the isolated regions

The original image along with the identified baseline for each extracted text line is shown in Fig. 3(a) and Fig. 3(b) shows image with separated secondary strokes.

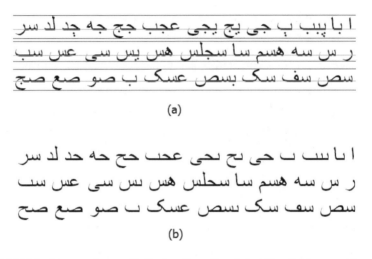

**Fig. 3.** (a) Original image along with the baseline of each text line (b) Image with separated secondary strokes

### 3.1.2  Primary Stroke Separation
Each line of text is scanned vertically left to right starting from the first character and ending at the last character; if the width of any white space between sub-words is found greater than (1/3) of threshold ($\tau$), a termination mark is placed that will be used

**Fig. 4.** Image with termination marks of primary strokes of sub-word

in later stages for the separation of primary stroke of sub-word. A threshold, $\tau$ *is* the maximum of the widths of the white space between the sub-words in the particular line of text. The resultant image is presented in Fig. 4.

## 3.2   Character Segmentation

This step does the job of breaking the connections between root characters of the sub-word, i.e. to identify the segmentation points. For this it computes the vertical projection graph for each line, which is then processed to generate a string indicating relative variations in pixels. The following Fig. 5 shows the string that is generated from the vertical projection of first line of text of the separated secondary strokes image in Fig. 3(b).

Here, in Fig. 5, the '+'sign indicates that the slope of the pixels is increasing, i.e. the number of black pixels in previous column is less than that of the current column, '-' sign indicates that the slope of the pixels in decreasing i.e. the number of black pixels in current column is less than that of the previous column, '?' indicates no pixel i.e. the number of black pixels in current column is zero, 'x' indicates the peak is less than baseline yet is not on baseline and '.' indicates a zero slope i.e. the number of black pixels in previous column is equal to the number of black pixels in current column.

```
+???xxxxxxxxxxxxx++---.....++-----..++--++-????????????++.-..+.----
..+..--????????????++.-..+.-----..+++....-.--.----.????????????+++.++++--
.....+++....-.--.----.????????????+++...............-....+++....-.--.----
.????????????++--..-...............++--...+++....-.--.----...++.+.....-.--
..????????????????+xxxxxxxxxx++++.........+++....-.--.----.......++-
????????????+++...............-......++-
????????????+xxxxxxxxxx++++.........+++....-.--.----.????????????++--..-
..............++-????????????++--..-...............++--.....++--......++-
?????????+++----.........++-????????????+++-????????
```

**Fig. 5.** The string generated from the vertical projection of first line of text in Fig.2

The string is then scanned from left to right and wherever segmentation conditions fulfilled it cuts the image. The conditions that result in segmentation point are given below:

**Condition 1:** A '+'sign followed by a negative sign. In this case the point next to the position where the negative sign last appeared is the segmentation point.

"xxx++--.....+++...----....."

**Condition 2:** A 'x' sign is followed by '+', which is then followed by more than two '.' signs. In this case the line next to the two dots is a segmentation point.

"??+xxxxxxxxxxx++++.........+"

The graph of the vertical projection of the text line of the sample image and the positions where the segmentation points are marked are indicated by vertical lines in Fig. 6.

**Fig. 6.** Image with segmentation points for characters in sub-words

Finally, the extracted regions termed as secondary strokes are linked with their related segmented root characters. This linkage can be performed by comparing the ranges of each secondary stroke and the segmented root characters.

## 4   Conclusion

A complete system for the segmentation of offline printed Sindhi text into characters has been presented. It can be used for the segmentation of text written in other Arabic style scripting languages such as Arabic, Persian, Urdu and Pashto etc. This method can be a part of character recognition system as one of its steps is the segmentation of text into characters. The new method has been tested on images of text in many fonts and sizes, and the results show its success on their segmentation. It is observed that the characters like seen (س) and bay (ب) are segmented into more than one character, so it is required that such characters be given an extra attention at the later stages of the complete recognition system.

## 5   Future Work

As the future work, we are looking for the set of features that must be efficient and robust enough to discriminate the segmented characters even by using a simple k-NN classifier by considering Euclidean or Manhattan distance between the features of the characters.

**Acknowledgement.** The authors would like to thank and acknowledge the Centre for Research in Ubiquitous Computing (CRUC) at FAST-NU, Karachi campus, where this entire work has been carried and Higher Education Commission (HEC), Pakistan who sponsored this work through its indigenous PhD program.

# References

1. Abdullah, I., Al-Shoshan: Arabic OCR Based on Image Invariants. In: Proceedings of the Geometric Modeling and Imaging – New Trends, pp. 150–154 (2006)
2. Nouh, A., Sultan, A., Tolba, R.: An approach for Arabic characters Recognition. J. Engng Sci. Univ. Riyadh. 6, 185–191 (1980)
3. Badr, B.A., Mahmoud, S.A.: A survey and bibliography of Arabic optical text recognition. Signal Processing 41, 49–76 (1995)
4. Zheng, L.: Machine Printed Arabic Character Recognition Using S-GCM. In: Proceedings of 18th International Conference on Pattern Recognition, vol. 2, pp. 893–896 (2006)
5. Mandana, K., Amin, A.: Pre-processing and Structural Feature Extraction for a Multi-Fonts Arabic/Persian OCR. In: Proceedings of 5th Intl. Conference on Document Analysis and Recognition (1999)
6. Somaya, A.: Recognition of Off-Line Handwritten Arabic Words Using Neural Network. In: Proceedings of the Geometric Modeling and Imaging – New Trends, pp. 141–144 (2006)
7. Elgammal, A., Ismail, M.A.: A graph-based segmentation and feature extraction framework for Arabic text recognition. In: Proceedings of 6th Intl. Conference on Document Analysis and Recognition, pp. 622–627 (2001)
8. Pakker, K.R., Miled, H., Lecourtier, Y.: A new approach for Latin/Arabic character segmentation. In: Proceedings of 3rd Intl. Conference on Document Analysis and Recognition, vol. 2, pp. 874–878 (1995)
9. Najoua, B.A., Noureddine, E.: A robust approach for Arabic printed character segmentation. In: Proceedings of 3rd Intl. Conference on Document Analysis and Recognition, vol. 2, pp. 865–868 (1995)
10. Motawa, D., Amin, A., Sabourin, R.: Segmentation of Arabic cursive script. In: Proceedings of the 4th International conference on Document Analysis and Recognition, pp. 625–628 (1997)
11. Sarfraz, M., Nawaz, S.N., Al-Khuraidly, A.: Offline Arabic text recognition system. In: Proceedings of the Int. Conference on Geometric Modeling and Graphics, pp. 30–35 (2003)
12. Ymin, A., Aoki, Y.: On the segmentation of multi-font printed Uygur scripts. In: Proceedings of Intl. Conference on Pattern Recognition, vol. 3, pp. 215–220 (1996)
13. El-Khaly, F., Sid-Ahmed, M.A.: Machine recognition of optically captured machine printed Arabic text. Proceedings of Pattern Recognition 23, 1207–1214 (1990)
14. Margner, V.: SARAT - A system for the recognition of Arabic printed text. In: Proceedings of 11th Intl. Conference on Pattern Recognition, pp. 561–564 (1992)
15. Sari, T., Souici, L., Sellami, M.: Off-line handwritten Arabic character segmentation algorithm: ACSA. In: Proceedings of Intl. Workshop on Frontiers of Handwriting Recognition, pp. 452–456 (2002)
16. Mehran, R., Pirsiavash, H., Razzazi, F.: A Front-end OCR for Omni-font Persian/Arabic Cursive Printed Documents. In: Proceedings of the Digital Imaging Computing: Techniques and Applications, pp. 56–60 (2005)
17. Wang, Z., Lu, Y., Tan, C.L.: Word extraction using area Voronoi diagram. In: Proceedings of the Conference on Computer Vision and Pattern Recognition Workshop, vol. 3, pp. 31–36 (2003)
18. Gonzalez, R.C., Woods, R.E., Eddins, S.L.: Digital image processing using MatLab. 3rd Indian Reprint (2005)

# Improved Texture Description with Features Based on Fourier Transform

Ahsan Ahmad Ursani[1], Kidiyo Kpalma[2], Joseph Ronsin[3],
Aftab Ahmad Memon[4], and Bhawani Shankar Chowdhry[5]

[1] INSA / IETR (Current Institute), 20 Avenue de Buttes des Coësmes, 35000 Rennes, France
and Mehran UET, Jamshoro Pakistan
Aursani@ens.insa-rennes.fr
[2,3] INSA / IETR, 20 Avenue de Buttes des Coësmes, 35000 Rennes, France
{kkpalma, jronsin}@insa-rnnes.fr
[4] Mehran University of Engineering & Technology, Jamshoro, Pakistan
memon_aftab@yahoo.com
[5] Mehran University of Engineering & Technology, Jamshoro, Pakistan
& School of ECS, University of Southampton, UK
bsc06v@ecs.soton.ac.uk

**Abstract.** This paper presents an improved version of the features based on Discrete Fourier Transform (DFT) for texture description that demonstrate robustness against rotation. The features have been tested on cropped parts of textures from the Brodatz collection and their rotated versions. The results show improved performance for both recognition and retrieval, compared to texture features based on the Gabor filter and older versions of the DFT-based features.

**Keywords:** Texture description, texture feature extraction, Fourier transform, Gabor filters.

## 1 Introduction

Apart from various other applications, the texture features are an essential part of computer graphics [1], image retrieval/computer vision systems [2], Medical Imaging [3], and the land-cover classification in remote sensing systems [4]. In all of these applications, since it becomes inevitable to encounter the rotated versions of the textures, it is highly necessary for texture features to be rotation invariant. Although many works on rotation invariant texture recognition have been published [5]-[8], but so far, practically there hasn't been any effort of estimating the claimed rotation invariance. Therefore, a measure for rotation variance is proposed and presented in section 4.1.

The texture features can be categorized into two major classes; one that extracts the global properties of the texture images and the other that extracts the local neighbourhood properties around the pixels of the textured image [9]. The first category includes features based on Gabor or other wavelets that use large filter banks and the second one includes the texture features based on Gaussian Markov random fields, the local binary pattern operator, the spatial grey level co-occurrence matrix and the 1D

D.M.A. Hussain et al. (Eds.): IMTIC 2008, CCIS 20, pp. 19–28, 2008.
© Springer-Verlag Berlin Heidelberg 2008

DFT [5]. The two categories are investigated and compared in [5], [9]-[11], concluding that the features from second category outperform the features from the first category. Considering this finding, we investigate the role that the smallest possible, i.e. 2×2 pixel neighbourhoods may play in the texture description. To do so, we append the texture feature vector introduced in [5] by introducing in it the texture features extracted from 2×2 pixel neighbourhoods.

## 2 Method

The 1D DFT features [5] were extracted from the 3×3-pixel neighbourhood in the spatial domain, as shown in fig. 1. The work mentioned in [5] takes 1D DFT of the 8-pixel sequence around each pixel in the image. Later, to obtain the overall picture of the texture, the sets of magnitudes of the DFT coefficients are quantized into bins to compute what is called as Local Fourier Histograms (LFHs). These LFHs are used as a texture descriptor. More is explained in the following subsection.

Though larger windows can also be considered for texture feature extraction, but considering obvious disadvantages inherent in the larger neighbourhoods for the application of texture segmentation, we suggest here more features extracted from within the same neighbourhood of 9 pixels; since larger neighbourhoods would cause more errors around border of the two textures in case of texture segmentation. It should be noted that the larger window sizes were also tried but yielded worse results.

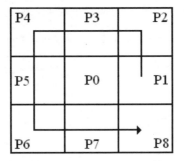

**Fig. 1.** The 9-pixel neighbourhood

### 2.1 Extracting DFT Features

Moving 3×3-pixel window across a texture image, DFT of the 8-pixel sequence, P1 through P8, shown in figure 1 is computed around each pixel in the image. Later, absolute value of first five coefficients i.e. $|X_0|$ through $|X_4|$ are saved in their respective arrays. The Fourier transform in a single dimension is computed as in (1), where $X_k$ represents the $k^{th}$ Fourier coefficient, $N$ represents length of the sequence in spatial domain, i.e. 8, $x_n$ represents the $n^{th}$ value in the spatial domain ($x_0$ is P1 and $x_7$ is P8), and $0 \leq k \leq 4$.

$$X_k = \sum_{n=0}^{N-1} x_n e^{-\frac{2\pi i}{N}kn} \tag{1}$$

From the computed DFT, the absolute values of the first 5 coefficients, i.e. $|X_0|$ through $|X_4|$ are used for the texture description. Altogether, the computed coefficients are normalized to take values from 0 to 255. After normalization, $|X_0|$ is linearly quantized into 8 bins and $|X_1|$ through $|X_4|$ are linearly quantized into 16 bins. For describing the textures, all the 8 bins of $|X_0|$ are made part of the resulting texture models, but only first 8 bins of the remaining coefficients are used as features. This is because all the bins following bin 8 are always zero, which is evident from the histograms shown in fig. 2. It can be seen that the histograms of the coefficients other than $X_0$ do not go beyond 127 on the x-axis. In this way, we have 8×5, i.e. 40 features in the texture descriptor.

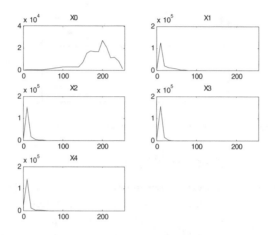

**Fig. 2.** Histograms of the coefficients $X_0$ through $X_4$ for image D1

## 2.2   The Augmented Feature Set

The features obtained in section 2.1 are a representation of local properties of the 3×3 pixel neighbourhoods [5]. But since these do not involve the central pixel itself, i.e. the pixel under process, these lack the relationship of the central pixel with the neighbourhood. Moreover, the smaller neighbourhoods can also provide spatial dependencies of the texture that might complement those represented by the features obtained in the previous section, especially in case of microscopic textures. Fig. 3 shows how we extract more features from the same 9-pixel neighbourhood. As shown, four 4-pixel sequences are used to extract 32 more features. As shown, the four 4-pixel sequences are {P0 P1 P2 P3}, {P0 P3 P4 P5}, {P0 P5 P6 P7}, and {P0 P7 P8 P1}. DFTs of the four 4-pixel sequences are computed and amplitude of only second DFT coefficient, i.e. $|X_1|$ from each sequence is used as a feature. Altogether, the computed coefficients are normalized to take values from 0 to 255 and then quantized into 8 bins, giving 4×8, i.e. 32 texture features. Hence we have 40+32, i.e. 72 features in all.

**Computational Cost.** Although the augmented feature set is larger, but computationally, it is only a very slightly more expensive than the smaller set and much less expensive than the features based on Gabor or other wavelets. Expression (1) used to compute the 40 DFT features can be explained as 40 sums of products, since k takes 5 values and n takes 8 values. On the other hand, the new 32 features introduced by us cost only 4 more sums of products, since k takes 1 value and n takes 4 values. This causes merely 10% increase in the cost of the features.

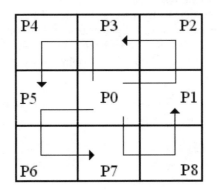

**Fig 3.** The four 4-pixel sequences

## 2.3  Texture Features Based on Gabor Filters

Although there are several works reporting the use of texture descriptors based on Gabor filters, we present experimental results with the descriptor first presented in [11], again in [12] along with a method of rotation invariant similarity measurement and compared in [10]. These descriptors are designed quite carefully to avoid filter outputs carrying redundant information. Equations (2) through (10) are the expressions as they appear in [11]-[12].

$$G_{mn}(x, y) = \sum_{s=0}^{S} \sum_{t=0}^{T} I(x - s, y - t) \psi_{m,n}^{*}(s,t) \ , \tag{2}$$

where $S$ and $T$ specify the filter mask size, and * represents operation of complex conjugate.

$$\psi(x, y, W) = \frac{1}{2\pi\sigma_x\sigma_y} \exp\left[-\frac{1}{2}\left(\frac{x^2}{\sigma_x^2} + \frac{y^2}{\sigma_y^2}\right)\right].\exp(j2\pi Wx) \ , \tag{3}$$

where $\psi$ is the mother Gabor wavelet, $\sigma_x$ and $\sigma_y$ are the functions of scale $(m)$, described in (9) and (10), and $W$ is the modulation frequency described in (8). The children wavelets are defined as

$$\Psi_{m,n}(x, y, W_m) = a^{-m} \Psi(\tilde{x}, \tilde{y}, W_m) \ , \tag{4}$$

where $\tilde{x}$ and $\tilde{y}$ are given as

$$\tilde{x} = a^{-m}(x\cos\theta + y\sin\theta) \ , \tag{5}$$

and

$$\tilde{y} = a^{-m}\left(- x \sin s\theta + y \cos \theta\right) , \tag{6}$$

respectively, where $\theta = n\pi/N$, $n$ specifies the orientation of the wavelet, $N$ is the total number of orientations, $a>1$ and is given by

$$a = (U_h / U_l)^{\frac{1}{M-1}} , \tag{7}$$

where the values of $U_l$ and $U_h$ are the lower and upper centre frequencies, whose values are proposed in [11]-[12] as 0.05 and 0.4 respectively. The modulation frequency $W_m$ is given as

$$W_m = a^m U_l . \tag{8}$$

The $\sigma_x$ and $\sigma_y$ of the children wavelets are given by

$$\sigma_{x,m} = \frac{(a+1)\sqrt{2\ln 2}}{2\pi a^m (a-1)U_l} , \tag{9}$$

$$\sigma_{y,m} = \frac{1}{2\pi \tan\left(\dfrac{\pi}{2N}\right)\sqrt{\dfrac{U_h^2}{2\ln 2} - \left(\dfrac{1}{2\pi\sigma_{x,m}}\right)^2}} , \tag{10}$$

The mean ($\mu_{mn}$) and standard deviation ($\sigma_{mn}$) of the magnitude of the filter outputs ($G_{mn}$) from (2) are used as texture features. Five scales ($M$=5) and six orientations ($N$=6), hence a total of 30 filters were proposed in [12]. Therefore, the resulting texture descriptor has 60 values in all. The rotation invariance is achieved by circularly shifting the feature vector so that the features (mean and variance) from the filter giving highest output energy are placed first. This was also proposed in [12]. However, since our test images are only 128×128 in size against 200×200 in [12], we used the filter size of 31×31 instead of 61×61 proposed in [11] and [12].

It should be noted that the DFT features come from the DFT coefficients that are quantized in to 8 bins each. Each bin is used as a feature, resulting in a feature vector of 40 values in conventional feature set and 72 values in the augmented feature set. On the other hand, the Gabor features come from 30 Gabor filters whose means and standard deviations are used as texture features, resulting in a feature vector of 60 values.

## 3 Materials

The results are presented for two sets of images from Brodatz collection. The first one consists of 107 texture images, i.e. all except D14, D43, D44, D45, and D59 that were too irregular to be considered as a single texture. Second one is a subset of 32 images from this set, listed and shown in fig. 4. Realizing that the Brodatz album contains many variants of many images, with different lighting conditions, with different zoom, etc., we selected only one image of each type including paper, woven wire, canvas, stone, cloth, brick wall, flower, water, skin, etc. Again, 80 siblings of each one of the 32 images, i.e. 2560 images, were used as a query image.

The DFT based texture features were extracted from all the Brodatz texture images and were made part of the database. Later, 16 partly (approximately 25%) overlapping square images having dimensions of 200×200 were cropped from each database image. This gave us 1712 texture images to be used as query images. Further, each one of the 1712 images was rotated by principal angles of 30°, 45°, 60° and 90°. Square images with the dimensions of 128×128 were cropped from the rotated ones and the un-rotated alike. This made us available a total of 1712×5, i.e. 8560 images with orientations of 0°, 30°, 45°, 60° and 90° to be used as the query images. In this way, we had 16×5 i.e. 80 siblings of each and every image in the database.

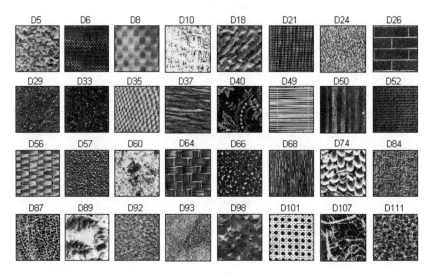

**Fig. 4.** The 32 Brodatz texture images selected for experimentation

# 4   Experiments and Results

We experiment with the three feature sets presented in section 2.1 through section 2.3 on the two datasets explained in section 3. The Euclidean distance was used as the dissimilarity measure in all the experiments.

We have performed experiments separately for recognition and for the retrieval. Recognition as well as retrieval results are shown for the three sets of Brodatz images.

## 4.1   Recognition

Every image in the database has 80 siblings, 16 sub-images each in 5 orientations, i.e. 0°, 30°, 45°, 60° and 90°. The results are not presented only for overall accuracy, but are presented for each orientation as well. This allows us estimating the rotation variance or otherwise of the feature sets. We define *rotation variance* as in (11).

$$RV = \frac{std(\{A_0, A_{30}, A_{45}, A_{60}, A_{90}\})}{OA} * 100 \ , \tag{11}$$

where $RV$ represents *rotation variance*, *std* represents standard deviation, OA represents overall percent accuracy of recognition, and $A_0$ through $A_{90}$ represent the percent accuracy of recognitions in the respective angles of rotation.

**Full Brodatz Album.** Each one of the 8560 images is taken as a query image that is to be recognized as one of the 107 images. Table 1 shows results obtained for set of all the 107 Brodatz images. The recognition was performed using closest mean methods. As shown, the best results are obtained with the largest query image size and the worst are obtained with the smallest query image size. The recognition results are also presented orientation-wise. The set of 72 DFT features exhibits less rotation variance as compared to the set of 40 DFT features. This shows that the augmented feature-set doesn't perform better only in terms of percent accuracy but also in terms of the rotation invariance. The results demonstrate clear superiority of the 72 DFT features set over the 40 DFT features set and the Gabor features.

**Table 1.** Overall recognition results for all the 107 Brodatz images

| Out of 8560 images | | | | |
|---|---|---|---|---|
| Feature Set → | | Gabor | DFT 40 | DFT 72 |
| Rotation | 0 | 54.3% | 63.0% | 64.3% |
| | 30 | 50.9% | 56.1% | 57.8% |
| | 45 | 44.5% | 56.1% | 60.9% |
| | 60 | 50.8% | 56.8% | 58.6% |
| | 90 | 54.4% | 63.4% | 63.0% |
| Total | | **51.0%** | **59.4%** | **60.9%** |
| Rotation variance | | **7.94** | **6.46** | **4.53** |

**Table 2.** Overall recognition results for the 32 Brodatz images

| Out of 2560 images | | | | |
|---|---|---|---|---|
| Feature Set → | | Gabor | DFT 40 | DFT 72 |
| Orientation | 0° | 80.9% | 87.1% | 90.0% |
| | 30° | 79.1% | 86.7% | 89.3% |
| | 45° | 75.0% | 86.1% | 90.6% |
| | 60° | 76.8% | 85.4% | 89.8% |
| | 90° | 79.7% | 86.9% | 89.6% |
| Total | | **78.3%** | **86.4%** | **89.9%** |
| Rotation variance | | **3.02** | **0.82** | **0.56** |

**Subset of 32 Images from Brodatz Album.** Table 2 shows the overall and the orientation-wise accuracy. Again, the set of 72 DFT features clearly outperforms the set of 40 DFT features and Gabor features in terms of both, the overall recognition accuracy as well as the rotation invariance. The accuracy of the 40 DFT features is 86.4% and that of 72 DFT features is 89.9% or 3.5% more than that of the former. On the other hand, the rotation variance has decreased from 0.82 to 0.56 that represents 31.7% decrease in the rotation variance.

## 4.2 Retrieval

In case of retrieval, we give full length images as the query images, and find the closest $N$ matches from a database that has 80 siblings of each query image, where $N$ takes the values from 16 to 320, with the step of 16. The results are presented in the form of precision and recall curves for 20 values of retrieved images. The recall is calculated as

$$recall = 100 * \frac{A}{A+B} ,$$ (12)

where $A$ represents the number of relevant images retrieved and $B$ represents the number of relevant images not retrieved. The precision is calculated as

$$precision = 100 * \frac{A}{A+C} ,$$ (13)

where $C$ is the number of irrelevant images retrieved. It should be noted that the precision and recall curves show the consolidated retrieval results for the complete sets rather than for individual images.

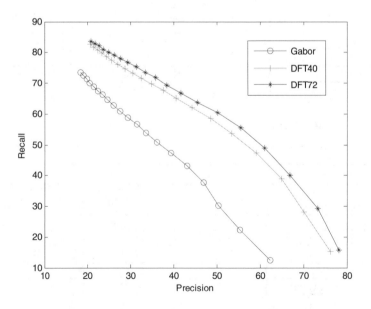

**Fig. 5.** Retrieval results for all the 107 Brodatz images

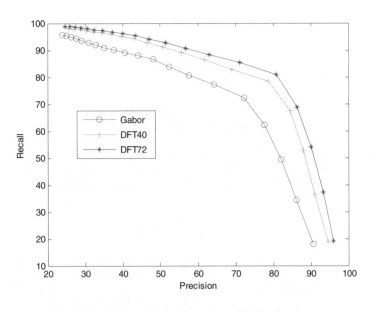

**Fig. 6.** Retrieval results for the subset of 32 Brodatz images

**Full Brodatz Album.** This section presents the results of the retrieving non-calibrated set of all the 107 Brodatz images. The graph of figure 5 illustrates the results obtained with the three feature sets, revealing that for a given value of precision, the set of 72 DFT features always provides higher recall than the sets of 40 DFT features and the Gabor features. Similarly, for a given value of recall, the augmented set provides higher precision than the sets of 40 DFT features and Gabor features. The curve representing the set of 72 DFT features always remains on the upper-right side of the other two curves, evidencing in favour of the augmented feature set.

**Subset of 32 Brodatz Images.** This section presents the results of the retrieving the set of the selected 32 Brodatz images. In the graph of figure 5 are the precision vs. recall curves for the three texture feature sets. The curves clearly decide in favour of the set of 72 DFT features which shows better recall than 40 DFT features and the Gabor features for a given precision and a better precision for a given recall.

## 5  Conclusion

The set of 72 DFT based features are superior to the set of 40 DFT based features [5] and those based on Gabor filters [10]-[12] not only with respect to the overall recognition and retrieval performance, but also with respect to the rotation invariance. The augmented feature set is slightly more expensive than the smaller set and much less expensive than the features based on Gabor or other wavelets, but it gives a gain of 3.5% (see table 2) in the accuracy that is really a big gain since a percent increase in the accuracy that is already 86% is much more important than a percent increase in the accuracy that is 50%.

## Acknowledgement

The first author (Ahsan Ahmad Ursani) is thankful to his mother institute of the paper is the Mehran University of Engineering & Technology (MUET), where he is assistant professor. Currently he is associated with INSA de Rennes, France, to pursue his PhD research sponsored by the MUET. One of the coauthors (Bhawani Shankar Chowdhry) is grateful to the Higher Education Commission (HEC), Islamabad, for sponsoring his postdoctoral fellowship at the School of ECS, University of Southampton.

## References

1. De Bonet, J.S.: Multiresolution sampling procedure for analysis and synthesis of texture images. In: Proceedings of the 24th annual conference on Computer graphics and interactive techniques, pp. 361–368 (1997)
2. Liu, Y., Zhou, X.F.: A Simple Texture Descriptor for Texture Retrieval. In: International Conference on Communication Technologies, vol. 2, pp. 1662–1665 (2003)
3. Kim, J.K., Park, H.W.: Statistical Textural Features for Detection of Microcalcifications in Digitized Mammograms. IEEE Transactions on Medical Imaging 18(3), 231–238 (1999)
4. Hongyu, Y., Bicheng, L., Wen, C.: Remote Sensing Imagery Retrieval Based-On Gabor Texture Feature Classification. In: Proceedings of 7th International Conference on Signal Processing (ICSP 2004), vol. 1, pp. 733–736 (2004)
5. Zhou, F., Feng, J.-F., Shi, Q.-Y.: Texture Feature Based on Local Fourier Transform. In: Proceedings of International Conference on Image Processing, vol. 2, pp. 610–613 (2001)
6. Pun, C.-M.: Rotation-invariant texture feature for image retrieval. Computer Vision and Image Understanding 89, 24–43 (2003)
7. Mayorga, M.A., Ludeman, L.C.: Shift and Rotation Invariant Texture Recognition with Neural Nets. In: IEEE Internation Conference on Neural Networks (1994)
8. Do, M.N., Veterli, M.: Rotation Invariant Texture Characterization and Retrieval using Steerable Wavelet-Domain Hidden Morkov Models. IEEE Transactions on Multimedia 4(4), 517–527 (2002)
9. Varma, M., Zisserman, A.: Texture Classification: Are Filter Banks Necessary? In: Proc. Conf. Computer Vision and Pattern Recognition, vol. 2, pp. 691–698 (2003)
10. Toyoda, T., Hasegawa, O.: Texture Classification using Extended Higher Order Local Autocorrelation Features. In: Proceedings of the 4th Int. Workshop on Texture Analysis and Synthesis, pp. 131–136 (2005)
11. Manjunath, B.S., Ma, W.Y.: Texture Features for Browsing and Retrieval of Image data. IEEE Transactions on Pattern Analysis and Machine Intelligence 18(8), 837–842 (1996)
12. Zhang, D., Wong, A., Indrawan, M., Lu, G.: Content-based Image Retrieval Using Gabor Texture Features. In: IEEE Intern. Symposium on Multimedia Information Processing (2000)

# Optimization of Projections for Parallel-Ray Transmission Tomography Using Genetic Algorithm

Shahzad Ahmad Qureshi[1], Sikander M. Mirza[2], and M. Arif[3]

[1] Department of Computer & Information Sciences, Pakistan Institute of Engineering & Applied Sciences (PIEAS), P.O. Nilore, Islamabad 45650, Pakistan
[2] Department of Physics and Applied Mathematics, Pakistan Institute of Engineering & Applied Sciences (PIEAS), P.O. Nilore, Islamabad 45650, Pakistan
[3] Department of Electrical Engineering, Pakistan Institute of Engineering & Applied Sciences (PIEAS), P.O. Nilore, Islamabad 45650, Pakistan
{SAQureshi,Sikander,fac097}@pieas.edu.pk

**Abstract.** In this work, a Hybrid Continuous Genetic Algorithm (HCGA) based methodology has been used for the optimization of number of projections for parallel-beam transmission tomography. Image quality has been measured using root-mean-squared error, Euclidean error and peak signal-to-noise ratio. The sensitivity of the reconstructed image quality has been analyzed with the number of projections used for the estimation of the inverse Radon transform. The number of projections has resulted in the maximization of image quality while minimizing the radiation hazard involved. The results have been compared with the intensity levels of the original phantom and the image reconstructed by the Filtered Back-Projection (FBP) technique, by using Matlab ® functions *radon* and *iradon*. For the 8 × 8 Head and Lung phantoms, HCGA and FBP have resulted in PSNR values of 40.47 & 8.28 dB and 26.38 & 12.98 dB respectively with the optimum number of projections.

**Keywords:** Genetic Algorithm, Inverse Radon Transform, Filtered Back-Projection, Transmission Tomography.

## 1 Introduction

The inversion of sinogram to estimate the cross-sectional view of an object forms the basis for computed tomography (CT). The projections obtained from the Data Acquisition System (DAS) are used to carry out the estimation of inverse Radon transform (IRT) in discrete domain. Various medical imaging techniques are classified into deterministic and stochastic categories. The Filtered Back-Projection (FBP), a deterministic technique, is commercially used in many applications. Other deterministic techniques include the Direct Fourier (DF) method and the Algebraic Reconstruction technique (ART) [1, 2]. These techniques loose their efficacy in case the projections data is missing or the data is noise contaminated and lose the resulting image quality. Consequently, many researchers worked towards the estimation of d-IRT (discrete-IRT) by using stochastic techniques. Metropolis et al. used the principle of thermodynamic equilibrium in numerical computation for global optimization [3]. Webb

D.M.A. Hussain et al. (Eds.): IMTIC 2008, CCIS 20, pp. 29–39, 2008.

discussed SA in optimizing radiation therapy planning from practical viewpoint [4-6]. Murino and Trucco put forward a technique based on Markov random fields methodology and used SA to minimize the energy functional to get the optimal estimates of underwater acoustic images [7]. Qureshi et al. introduced a Simulated Annealing based algorithm for image reconstruction where an IRT estimate is randomly evolved and the error minimization is improved by tuning various generic and problem specific parameters [8].

The stochastic technique of Genetic Algorithm (GA), originally developed by Holland, was based on Darwinian theory of biological evolution [9, 10]. Pastorino et al. used GAs for image reconstruction to determine the distribution of dielectric permittivity of two-dimensional object using scattered electric-field measurements through microwave inverse scattering technique. The major limitations identified were the design and realization of efficient illumination [11]. Another group of researchers introduced GA based tomography using electrical resistivity measurements distributed in the object. The poor resolution of reconstructed images was attributed mainly to higher number of degrees of freedom [12]. Franconi and Jennison proposed a hybrid algorithm for finding the maximum a posteriori (MAP) estimation of a binary image in Baysian image analysis. They used crossover to merge subsections developed by SA. The partitioning strategy and the use of various deterministic techniques, and the way they may be applied to combine solutions, in place of SA still need to be explored [13]. The representation as float was found superior in comparison to both binary GA as well as in SA in terms of quality and efficiency [14, 15]. In order to improve the efficiency of convergence, Caorsi et al. applied for the first time a real-coded GA (RCGA) for the detection and reconstruction of dielectric objects for short-range microwave imaging applications [16]. Recently, Qureshi et al. have introduced Hybrid Continuous Genetic Algorithm (HCGA) for the estimation of d-IRT. They used continuous GAs, hybridized with a preprocessed population initialization [17]. The number of projections, a DAS dependent parameter, is linked with radiation dose reduction and image reconstruction through incomplete projection data or when the data is noise contaminated. In this work, thorough analysis for the optimization of number of projections has been carried out as a single dominant parameter affecting the image quality with the available input data for various sizes of the Head and the Lung phantoms.

## 2 Mathematical Model

The coordinate system is shown in Fig. 1 for Head phantom f(x, y) [18, 19]. A projection along ($\theta$, t)-line is apprehended to be a set of line integrals P$\theta$ (t) for $\theta$ views. This form the Radon transform of the cross section. The view fIRT(x, y) may be generated from the known projections. In the case of FBP technique, the Radon transform is convolved with a filter |w| to simulate the template [1]:

$$f_{IRT}(x, y) = \int_0^\pi \int_{-\infty}^\infty F_{1D}\left(P_\theta(t)\right) e^{j2\pi wt} |w| \, dw \, d\theta. \tag{1}$$

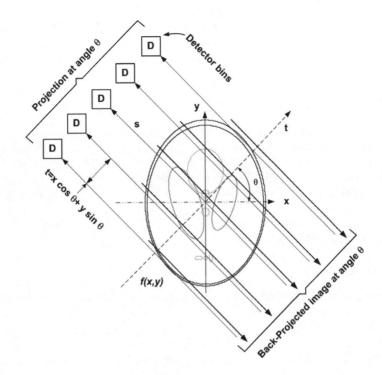

**Fig. 1.** Formation of a Projection at a distance $t$, from the centre through the Head phantom in rectangular coordinates $(x,y)$. Back-projected image at angle $\theta$ is shown as the inverse Radon transform for $P_\theta(t)$.

The Radon transform $P_{RT}$ of $f_{IRT}(x, y)$ is given by:

$$P_{RT}(t,\theta) = R_a\left(f_{IRT}(x,y)\right),$$ (2)

where $R_a$ is the Radon transform operator.

The fitness function **F** is based on the relative difference between the postulated and measured projections for $\theta_n$ views and $t_b$ as follows:

$$\mathbf{F} = \left(1+E\right)^{-1},$$ (3)

where

$$E = \left[\frac{1}{\theta_n t_b}\sum_{i=0}^{\theta_n}\sum_{j=0}^{t_b}\left\{R_a\left(f_{IRT}(x,y)\right) - \sum_{x=0}^{R_{Max}-1}\sum_{y=0}^{C_{Max}-1} f(x,y)\zeta(\theta_i)\right\}^2\right]^{1/2},$$ (4)

and

$$\zeta(\theta) = \begin{cases} 1 & \textit{iff} \quad t = x\cos\theta + y\sin\theta. \\ 0 & \textit{otherwise.} \end{cases}$$ (5)

*E* represents root-mean-squared error (RMSE) of this system. HCGA uses selection, crossover and mutation operators to improve **F** of the population by going through the GA cycle [20]. Various implementation steps for this technique are given briefly in the sections below.

## 2.1 Initialization and Selection Procedure

The preprocessed template, or even the actual human data, is supplied to initialize the population *P*. The selection operation forms the mating pool (MP) with improved fitness of population. In this work, Mixed Selection Scheme (MSS) has been used [20]. It uses the Truncation Scheme (TS) followed by the Roulette Wheel Scheme (RWS) with the Elitism. The fraction of chromosomes selected using TS, is fed to RWS and the chromosomes are assigned sectors on the roulette wheel according to their fitness. This selection process with RWS is repeated *m*-1 times. The $m^{th}$ chromosome of the mating pool is provided with a copy of the *best-string*, as an elite entity, without any alteration.

## 2.2 Crossover

Conventional crossover (CO) involves exchanging genes to share valuable information between each pair of mating parents. It is generally applied with high value of probability $P_c$. Various CO schemes employed in these simulations consist of Single-Point-, Two-Point-, Multi-Point- (with 4, 6, 8 and any random integer $I \in [1, n-1]$, where *n* is the length of the chromosome) and Uniform- schemes. The details may be seen in [20].

## 2.3 Mutation

This operation induces diversity in the population with a small mutation probability $P_m$ [21]. Offset-Based Mutation (OBM) scheme has been used in this work. In this scheme, a fixed gray level offset $O_s$ defines the range for random value as the possible mutate in either direction. The offset value depends on the complexity of the reconstructed image. The algorithm used is given by:

$$a = N_p(i, j) - O_s \quad if \quad a < 0.0 \quad a = 0.0,$$
$$b = N_p(i, j) + O_s \quad if \quad b > 1.0 \quad b = 1.0, \tag{6}$$
$$if \quad R_f < P_m \quad N_p(i, j) = R_v(b - a) + a \; ; \quad j = 1, 2, ..., n, \quad i = 1, 2, ..., m,$$

where $R_v$ is a random value in the range [0, 1], to be converted to an allele, and $R_f$ is another random number generated in the range [0, 1] to check the occurrence of event. In Eq. (6), *a* or *b* are adjusted accordingly to avoid surpassing the values 0 or 1 at either end.

## 3  Simulation Results and Discussion

The Head phantom and the Lung phantom have been used for this simulation [18]. The International Commission on Radiation Units and Measurements defined a phantom in detail [19]. The constraints for uniform basis include zero noise level, evenly distributed projections and 256 grey levels scaled in the range [0, 1]. Matlab ® has been used to carry out various simulations of image reconstruction. The FBP technique has been used to prepare templates of various sizes. The details about the selection of various composite filters may be seen in [17]. The root-mean-squared error (RMSE), peak signal-to-noise ratio (PSNR) and Euclidean error (EE) have been measured as the image quality parameters [8, 22]. The details for computation of PSNR may be seen in [23]. The termination of HCGA is carried out after fixed number of generations is surpassed or predefined fitness level is achieved [20].

In the case of Head phantom, the simulation parameters consist of MSS with fraction $f = 0.5$, crossover probability $P_c = 0.8$, population size $N_p = 75$ and generations $G = 5000$. Double-Point CO Scheme has been used for $8 \times 8$ Head phantom whereas Multi-Point CO Scheme has been employed for $16 \times 16$ image size. The details for CO effect on HCGA may be seen in [20]. The mutation probabilities $P_m = 0.01$ & $0.002$ have been used with OBM scheme for $8 \times 8$ and $16 \times 16$ sized images respectively. The parameters for Lung consist of MSS with $f = 0.4$, $N_p = 50$, $G = 5000$, Multi-Point CO Scheme with $P_c = 0.8$, and $P_m = 0.01$ & $0.005$ for $8 \times 8$ and $16 \times 16$ sizes respectively.

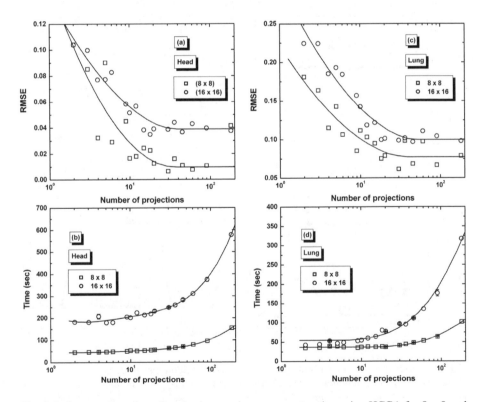

**Fig. 2.** Influence of number of projections on image reconstruction using HCGA for $8 \times 8$ and $16 \times 16$ image sized phantoms, (a) & (c) root-mean-squared error (RMSE) plotted against the number of projections for $8 \times 8$ & $16 \times 16$ Head and Lung phantoms respectively, and (b) & (d) time for convergence plotted against the number of generations for $8 \times 8$ & $16 \times 16$ Head and Lung phantoms respectively

Fig. 2(a) shows the RMSE plotted against the number of projections where no improvement in image quality is seen after a specific number of projections, $p = 30$, for $8 \times 8$ and $16 \times 16$ sizes where the error minimizes at a certain point. This is in consistency with the findings of other researchers [24]. In Fig. 2(b), execution time has been plotted against the projections for various sizes of the reconstructed phantoms. The

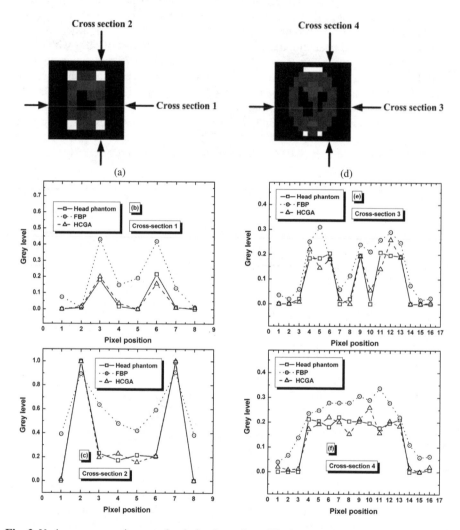

**Fig. 3.** Various cross-sections results during inversion of Radon transform for $8 \times 8$ & $16 \times 16$ Head phantoms by using Filtered Back-Projection (FBP) and Hybrid Continuous Genetic Algorithm (HCGA) techniques with $p = 30$, (a) cross-sections 1 & 2 for $8 \times 8$ Head phantom, (b) & (c) cross-sections 1 & 2 showing the grey levels plotted against the pixel positions for $8 \times 8$ Head phantom for original phantom (solid line), FBP (dotted line) and HCGA (dashed line), (d) cross-sections 3 & 4 for $16 \times 16$ Head phantom, (e) & (f) cross-sections 3 & 4 showing the grey levels plotted against the pixel positions for $16 \times 16$ Head phantom for original phantom (solid line), FBP (dotted line) and HCGA (dashed line)

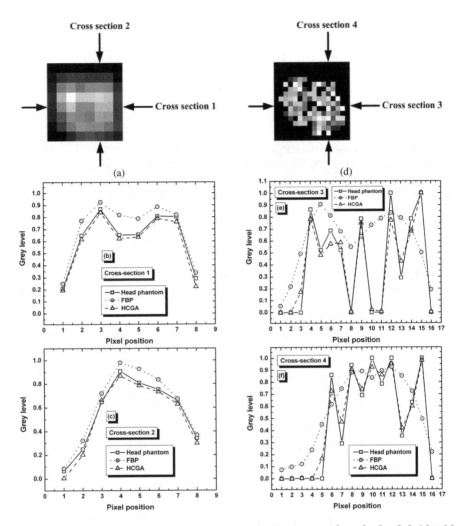

**Fig. 4.** Various cross-sections results during inversion of Radon transform for 8 × 8 & 16 × 16 Lung phantoms by using Filtered Back-Projection (FBP) and Hybrid Continuous Genetic Algorithm (HCGA) techniques with $p = 30$, (a) cross-sections 1 & 2 for 8 × 8 Lung phantom, (b) & (c) cross-sections 1 & 2 showing the grey levels plotted against the pixel positions for 8 × 8 Lung phantom for original phantom (solid line), FBP (dotted line) and HCGA (dashed line), (d) cross-sections 3 & 4 for 16 × 16 Lung phantom, (e) & (f) cross-sections 3 & 4 showing the grey levels plotted against the pixel positions for 16 × 16 Lung phantom for original phantom (solid line), FBP (dotted line) and HCGA (dashed line)

execution time rises with the number of projections and the image size. The abrupt behavior in the initial part may be attributed to the unavoidable statistical fluctuations due to the stochastic nature of the process. The Lung phantom follows the same trend as the Head phantom as shown in Fig. 2(c) & (d). The simulation results of the Head and the Lung phantoms are listed in Table 1 for 8 × 8 and 16 × 16 image-sized phantoms by using RMSE, EE and PSNR as image quality measures. The PSNR value has been found to be 40.47 & 8.28 and 26.38 & 12.98 dB for the 8 × 8 Head and Lung phantoms using HCGA and FBP respectively with the use of optimum number of projections. The same trend is followed for the larger sized image as well.

Fig. 3(a) shows the cross-sections 1 & 2 marked for 8 × 8 Head phantom. The inversion of Radon transform has been estimated by using FBP technique with Matlab ® functions, *radon* and *iradon*, for $p = 30$. The three tracks showing the pixel intensities of the cross-sections 1 & 2 are shown in Fig. 3(b) & (c) for original Head phantom, FBP ($p = 30$) and HCGA ($p = 30$). The results of HCGA with 30 projections have been found more encouraging than FBP with the same number of projections. The FBP is a deterministic technique and its results are strongly dependent on the number of projections used for image reconstruction. The details may be seen in [25]. Similarly Fig. 3(d) shows the cross-sections 3 & 4 marked for 16 × 16 Head phantom. Fig. 3(e) & (f) show the plots of grey levels against the pixel positions for 16 × 16 image-sized cross-sections. The same trend has been found as in the previous case of

**Table 1.** Image reconstruction by using Hybrid Continuous Genetic Algorithm (HCGA) and Filtered Back-Projection (FBP) for 8 × 8 and 16 × 16 Head and Lung phantoms for number of projections $p = 30$ with image quality measures selected as root mean-squared error (RMSE), Euclidean error (EE) and peak signal-to-noise ratio (PSNR)

| Type | Size | Technique | Ref. | RMSE | EE | PSNR (dB) |
|---|---|---|---|---|---|---|
| Head phantom | 8 × 8 | FBP | Matlab® | 0.3855 | 1.5663 | 8.28 |
| | | HCGA | [20] | 0.0095 | 0.0385 | 40.47 |
| | 16 × 16 | FBP | Matlab® | 0.1691 | 0.9779 | 15.44 |
| | | HCGA | [20] | 0.0211 | 0.1223 | 33.49 |
| Lung phantom | 8 × 8 | FBP | Matlab® | 0.2244 | 0.6378 | 12.98 |
| | | HCGA | [20] | 0.0480 | 0.1280 | 26.38 |
| | 16 × 16 | FBP | Matlab® | 0.2312 | 0.6497 | 12.72 |
| | | HCGA | [20] | 0.0796 | 0.2106 | 23.32 |

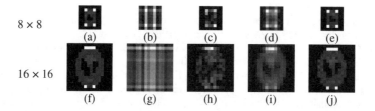

8 × 8     (a)     (b)     (c)     (d)     (e)

16 × 16     (f)     (g)     (h)     (i)     (j)

**Fig. 5.** Reconstructed images for 8 × 8 and 16 × 16 sized Head phantom with Filtered Back-Projection (*radon* and *iradon* functions) and Hybrid Continuous Genetic Algorithm techniques using 2 and 30 projections, 8 × 8 image sized (a) original phantom, (b) & (c) FBP and HCGA reconstructions by using 2 projections, and (d) & (e) FBP and HCGA reconstructions by using 30 projections, and 16 × 16 image sized (f) original phantom, (g) & (h) FBP and HCGA reconstructions by using 2 projections, and (i) & (j) FBP and HCGA reconstructions by using 30 projections

8 × 8

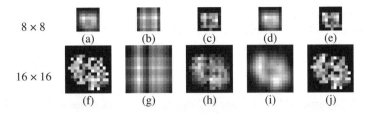

(a)    (b)    (c)    (d)    (e)

16 × 16

(f)    (g)    (h)    (i)    (j)

**Fig. 6.** Reconstructed images for 8 × 8 and 16 × 16 sized Lung phantom with Filtered Back-Projection (*radon* and *iradon* functions) and Hybrid Continuous Genetic Algorithm techniques using 2 and 30 projections, 8 × 8 image sized (a) original phantom, (b) & (c) FBP and HCGA reconstructions by using 2 projections, and (d) & (e) FBP and HCGA reconstructions by using 30 projections, and 16 × 16 image sized (f) original phantom, (g) & (h) FBP and HCGA reconstructions by using 2 projections, and (i) & (j) FBP and HCGA reconstructions by using 30 projections

8 × 8 size phantoms. Fig. 4(a) & (d) show the cross-sections 1 to 4 marked for the 8 × 8 & 16 × 16 Lung phantoms. These cross-sections in Fig. 4(b), (c), (e) & (f) follow the same trend as in the case of Head phantom for optimum number of projections found to be $p = 30$.

Fig. 5(a) & (f) show the original Head phantom for 8 × 8 & 16 × 16 image sizes. Fig. 5(b) & (c) show the results of FBP and HCGA reconstructions with $p = 2$ and Fig. 5(d) & (e) show the results for $p = 30$. Similarly the reconstructed images have been shown in Fig. 5(g) & (h) and (i) & (j) for $p = 2$ & 30 for 16 × 16 Head phantom using FBP and HCGA respectively. Fig. 6(a) & (f) show original 8 × 8 & 16 × 16 Lung phantoms. FBP reconstructions for $p = 2$ & 30, with *radon* and *iradon* functions of Matlab ®, are shown in Fig. 6(b) & (d) for 8 × 8 sized images. Fig. 6(c) & (e) show the results for HCGA with $p = 2$ & 30 for 8 × 8 sized images respectively. Fig. 6(g) & (i) show the FBP reconstructions under the prevailing conditions for the 16 × 16 Lung phantom for $p = 2$ & 30 respectively. Similarly Fig. 6(h) & (j) show HCGA results for $p = 2$ & 30 under conditions defined for the previous simulations. Fig. 5 & 6 show relatively better results for HCGA than FBP with $p = 30$ for the sizes under consideration. As far as the shape complexity is concerned, the complex shapes, like the Lung phantom, has been found requiring more generations in comparison to the simpler shapes, like the Head phantom, as shown in Fig. 2(a) and (c). In the case of the Head phantom, RMSE has been found relatively lower than found for the Lung phantom.

## 4   Conclusions

The optimum number of projections has found to be $p = 30$. The use of smaller number of projections results in a lower radiation dose while yielding a higher quality reconstructed image even with incomplete data set and/or noisy projections data. The comparison of projections at which maximization of PSNR occurs indicates that the image quality is also dependent on the shape-complexity of the cross-section involved since the Lung phantom has been found requiring more generations to acquire the same PSNR as shown in Table. 1.

**Acknowledgments.** Shahzad A. Qureshi gratefully acknowledges the funding by Higher Education Commission, Pakistan under PIN No. 041 103076 Cu-027.

# References

1. Kak, A.C., Slaney, M.: Principles of Computerized Tomographic Imaging. IEEE Press, New York (1988)
2. Mersereau, R.M.: Direct Fourier transform techniques in 3-D image reconstruction. Comput. Biol. Med. 6, 247–258 (1976)
3. Metropolis, N., Rosenbluth, A.W., Teller, A.H., Teller, E.: Equations of state calculations by fast computing machines. Journal of Chemical Physics 21, 1087–1092 (1953)
4. Press, W.H., Teukolsky, S.A., Vetterling, W.T., Flannery, B.P.: Numerical recipes in C: The art of scientific computing. Cambridge University Press, New York (2002)
5. Webb, S.: Optimizing radiation therapy inverse treatment planning using the simulated annealing technique. International Journal of Imaging Systems and Technology 6(1), 71–79 (2005)
6. Trosset, M.W.: What is simulated annealing? Optimization and Engineering 2, 201–213 (2001)
7. Murino, V., Trucco, A.: Markov-based methodology for the restoration of underwater acoustic images. International Journal of Imaging Systems and Technology 8(4), 386–395 (1998)
8. Qureshi, S.A., Mirza, S.M.: Inverse Radon transform-based image reconstruction modalities. In: National Conference on Information Technology and Applications, pp. 13–25. Balochistan University of Information Technology and Management Sciences, Quetta (2005)
9. Holland, J.H.: Outline for logical theory of adaptive systems. J. ACM 3, 297–314 (1962)
10. Holland, J.H.: Adaptation in natural and artificial systems. University of Michigan Press, Ann Arbor (1975)
11. Pastorino, M., Massa, A., Caorsi, S.: A microwave inverse scattering technique for image reconstruction based on a genetic algorithm. IEEE transactions on instrumentation and measurement 49(3), 573–578 (2000)
12. Cheng, K.-S., Chen, B.-H., Tong, H.-S.: Electrical Impedance image reconstruction using the genetic algorithm. In: 18th Annual International Conference of the IEEE Engineering in Medicine and Biology Society, Amsterdam, pp. 768–769 (1996)
13. Franconi, L., Jennison, C.: Comparison of a genetic algorithm and simulated annealing in an application to statistical image reconstruction. Statistics and Computing 7, 193–207 (1997)
14. Christopher, R.H., Jeffery, A.J., Michael, G.K.: A genetic algorithm for function optimization: A Matlab implementation. Technical Report, Raleigh: NSCU, pp. 1–14 (1995)
15. Michalewicz, Z.: Genetic algorithms + Data structures = Evolution programs. Springer, New York (1994)
16. Caorsi, S., Massa, A., Pastrino, M.: A computational technique based on a real-coded genetic algorithm for microwave imaging purposes. IEEE transactions on geoscience and remote sensing 38(4), 1697–1708 (2000)
17. Qureshi, S.A., Mirza, S.M., Arif, M.: Quality of inverse Radon transform-based image reconstruction using various frequency domain filters in parallel beam transmission tomography. Science International 18(3), 181–186 (2006)
18. Shepp, L.A., Logan, B.F.: The Fourier reconstruction of a head section. IEEE Trans. Nucl. Sci. NS-21, 21–43 (1974)

19. White, D.R., Wambersie, A.: Tissue Substitutes in Radiation Dosimetry and Measurement. Technical Report, International Commission on Radiation Units and Measurements, pp. 1–132 (1999)
20. Qureshi, S.A., Mirza, S.M., Arif, M.: A Template Based Continuous Genetic Algorithm for Image Reconstruction. In: 11th IEEE INMIC 2007, Comsats Institute of Information Technology, Lahore (in press)
21. Bessaou, M., Siarry, P.: A genetic algorithm with real-value coding to optimize multimodal continuous functions. Struct. Multidisc. Optim. 23, 63–74 (2001)
22. Inters, X., Ntziachrostos, V., Culver, J., Yodh, A., Chance, B.: Projection access order in algebraic reconstruction technique for diffuse optical tomography. Physics in Medicine and Biology 47, N1–N10 (2001)
23. Brankov, J.G., Yang, Y., Wernick, M.N.: Tomographic Image Reconstruction Based on a Content-Adaptive Mesh Model. IEEE Transactions on medical imaging 23(2), 202–212 (2004)
24. Van Daatselaar, A.N., Tyndall, D.A., Verheij, H., van der Stelt, P.F.: Minimum number of basis projections for caries detection with local CT. Dentomaxillofacial Radiology 33, 355–360 (2004)
25. Qureshi, S.A., Mirza, S.M., Arif, M.: Effect of number of projections on inverse Radon transform-based image reconstruction by using filtered back-projection for parallel beam transmission tomography. Science International 19(1), 5–10 (2007)

# Seasonal to Inter-annual Climate Prediction Using Data Mining KNN Technique

Zahoor Jan[1], M. Abrar[2], Shariq Bashir[3], and Anwar M. Mirza[1]

[1] FAST-National University of Computer and Emerging Sciences, A. K. Brohi Road,
H-11/4, Islamabad, Pakistan
zahoor_jan2003@yahoo.com, anwar.m.mirza@nu.edu.pk
[2] NWFP Agricultural University Peshawar, Pakistan
gulabson2@gmail.com
[3] Vienna University of Technology, Austria
shariqadel@yahoo.com

**Abstract.** The impact of seasonal to inter-annual climate prediction on society, business, agriculture and almost all aspects of human life, force the scientist to give proper attention to the matter. The last few years show tremendous achievements in this field. All systems and techniques developed so far, use the Sea Surface Temperature (SST) as the main factor, among other seasonal climatic attributes. Statistical and mathematical models are then used for further climate predictions. In this paper, we develop a system that uses the historical weather data of a region (rain, wind speed, dew point, temperature, etc.), and apply the data-mining algorithm "**K-Nearest Neighbor (KNN)**" for classification of these historical data into a specific time span. The k nearest time spans (k nearest neighbors) are then taken to predict the weather. Our experiments show that the system generates accurate results within reasonable time for months in advance.

**Keywords:** climate prediction, weather prediction, data mining, k-Nearest Neighbor (KNN).

## 1 Introduction

Seasonal to inter annual (S2I) climate prediction is the recent development of meteorology with the collaboration of oceanography and climatology all over the world. Weather and climate affects human society in all dimensions. In agriculture, it increases or decreases crop production [1], [2]. In water management [3] rain, the most important factor for water resources, an element of weather. Energy sources, e.g. natural gas and electricity are greatly depends on weather conditions. The day-to-day weather prediction is used for decades to forecast few days in advance, but recent developments move the trend from few days to inter annual forecast [4]. The S2I forecast is to forecast climate from months to year in advance. Climate is changing from year to year , e.g. rain/ dry, cold/warm seasons significantly influence society as well as economy. Technological improvements increase the understanding in meteorology that how the different cycles, ENSO (El Niño Southern Oscillation, i.e. the warm and cold and vice versa phenomena of ocean) over the Pacific Ocean and Sea Surface Temperature (SST), affects the

D.M.A. Hussain et al. (Eds.): IMTIC 2008, CCIS 20, pp. 40–51, 2008.

climate of regions world widely. Many countries United State of America (National Ocean and Atmospheric Administration – NOAA), England (MetOffice – Meteorology Department and London Weather Center), Sri Lanka (Department of Meteorology, Sri Lanka), India (India Meteorology Department and National Center for Medium Range Weather Forecasting – NCMRWF), and Bangladesh (Bangladesh Meteorological Department) etc have started the utilization of Seasonal Climate forecast Systems. In 1995, the International Research Institute for climate prediction (IRI) was formed by a group of scientists to overcome the shortcomings of previous work. The main goal and purpose of this institute was to improve performance of existing systems, develop new accurate and sophisticated systems for S2I. The IRI with the collaboration of other international institute for climate prediction developed new models for forecasting. IRI developed dynamical models based on the Atmosphere General Circulation Model (AGCM), e.g. ECHAM3 (Max Planck Institute), MRF9 (National Center for Environmental Prediction – NECP), CCA3 (National Center for Atmospheric Research – NCAR). Other models are Canonical Correlation Analysis (CCA) [5], [6], Nonlinear Canonical Correlation Analysis (NCCA) [7], Tropical Atmosphere Ocean Array TAOA [8], Global Forecast System, Climate Forecast Model[1] are statistical, Numerical and dynamical (Two-tiered). These models use SST as main attribute for forecasting among other climatic attributes. The sources of these attributes [9] are ENSO teleconnections (effects the global climate), Indian and Atlantic Ocean (effect the regional climate). None of these models is accurate for all situations and regions. These systems also use the geographical (longitude and latitude) location to identify the different regions instead of specific country or city, e.g. The Tropical region (the area between 23.5° N and 23.5° S along the equator).

The main purpose of this paper is how to use a data mining technique, "K-Nearest Neighbor (KNN)", how to develop a system that uses numeric historical data (instead of geographical Location) to forecast the climate of a specific region, city or country months in advance.

Data Mining is the recent development in the field of VLDB/ Data warehousing. It is used to discover the hidden interesting patterns in huge databases that are impossible otherwise. Data Mining can be classified by its techniques into three basic types, i.e. Association Rules Mining, Cluster analysis and Classification/Prediction [10]. KNN is classification algorithm that is based on Euclidean distance formula, which is used to find out the closeness between unknown samples with the known classes. The unknown sample is then mapped to the most common class in its k nearest neighbors. Rest of the paper is organized as follows, section two narrate the related work, section three describes the general aspect of KNN algorithm section 4 shows the KNN application in forecast system, section 5 is about the experiment and result, section 6 concludes the work and for the future extension.

## 2  Related Work

A number of tools are available for climate Prediction. All the initial efforts use statistical models. Most of these techniques predict the SST based on ENSO phenomena

---

[1] GFS, CFS and large collection of regional and global climate forecasting system are the developed and run by NOAA. http://www.noaa.gov, http://www.nco.ncep.noaa.gov/pmb

[6]. NINO3 uses simple average for a specific latitude and longitude (5°S-5°N; 150°E-90°W), etc. Canonical correlation analysis [15, 16] is another statistical model that takes data from different oceans, i.e. Indian, Atlantic, Pacific etc) and forecast the SST monthly anomalies (notable changes from routine measurement, very high or very low SST).

Data Mining is recently applied that how the climate effects variation in vegetation [17]. The *Regression Tree* technique is used to find this relation and predicts the future effects of climate on vegetation variability. The Independent Component Analysis is incorporated in Data Mining [18] to find the independent component match in spatio-temporal data specifically for North Atlantic Oscillation (NAO). The Neural Networks using nonlinear canonical correlation analysis [7] are used to find the relationship between Sea Level Pressure (SLP) and Sea Surface Temperature (SST) that how SST is effected by SLP and changes the climate of specific regions.

## 3   K Nearest Neighbor (KNN)

KNN [11] is widely and extensively used for supervised classification, estimation and prediction [10], [12]. It classify the unknown sample **s** to a predefine class $c_i \in C$, $1 < i \leq n$, based on previously classified samples (training data). It is called *Lazy Learner* because it performs the learning process at the time when new sample is to be classified, instead of its counterpart *Eager Learner*, which pre-classifies the training data before the new sample is to be classified. The KNN therefore requires more computation than eager learner techniques. The KNN is however beneficial to dynamic data, the data that changes/updates rapidly.

When new sample **s** is to be classified, the KNN measures its distance with all samples in training data. The *Euclidean distance* is the most common technique for distance measurement. All the distance values are then arranged such that $d_i \leq d_{i+1}$, $i = 1,2,3,....n$. The k samples with the smallest distance to the new sample are known k-nearest neighbors and are used to classify the new sample **s** to the existing class $c_i \in C$, $1 < i \leq n$. The decision of classification depends on the nature of the data. If the data is of categorical nature then simple voting or weighted voting is used for classification. In case of continuous/quantitative data, the average, median or geometric mean is used. The new sample **s** is then classified to $c_i \in C$, $1 < i \leq n$. The process is simplified below.

```
KNN Algorithms:

Step 1: Measure the distance between the new sample s
and training data.
```

$$\left( Euclidean\ Distance,\ D(x_s, y_s) = \sqrt{\sum_{i=0}^{n}(x_i - y_i)^2} \right)$$

```
Step 2: Sort the distance values as d_i ≤ d_{i+1}, select k
smallest samples

Step 3: Apply voting or means according to the applica-
tion
```

**Table 1.** The training dataset with class labels

| x | y | Label |
|---|---|-------|
| 2 | 3 | RED |
| 7 | 8 | RED |
| 5 | 7 | BLUE |
| 5 | 5 | RED |
| 4 | 4 | BLUE |
| 1 | 8 | BLUE |

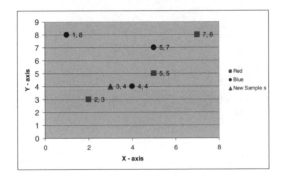

**Fig. 1(a).** The Plotted data in XY plane

Example: A typical example of KNN is classification. Suppose we have the following training data with class C = {RED, BLUE}.

The new test sample is s = (3,4), what will be the class of s for k = 3, k = 5? The data is plotted on XY plane in fig 1(a).

The KNN show the nearest neighbors with k = 3 And k = 5 with respect to new sample *s*. The distance calculation between the training Dataset and new sample s is shown in Table 2.

**Table 2.** The distance calculation between the training Dataset and new sample **s**

| X | Y | Euclidean Simplest form | Distance | Label |
|---|---|-------------------------|----------|-------|
| 2 | 3 | $\sqrt{(3-2)^2 + (4-3)^2}$ | 1.41 | RED |
| 7 | 8 | $\sqrt{(3-7)^2 + (4-8)^2}$ | 5.66 | RED |
| 5 | 7 | $\sqrt{(3-5)^2 + (4-7)^2}$ | 3.61 | BLUE |
| 5 | 5 | $\sqrt{(3-5)^2 + (4-5)^2}$ | 2.24 | RED |
| 4 | 4 | $\sqrt{(3-4)^2 + (4-4)^2}$ | 1.00 | BLUE |
| 1 | 8 | $\sqrt{(3-1)^2 + (4-8)^2}$ | 4.47 | BLUE |

**Table 3.** The sorted list based on distance from Table 2

| S No | Distance | Label |
|------|----------|-------|
| 5 | 1.00 | BLUE |
| 1 | 1.41 | RED |
| 4 | 2.24 | RED |
| 3 | 3.61 | BLUE |
| 6 | 4.47 | BLUE |
| 2 | 5.66 | RED |

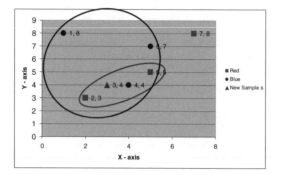

**Fig. 1. (b)** The classification for k = 3 and k = 5

In simple voting based on KNN where k = 3 the number of RED label is greater than the number of BLUE label so the new sample s will be classified as RED while for k = 5, the number of BLUE label is greater than the number of RED label so the s will be classified as BLUE. The process is shown in fig. 1(b)

## 4   Forecasting Using KNN

The dataset for the system was collected from the Pakistan Meteorological Department (PMD) and the National Climatic Data Center (NCDC), which consist of 10 years of historic data with a rich collection of attributes, i.e. temperature, Max Temp, Min Temp, Wind Speed, Max Wind Speed, Min Wind Speed, Dew Point, Sea Level, Snow Depth, Fog, etc. The dataset was noisy with some missing values. The data mining techniques (Means, bin means) were used for data cleansing. The attributes with textual values were replaced with numeric values, i.e. (YES/NO with 1/0, etc.). The final dataset (consisting of 10 years, 17 attributes and 40000 records for 10 cities) was brought in a consistent state and stored in MS ACCESS format.

*Initialization and Prediction range setting:* four basic values need to be initialized. *Current date*, the date from which onward the prediction is sought; *No-of-days-to-forecast*; *k – value* required by KNN; and *Attribute selection*, for which the prediction should be made. Based on this information, the data is extracted from the database and loaded into main memory in appropriate data structures (arrays). We have the following information in main memory.

*Actual Record*, for comparison with predicted values
*Previous Records*, these records are used as a base sequence during Distance Measurement (explained later in this section)
*All Records*, all records in a database for selected attributes and selected regions/City.

*Applying Euclidean Distance (ED)*: Once the data is ready in memory, the data is divided into sequences. Each sequence size is *No of days to forecast* x *selected Attribute*. Thus, the *previous records* become a sequence by itself and will be used as a *base sequence* in ED. The dataset consists of all record for selected attributes are divided into sequences. The number of total sequences is *Total values in all records / total elements in base sequence*. Each sequence is taken and ED is applied element by element with base sequence. The process is summarized

```
Algorithm ED(No_of_Sequences, Size_of_Sequence)
While i < Total_element Do
  For j = 0 to Size_of_Sequence
    Sum = Sum + (Power(BaseSequence[j] - AllSequence[i]),2)
    i++
  End For
Distance = Sqrt(Sum)
End While.
```

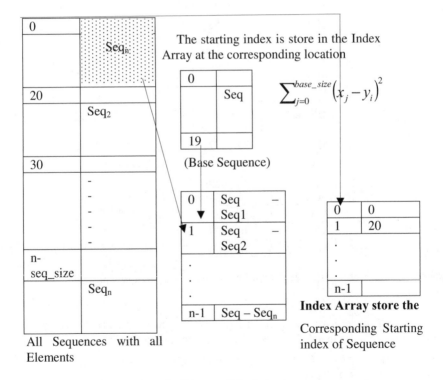

**Fig. 2.** Sequence wise Distance Calculation array management

Example: if *current date* is 20-Jan-1995 and *days to forecast* is 10 with 2 attribute selected for forecast and we have 507 records, then the previous record consist of 10 days back from the current date, i.e. 11-Jan-1995 to 20-Jan-1995, and the size of each sequence will be (10x2) = 20 elements. The total number of sequences will be 507/20 = 25.

The outer loop in the algorithm will take care of the sequences while the inner loop will calculate the formula

$$\sum_{j=0}^{total\_elements} \left(x_j - y_i\right)^2, 0 \leq i < size\_of\_Sequence$$

where $x_j$ is element in *base sequence* and $y_i$ is an element in all sequences. The square root of the sum is taken sequences wise for all sequences and store in *Distance Array* along with starting index in *index Array* of each sequence at the corresponding element. (Fig. 2).

**Table 4.** The k nearest sequences          **Table 5.** Final Predicted values

| Sequences | Days | Attributes | Values |
|---|---|---|---|
| Seq 1 | 1 | Max | 52.2546 |
| | | Min | 35.2457 |
| | 2 | Max | 50.3658 |
| | | Min | 37.2657 |
| | 3 | Max | 48.2689 |
| | | Min | 36.9882 |
| | 4 | Max | 49.8754 |
| | | Min | 32.3655 |
| Seq 2 | 1 | Max | 4.5 |
| | | Min | 34.4587 |
| | 2 | Max | 55.6589 |
| | | Min | 38.4556 |
| | 3 | Max | 51.2154 |
| | | Min | 36.3636 |
| | 4 | Max | 48.1458 |
| | | Min | 32.1554 |

| Days | Attributes | Values |
|---|---|---|
| 1 | Max | 51.50313 |
| | Min | 33.41967 |
| 2 | Max | 50.86453 |
| | Min | 37.01423 |
| 3 | Max | 49.20323 |
| | Min | 36.53553 |
| 4 | Max | 49.4933 |
| | Min | 34.1263 |

Day1.Max = Seq1.Day1.Max + Seq2.Day1.Max;

Day1.Min = Seq1.Day1.Min + Seq2.Day1.Min;

**Fig. 3.** The top k sequences that are used for prediction; the mean is applied attribute wise and day wise

*Sorting of Distance array and index array:* The Distance array and Index array is arrange into such order that $d_i \leq d_{i+1}$ the following algorithm is used to sort the both arrays.

```
Algorithm Sort(DistanceArray, IndexArray, Size)
For i = 0 to size - 1
  For j = 0 to (size - i) - 1
    If Distance[i] > DistanceArray[i+1] then
      Swap DistanceArray[i] and DistanceArray[i+1]
```

```
    Swap IndexArray[i] and IndexArray[i+1]
  End If
 End For
End For.
```

*Applying KNN mean:* Once the arrays are sorted the top – k sequences (the sequences with the smallest differences) are chosen for KNN means and their corresponding values are retrieved using the indices store in IndexArray. The sum is taken, attributes wise and day wise, of all sequences and then its mean is calculated, that is the predicted value for the specific day and attribute.

*Example:* The final prediction for 2 selected attributes (Max Temperature, min temperature), days to forecast = 4 and k = 2, is shown in Fig. 3.

## 5  Experiment

The experiments were performed on two datasets. One mentioned in section 3 had over 40000 records, and a second had 80000+ records. Each had the same number of attributes and belongs to same regions. The accuracy of the results was checked for deferent values of k and different time spans for both dataset. The performance of the system is evaluated on Windows XP, RAM 256 with 733 MHz. The execution time for KNN and prediction was measure from one to seventeen attributes. Fig. 4 show the KNN and Prediction for all attribute. The average time is less than 1 second when seventeen attributes are predicted at the same time.

Fig. 5 shows the load time for database into main memory. Although MS Access provides less overhead on loading the database, still it is long for attribute greater than 5. Fig. 6 show the accuracy of prediction with actual record for Sea Level Pressure (SLP) for 40K dataset with k = 5. The predicted records were checked against the actual records in the database for accuracy, i.e. prediction for 20 days from 15-Feb-2000, and the results were compared to the actual record of the specified time span in the database. Fig. 7 compares the database load time and prediction time for different values of k. the experiment shows that the load time is not effected by deferent values of k, if number of attribute remain constant. The prediction time although varies but remain less than 1 second for all values of k.

The value of K is important and should be selected carefully. It depends on the nature and volume of dataset. In our experiment k = 5 to k = 10 gives accurate results for small (40k) dataset while for large (80k) dataset the results are accurate for k = 35 to k = 40. Figure 8 shows the graphs for different values of k. The attributes are same in all figures to show the difference for both datasets.

The reason for these different values of K is, for large dataset we take the sequences with larger difference that in turn affects the predicted value which is mean of these selected sequences. Thus for small dataset we select small no of nearest neighbors while in large dataset, the small no of sequences are not enough to predict the accurate result and we require to select larger value for k to get accurate results. All odd figures (a, c, e, etc.) belong to 40k dataset while the even figures (b, d, f, etc.) are for 80k. The value of k is same for two dataset in consecutive figures.

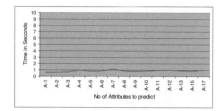

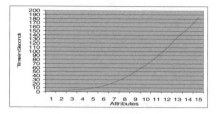

**Fig. 4.** Shows the prediction time for various values of k

**Fig. 5.** Database load time for all attributes

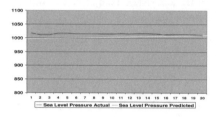

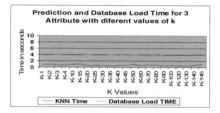

**Fig. 6.** Compare the actual and predicted record for SPL

**Fig. 7.** Database load time and prediction time All Attributes simultaneously

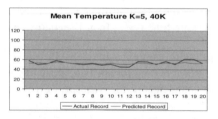

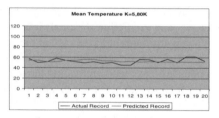

**Fig. 8a.** Mean Temperature for K=5

**Fig. 8b.** Mean Temperature for K=5

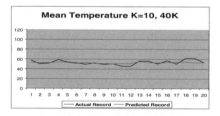

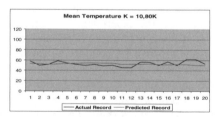

**Fig. 8c.** Mean Temperature for K=10

**Fig. 8d.** Mean Temperature for K=10

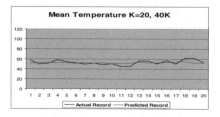

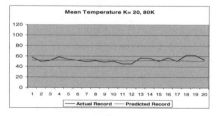

**Fig. 8e.** Mean Temperature for K=80

**Fig. 8f.** Mean Temperature for K=20

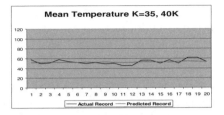

Fig. 8g. Mean Temperature for K=35

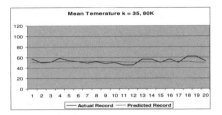

**Fig. 8h.** Mean Temperature for K=35

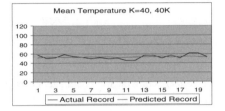

Fig. 8k. Mean Temperature for K=40

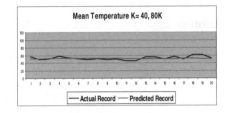

**Fig. 8j.** Mean Temperature for K=40

The accuracy is 96.66% for attributes having Boolean values, i.e. Fog, Hail, snow ice, Thunder etc. and for larger dataset the accuracy is even above 98.33%. Fig. 9a. shows prediction for small (40K) dataset and Fig. 9b. shows the same results for large (80K) dataset. The Figures show the result is 100% for hail and snow ice in both dataset. While for fog the results of 80K is better than 40k. The reason is discrete values of attributes. The results generated by the system are real numbers and were rounded to convert it to yes/no. the values 0.0 to 0.49 were rounded to Zero (No) and 0.5 and above are considered 1 (Yes).

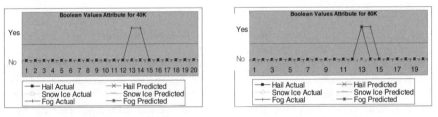

**Fig. 9a.** Boolean values attribute k=10                **Fig. 9b.** Boolean values attribute k=40

## 6  Conclusion and Future Work

The CP-KNN can predict up to seventeen climatic attributes, i.e. Mean Temperature, Max Temperature, Min Temperature, SST, SLP, Gust etc. at the same time. None of the previous developed systems can predict such a huge set of attribute at the same time with such level of accuracy. Recently Climate Prediction Tool (CPT) [from NOAA] has developed that works with multiple attributes but this new version is only used in research labs and not publicly available. The predicted result of CP-KNN is easier to understand as these results are in form YES/NO (for Boolean Attributes) and numeric values. Thus this new system can be used by non-professionals related to any field, e.g. Agriculture, Irrigation, retailers and those specially related to weather sensitive businesses [13].

CP-KNN cannot incorporate to reflect the global changes (ENSO events) but will work correctly with the areas not prone to these global effects. However as these events has some known pattern is advance, e.g. SLP, SST and wind speed etc [6, 14]. It can be modeled using data mining pattern recognition techniques.

# References

1. Hansen, J.W., Sivakumar, M.V.K.: Advances in applying climate prediction to agriculture. Climate Research 33, 1–2 (2006)
2. Sayuti, R., Karyadi, W., Yasin, I., Abawi, Y.: Factors affecting the use of climate forecasts in agriculture: a case study of Lombok Island, Indonesia. ACIAR Technical Reports Series, No. 59, pp. 15-21 (2004)
3. Murray-Ruest, H., Lashari, B., Memon, Y.: Water distribution equity in Sindh Province, Pakistan, Pakistan Country Series No. 1, Working Paper 9, International Water Management Institute, Lahore, Pakistan (2000)
4. Stern, P.C., Easterling, W.E.: Making Climate Forecasts Matter. National Academy Press (1999)
5. Landman, W.A., Mason, S.J.: Change in the association between Indian Ocean sea-surface temperatures and summer rainfall over South Africa and Namibia. International Journal of Climatology 19, 1477–1492 (1999)
6. Landman, W.A.: A canonical correlation analysis model to predict South African summer rainfall. NOAA Experimental Long-Lead Forecast Bulletin 4(4), 23–24 (1995)
7. Hsieh, W.W.: Nonlinear Canonical Correlation Analysis of the Tropical Pacific Climate Variability Using a Neural Network Approach. Journal of Climate 14(12), 2528–2539 (2001)
8. Hays, S.P., Mangum, L.J., Picaut, J., Sumi, A., Takeuchi, K.: TOGA-TAO: A moored array for real time measurement in the tropical Pacific ocean. Bulletin of the American Meteorological Society 72(3), 339–347 (1991)
9. Mason, S.E., Goddard, L., Zebiak, S.J., Ropelewski, C.F., Basher, R., Cane, M.A.: Current Approaches to Seasonal to Interannual Climate Predictions. International Journal of Climatology 21, 1111–1152 (2001)
10. Han, J., Kamber, M.: Data Mining Concepts and Techniques. Elsevier Science and Technology, Amsterdam (2006)
11. Fix, E., Hodges, J.L.: Discriminatory Analysis - Nonparametric Discrimination: Consistency Properties. USAF school of Aviation Medicine, Randolph Field Texas (1951)
12. Larose, D.T.: Discovering Knowledge in Data: An Introduction to Data Mining. Wiley, Chichester (2005)
13. Lettre, J.: Business Planning, Decisionmaking and Ethical Aspects of Seasonal Climate Forecasting (1999), http://members.aol.com/gml1000/busclim.html
14. Mason, S.J., Goddard, L., Graham, N.E., Yulaeva, E., Sun, L., Arkin, P.A.: The IRI seasonal climate prediction system and the 1997/1998 El Niño event. Bulletin of the American Meteorological Society 80, 1853–1873 (1999)

15. Landman, W.A., Mason, S.J.: Forecasts of Near-Global Sea Surface Temperatures Using Canonical Correlation Analysis. Journal of Climate 14(18), 3819–3833 (2001)
16. Rogel, P., Maisonnave, E.: Using Jason-1 and Topex/Poseidon data for seasonal climate prediction studies. AVISO Altimetry Newsletter 8, 115–116 (2002)
17. White, A.B., Kumar, P., Tcheng, D.: A data mining approach for understanding control on climate induced inter-annual vegetation variability over the United State. Remote sensing of Environments 98, 1–20 (2005)
18. Basak, J., Sudarshan, A., Trivedi, D., Santhanam, M.S.: Weather Data Mining using Component Analysis. Journal of Machine Learning Research 5, 239–253 (2004)

# Application of a Proposed Efficient Smoothing Algorithm on Directional Vectors for Isolated Handwritten Characters

Zafar M. Faisal[1], M. Dzulkifli[2], Abdul Rab[3], and Otman M. Razib[2]

[1] Informatics Complex (ICCC), H-8/1, Islamabad, Pakistan
[2] FSKSM, University Malaysia
[3] Islamia College Civil Line Lahore, Pakistan
hmfzafar@iccc.org.pk

**Abstract.** This paper describes an online isolated character recognition system using advanced techniques of pattern smoothing and direction feature extraction. The composition of direction elements and their smoothing are directly performed on the online trajectory, and therefore are computationally efficient. We compare recognition performance when direction features are formulated using smoothed and unsmoothed direction vectors. In experiments, direction features from the original pattern yielded inferior performance, while primitive sub-character direction features, using smoothed direction encoded vectors, made a large difference. The average recognition rates obtained were about 75% using original direction vectors, and about 86% using the smoothed direction vectors.

**Keywords:** Online character recognition, direction vector encoding, smoothing, feature extraction, neural networks.

## 1 Introduction

Computer recognition of pen-input (online) handwritten characters is involved in various applications, like text editing, online form filling, note taking, pen interface, and so on. Handwriting using a pen needs reliable recognition rates to be accepted as an alternative input modality [1]. A great deal of research works have contributed to online character recognition since the 1960s and many effective methods have been proposed [2], [3]. The methods include dynamic time warping, stroke segment matching, hidden Markov models (HMMs), artificial neural networks (ANNs), etc. We used ANN for classification and recognition purposes as ANNs in general obtained better results than HMMs, when a similar feature set is applied [4]. We did not implement discriminatively trained dimensionality reduction and classification methods [5] since the aim of this work is to evaluate smoothing and feature extraction methods.

A variety of experimental set ups and paradigms are currently employed in the online handwriting recognition for taking the input data of a character and for extracting features of interest out of it. As elaborated in this study, the data acquired by these devices suffer from unavoidable noise and artifact problems, which critically affect the results. Noisy data deform the actual path traced by the digital device. It is common practice to reduce the noise by smoothing. The smoothing thus uncovers an

D.M.A. Hussain et al. (Eds.): IMTIC 2008, CCIS 20, pp. 52–63, 2008.
© Springer-Verlag Berlin Heidelberg 2008

underlying "skeleton" consisting of an alternation of arrests and motion segments whose dynamic features can subsequently be analyzed [6].

Every smoothing technique also has its drawbacks. Most notably it may distort the actual path by making it too smooth. If the true underlying process is actually smooth, the path will not be much distorted while the noise will be reduced; but an actual rough path will become smooth. To overcome this problem, the level of smoothing can be controlled in most methods, and therefore its choice becomes an important decision reflecting the goals of the analysis. Typically, a researcher applies one smoothing method, tuning it to suit best the data and the questions of interest [6].

This paper describes an online character recognition system for handwritten upper-case English characters and reports our results using direction feature with and without smoothing direction vectors. In general, handwriting recognition generates high-dimensional problems, and is difficult due to the wide variability of writing styles and the confusion between similar characters [7, 8, 9].

The input pattern trajectory, composed of the direction vectors of filtered pen-down points, is smoothed by an efficient proposed algorithm. We did not try sophisticated smoothing, trajectory resampling and normalization operations. In normalization, the coordinates are transformed such that the pattern size is standardized and the shape is regulated [10].

This paper is organized as follows. Section 2 describes the system overview which includes data acquisition process, filtering methods, 8-direction vector encoding, novel proposed method of smoothing and feature exaction. In Section 3 the experimental results are provided, with some analysis and discussion. In Section 4, concluding remarks and plans for future work are provided.

## 2   System Overview

In this work, on-line data is converted into series of x, y coordinated first. Some pre-processing is applied to the data and then filtered data is encoded into direction

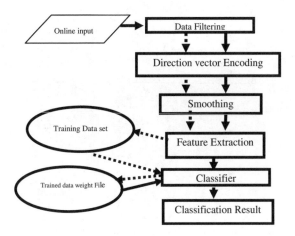

**Fig. 1.** Block diagram of the system

vector. A proposed smoothing algorithm smoothes the encoded sequence of direction vectors before extracting the features. These features are then fed to the input of back propagation neural networks (BPN) for training and recognition purposes. An overview of the whole system is presented in Fig. 1. In following sections, the techniques and algorithms to be involved for each processing block are briefly introduced.

## 2.1  Data Collection

The data used in this work was collected using tablet Summa Sketch III with a sampling rate of 80 points/second. Collecting the data requires writers to look at the monitor while writing on the tablet to see the graphically emulated inking. Thus, the letter data can be considered more natural. Instructions to the writers were reduced to a minimum to guarantee the most natural way of writing. The data obtained is stored in $(x, y)$ coordinates.

## 2.2  Data Filtering

Recognition results strongly depend on the quality of data provided to the recognizer. Ideally, the recognizer would get the data which contains only the features it needs for recognition. Strokes captured by a digitizing tablet tend to contain a small amount of noise, most likely introduced by the digitizing device, causing the sampled curves to be somewhat jagged. Removing irrelevant data decreases the amount of work necessary to locate the features and, more importantly, it reduces the possibility of incurring errors in classifying these features. Great care must be taken not to remove too much information, as this will affect the recognition [11]. It is perceived here that the filtered data ought to be smoothed and contain as few data points as possible, maintaining the appearance of the original data. In this research two filters: distance filter and cross product filter have been used. The distance filter removes the identical consecutive data points from input trajectory. The cross product filter approximates curved strokes by line segments. It selects three points from the input stroke (Fig. 2a).

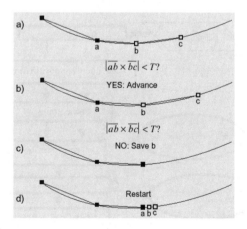

**Fig. 2.** Cross product filter algorithm. Adapted from [11].

One point denotes which input point last went to the output (point a), one tells where the algorithm is (point b) and one tells where the algorithm is going (point c). Point b is the candidate for output, and it is most likely to be accepted when it is at a corner. Having point *b* at a corner maximizes the area of triangle *abc*, which is one half the cross product of line segments *ab* and *bc*.

### 2.3   Direction Vector Encoding

Direction vector encoding (DVE) was introduced by Herbert Freeman [12]. It is a mean of describing line drawings in an approximate way using a number of vectors of quantised directions. Direction of each line segment is approximated to the nearest direction available within the vector encoding. When a line changes direction, another approximation is used. The result is a chain of vectors capable of representing any arbitrary line drawing. Handwriting can be perceived as a line drawing. This is particularly the case for dynamic data, which can be represented by polylines of various complexities. DVE well represents the angular variation of the pen path and is used in the segment and recognize approaches. At least eight direction encoding is necessary in order to represent horizontal, vertical and diagonal directions [11]. Thus the current recognition system uses eight vector directions (Fig. 3).

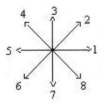

**Fig. 3.** Eight Vectors Encoding

All data samples were encoded using eight number of direction vectors. Fig. 4 presents results for some samples. The encoding was performed on filtered data. It can be seen eight directions encoding (Fig. 4b) provides quite a good approximation of the original data. Both vertical and horizontal directions are present making it easier to represent vertical strokes and horizontal ligatures. As the number of coding directions will increase, the encoding will provide better approximation of the original data at the expense of increased length and variability of the coding sequence.

**Fig. 4.** Examples of the vector direction encoding using eight number of vector directions: a) original characters, b) encoded characters

For eight direction encoding the following measures have been taken: Horizontal direction along x-axis (at angle 0) is considered as 1. Directions at angles 45, 90, 135,180, 225, 270, 315 are encoded as 2, 3, 4, 5, 6, 7, and 8 respectively. When a character is drawn, the value of angle $\theta$ is calculated by the relation:

$$\theta = \tan^{-1} \frac{y}{x} \tag{1}$$

As there are 4 quadrants, therefore, the values of $\theta$ in 1st and 3rd quadrants are negative. To code the angle space between two angle values, a threshold is fixed. The value of $\theta$ below this threshold is coded to lower nearest value and above it to upper nearest value. For instance, the threshold between 0° (0 rad.) and 45° (0.785 rad.) is taken as 22° (0.384 rad.). The $\theta$ value between 0° (0 rad.) and 22° (0.384 rad.) is encoded as '1' and between 22.1° and 45° is coded as '2' and so on. In terms of radian it will look like as:

Code = 1 when $-0.01 < \theta < -0.385$
Code = 2 when $-0.385 < \theta < -0.786$

The length of each vector is measured using Euclidean distance:

$$d = \sqrt{(x_2 - x_1)^2 + (y_2 - y_1)^2} \tag{2}$$

Also slope of a line with points $(x_1, y_1)$ and $(x_2, y_2)$ is given by

$$\tan \theta = \frac{y_2 - y_1}{x_2 - x_1} \tag{3}$$

or
$$y_2 - y_1 = (x_2 - x_1) \tan \theta$$

Putting this value of $y_2 - y_1$ in (2):

$$d = \sqrt{(x_2 - x_1)^2 + ((x_2 - x_1) \tan \theta)^2}$$

or
$$d = (x_2 - x_1)\sqrt{1 + (\tan \theta)^2}$$

or
$$x_2 - x_1 = \frac{d}{\sqrt{1 + (\tan \theta)^2}}$$

or
$$x_2 = x_1 + \frac{d}{\sqrt{1 + (\tan \theta)^2}} \tag{4}$$

Similarly
$$y_2 = y_1 + \frac{d}{\sqrt{1 + \frac{1}{(\tan \theta)^2}}} \tag{5}$$

As the values of $x_1$, $y_1$, $d$, and $\theta$ are calculated from the previous points $(x_1, y_1)$ and $(x_2, y_2)$, the new value of $x_2$ and $y_2$ is measure by (4) and (5). In this way, the exact length of the segment remains same i.e. Euclidean distance $d$ after being encoded.

## 2.4  Smoothing

Smoothing is a technique to suppress the quantization noise of the point sampling, which averages an input point with its neighbors with some weighted mask [1]. The goal of smoothing is to reduce or eliminate some of the variations in handwriting that may exist and not useful for pattern class discrimination. Strokes captured by a digitizing tablet tend to contain a small amount of noise, most likely introduced by the digitizing device, causing the sampled curves to be somewhat jagged.

While it is a common statistical practice to reduce the noise by smoothing, not all smoothing methods were created equal. The most common smoothing technique used to reduce system noise is the Moving Average. This method uses a sliding window, shifting along the time series of coordinates from one data point to the next. The value of each data point is replaced by the average of the $n$ points surrounding it, $n$ is called the width of the smoothing "window" [6, 13, 14, 15]. The wider the window is, the more data points are used in the evaluation of each data point, leading to 'stronger' smoothing. Fig. 5 provides a good example of Moving Average smoothing and the effect of different choices of window sizes [6].

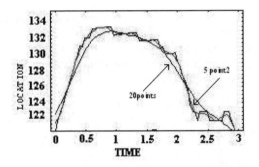

**Fig. 5.** Horizontal location vs. Time plot smoothed by Moving Average with window sizes of 5 and 20 points

The average used in this method can be the simple average, giving equal weights to each data point in the window (as is done in Fig. 5). A weighted average, which assigns large weight to data points in the center of the window, and gradually decreasing weights while getting further from the center, may alternatively be used. The result of such an operation is a smoothed time series, implying that noise is substantially reduced.

Another commonly used smoothing technique is the Local Polynomials method [16]. This method also uses a window of a pre-specified size around each data point in turn. However, instead of assigning each point the average of its surrounding neighbors (as the Moving Average does), this method uses a simple or weighted least squares fit of a low order polynomial (usually a line or a parabola) to the data points in each window. Once such a polynomial is found, the data point at the center of the window is replaced by the value of the polynomial at that point. As in the Moving Average, the window size affects, visually, the smoothed sequence of locations [6].

Some other forms of smoothing are often applied using spline filter [6], or a Yule-walk filter for low-pass filtering of the data [17] and Gaussian filter [1, 18] etc. But these filters sometimes involve complex mathematics and are applied on the initial data which is in the form of $x$, $y$ coordinates. To remove jitter from the handwritten text, Jaeger *et al.* [19] replaced every point $(x(t), y(t))$ in the trajectory by the mean value of its neighbors:

$$x'(t) = \frac{x(t - N) + \ldots + x(t - 1) + \alpha x(t) + x(t + 1) + \ldots + x(t + N)}{2N + \alpha} \tag{6}$$

and

$$y'(t) = \frac{y(t - N) + \ldots + y(t - 1) + \alpha y(t) + y(t + 1) + \ldots + y(t + N)}{2N + \alpha} \tag{7}$$

The parameter $\alpha$ is based on the angle subtended by the preceding and succeeding curve segment of $(x(t), y(t))$ and is empirically optimized. Babu *et al.* [15] used moving average window of fixed size for smoothing the online handwritten data. Using moving averages rather than raw totals has the effect of smoothing out peaks in the data [13].

In this research, a smoothing algorithm has been proposed which is quite simple in nature and applied on the direction encoded data of the input character. It involves simple mathematics and has enhanced overall results by about 10%. For a sequence $C = \langle p_1, p_2, \ldots, p_i, \ldots, p_n \rangle$ of direction code of a character, the proposed algorithm smoothes the encoded data in the following steps.

1. When both neighbors of an element in the sequence are same and it has difference of absolute $1^1$ with both, then change it equal to value of its neighbors e.g. 323323 to 333333. In symbolic form:

   if $p_i \mathbin{!=} p_{i+1} \wedge p_{i-1} = p_{i+1} \wedge |p_i - p_{i+1}| = 1$
   $$\Rightarrow p_i = p_{i+1}, i = 2,3,\ldots n-1$$

   See effects in Fig. 6 (a), (b), and (d).

2. Search for two consecutive sequence elements which are equal in value but different from their respective neighbors by a value of absolute $1^1$, when neighboring pair on both sides are same, then change these two elements equal to value of its neighbors e.g. 445544 to 444444. In symbolic form:

   if $p_i = p_{i+1} \wedge p_{i-1} = p_{i-2} = p_{i+2} = p_{i+3} \wedge |p_i - p_{i+1}| = 1$
   $$\Rightarrow p_i = p_{i+1} = p_{i+2}, i = 3,\ldots n-4$$

   See effects in Fig. 6 (e).

3. Search for the element of sequence which is different from its left side pair of neighbors by a value of absolute $1^1$, and also different from right side pair, when elements in that left neighboring pair are same, but different in right neighboring pair then change it equal to value of its left neighbors e.g. 5541723 to 5551723. In symbolic form:

   if $p_i \mathbin{!=} p_{i+1} \wedge p_{i-1} = p_{i-2} \wedge p_{i+2} \mathbin{!=} p_{i+3} \wedge |p_i - p_{i-1}| = 1$
   $$\Rightarrow p_i = p_{i-1}, i = 3,\ldots n-4$$

   See effects in Fig. 6 (a) and (f).

---

[1] In case of consecutive 1 and 8 in either order, this difference will be considered as absolute 7.

4. Search for the element of sequence which is different from its right side pair of neighbors by a value of absolute $1$[1], and also different from left side pair, when elements in that right neighboring pair are same, but different in left neighboring pair then changes it equal to value of its right neighbors e.g. 5551233 to 5551333.

$$\text{if } \begin{array}{l} p_i != p_{i-1} \Lambda p_{i-1} != p_{i-2} \Lambda p_{i+1} = p_{i+2} \Lambda |p_i - p_{i+1}| = 1 \\ \Rightarrow p_i = p_{i+1}, i = 3, \dots n-3 \end{array}$$

See effects in Fig. 6 (a), (b), (c), (d) and (f).

5. Search for the element of sequence which is different from its left side neighboring pair by a value of absolute $2$[2], and also different from left side pair, when elements in that left and right neighboring pairs are same inside the pair but different among the pairs, then change it equal to value of its left neighbors e.g. 2224777 to 2222777. In symbolic form: if

$$\begin{array}{l} p_i != p_{i+1} \Lambda p_{i-1} = p_{i-2} \Lambda p_{i+1} = p_{i+2} \Lambda p_{i-1} != p_{i+1} \Lambda |p_i - p_{i-1}| = 2 \\ \Rightarrow p_i = p_{i-1}, i = 3, \dots n-4 \end{array}$$

See effects in Fig. 6 (c).

6. Search for two consecutive sequence elements which are equal to each other but different from their respective neighbors, when neighboring pair on both sides are

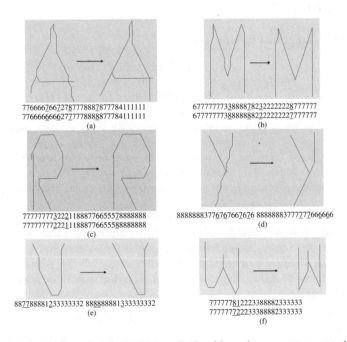

776666766727877788787777841111111
776666666627777788887778411111111
(a)

6777777733888878232222222287777777
677777773888888822222222277777777
(b)

77777777732221188877665557888888
77777777722211188877665558888888
(c)

8888888377676766676 88888883777777666666
(d)

88778888123333332 8888888813333333332
(e)

77777781222338882333333
77777772223388882333333
(f)

**Fig. 6.** Smoothing from left to right by Proposed Algorithm: characters representation graphically and numerically in 8-directional code

[2] In case of consecutive 2 and 8 in either order, this difference will be considered as absolute 6.

different by a value of absolute $1^3$, then change one of two elements equal to value of its neighbors e.g. 77733888 to 77773888. In symbolic form: if

$$p_i = p_{i+1} \Lambda p_{i-1} = p_{i-2} \Lambda p_{i+2} = p_{i+3} \Lambda |p_{i-2} - p_{i+2}| = 1$$
$$\Rightarrow p_i = p_{i-1}, i = 3,....n-4$$

See effects in Fig. 6 (c).

### 2.6 Feature Extraction

In direction features (also called sub-character primitives), the character shape is divided into equally spaced intervals and the line orientation code corresponding to each interval is computed. Usually 8 directional code is used. One advantage of the sub-character primitives is that they are more concise due to their encoded nature. Another is that they offer a higher-level view because they represent a region not a point.

DVE algorithm was applied to find out the vector direction code for the character, as stated earlier. Feature extraction scheme used in this work is based on the encoded direction vectors [20], as they carry useful information to process, which can be used for character recognition. They are produced by considering different combination of local as well as global traits of every character.

## 3 Experiments

### 3.1 Data Set and Model Parameters

The data used in this work was collected using tablet SummaSketch III. It has an electric pen with sensing writing board. An interface was developed to get the data from tablet and it was stored in UNIPEN [21] format. In the data set, the total number of handwritten characters is about 2000 characters used for training purposed, collected from 30 subjects. More than 1000 handwriting samples were used for testing purpose. Experiments were examined by extracting a feature vector of 22 elements from each character. Each subject was asked to write on tablet board (writing area). No restriction was imposed on the content or style of writing. The writers consisted of university students (from different countries), professors, and employees in private companies.

Recent experiments are based on the UNIPEN Train-R01/V07 database. Data was extracted from Section 1a and 1b of UNIPEN benchmark Data CD. About 1000 digits, lower case, and upper case characters were used for training and testing purposes. The experimental data was divided in two different categories; with smoothing and without smoothing, to deduce the results of our proposed smoothing algorithm.

### 3.2 Recognition Performance

For classification purpose, BPN has been used. Experiments were performed using data sets of 11, 22, 33, 44, 55 and 66 samples/character separately for direction features using smoothed and unsmoothed direction encoded vectors. Table 1 presents the

---

[3] In case of 1 and 8 in either side, this difference will be considered as absolute 7.

**Table 1.** Performance of BPN using input data with and without Smoothing

| Sample/ Character | Data without Smoothing | | | Data with Smoothing | | |
|---|---|---|---|---|---|---|
| | *CRs* | *FRs* | *RFs* | *CRs* | *FRs* | *RFs* |
| **11 Each** | 60.0% | 10.3% | 29.7% | 68.4% | 18.3% | 13.3% |
| **22 Each** | 64.5% | 15.3% | 20.2% | 70.2% | 24.8% | 5.0% |
| **33 Each** | 67.3% | 14.4% | 18.3% | 74.5% | 20.3% | 5.2% |
| **44 Each** | 70.4% | 14.1% | 15.5% | 80.1% | 12.6% | 7.3% |
| **55 Each** | 71.2% | 13.8% | 15.0% | 80.6% | 10.3% | 9.1% |
| **66 Each** | 75.0% | 12.0% | 13.0% | 86.1% | 9.0% | 4.9% |

**Table 2.** Performance of BPN using UNIPEN data

| Characters | Data without Smoothing | | | Data with Smoothing | | |
|---|---|---|---|---|---|---|
| | *CRs* | *FRs* | *RFs* | *CRs* | *FRs* | *RFs* |
| **1000** | 74.0% | 11.3% | 14.7% | 82.4% | 7.3% | 10.3% |

statistics; **CRs**, **FRs**, and **RFs** are abbreviation for Correct Recognitions, False Recognitions, and Recognition Failures respectively.

### 3.3 Performance Analysis

For developed BPN models, it was observed that learning became more difficult with the increase of training samples and, even after long time of training, models were unable to fully learn the training sets. Number of hidden PEs from 30 to 35 was found suitable. Value of learning rate < 1 appeared appropriate for training. Generally, the recognition performance of BPN models improved with increase in samples/character [20]. The recognition rate with smoothing was higher (up to 86.1%) than that of without smoothing (up to 75%) showing almost 11% improvements in recognition. Similar results (about 8% improvement) were obtained for bench mark UNIPEN data (more experiments are being carried out). This means that smoothing has ultimately improved the recognition rate. While in case of Jaeger *et al.* [19], smoothing has improved overall recognition rate by about 0.5%.

## 4   Conclusions

A variety of experimental setups and paradigms are currently employed in the online handwriting recognition for taking the input data of a character and for extracting features of interest out of it. Many of these features, however, especially those relating to the dynamics of movement, are critically sensitive to the noise and artifacts of trajectory. To remove the noise, etc., filtering and proposed smoothing algorithm have been implied. The composition of direction elements and their smoothing are directly performed on online trajectory, and therefore, are computationally efficient. Direction features for written character produced could then be stored for training purposes or can be supplied to a developed model for making a decision regarding the class of the character. The preliminary results are quite encouraging and indicate that this research is quite promising and proves to be worthy of further investigation.

## References

1. Oh, J.: An On-Line Handwriting Recognizer with Fisher Matching, Hypotheses Propagation Network and Context Constraint Models. PhD thesis, Department of Computer Science New York University, USA (2001)
2. Liu, C.L., Jaeger, S., Nakagawa, M.: Online recognition of Chinese characters: The state-of-the-art, IEEE Trans. Pattern Anal. Mach. Intell. 26(2), 198–213 (2004)
3. Liu, C., Zhou, X.: Online Japanese Character Recognition Using Trajectory-Based Normalization and Direction Feature Extraction. In: Proceedings of 10th International Workshop on Frontiers in Handwriting Recognition, Atlantia Congress Center, La Baule, France (2006)
4. Marcelo, N.K., Cinthia, O., de Freitas, A., Sabourin, R.: Methodology for the design of NN-based month-word recognizers written on Brazilian bank checks. Image and Vision Computing 25, 40–49 (2007)
5. Liu, C.L.: High accuracy handwritten Chinese character recognition using quadratic classifiers with discriminative feature extraction. In: Proc. 18th ICPR, Hong Kong (2006)
6. Hen, I., Sakov, A., Kafkafi, N., Golani, I., Benjamini, Y.: The Dynamics of Spatial Behavior: How can robust smoothing techniques help? J. Neuroscience Methods 133(1-2), 161–172 (2004)
7. Plamondon, R., Privitera, C.M.: The Segmentation of Cursive Handwriting: An Approach Based on Off-Line Recovery of the Motor-Temporal Information. IEEE Trans. Image Processing 8(1), 80–91 (1999)
8. Plamondon, R., Sargur, N.S.: On-Line and Off-Line Handwriting Recognition: A Comprehensive Survey. IEEE Transactions on PAMI 22(1), 63–84 (2000)
9. Xiaolin, L., Yeung, D.Y.: On-line Handwritten Alphanumeric Character Recognition Using Dominant Points in Strokes. Pattern Recognition 30(1), 31–44 (1997)
10. Liu, C.L., Nakashima, K., Sako, H., Fujisawa, H.: Handwritten digit recognition: investigation of normalization and feature extraction techniques. Pattern Recognition 37(2), 265–279 (2004)
11. Powalka, R.K.: An algorithm toolbox for on-line cursive script recognition. PhD thesis, Nottingham Trent University (1995)
12. Freeman, H.: Computer processing of line-drawing images. Computing Surveys 6(1), 57–97 (1974)
13. Souter, C., Churcher, G., Hayes, J., Hughes, J., Johnson, S.: Natural Language Identification using Corpus-Based Models. Hermes Journal of Linguistics 13, 183–204 (1994)
14. IPredict It: Forecasting Made Easy, http://www.ipredict.it/Methods/MovingAverage.aspx

15. Babu, V.J., Prasanth, L., Sharma, R.R., Bharath, A.: HMM-based Online Handwriting Recognition System for Telugu Symbols. In: Proceedings of Ninth International Conference on Document Analysis and Recognition (ICDAR 2007), pp. 63–67 (2007)
16. Fan, J., Gijbels, I.: Local Polynomial Modeling and Its Applications. Chapman and Hall, Boca Raton (1996)
17. Scott, D.C.: Online Handwriting Recognition Using Multiple Pattern Class Models. PhD Thesis, Dept. of Computer Science and Engineering, Michigan State University, USA (2000)
18. Joshi, N., Sita, G., Ramakrishnan, A.G., Madhvanath, S.: Tamil handwriting recognition using subspace and DTW based classifiers. In: Pal, N.R., Kasabov, N., Mudi, R.K., Pal, S., Parui, S.K. (eds.) ICONIP 2004. LNCS, vol. 3316, pp. 806–813. Springer, Heidelberg (2004)
19. Jaeger, S., Manke, S., Reichert, J., Waibel, A.: Online handwriting recognition: The NPen++ recognizer. International Journal on Document Analysis and Recognition 3, 169–180 (2001)
20. Zafar, M.F., Mohamad, D., Anwar, M.M.: Recognition of Online Isolated Handwritten Characters by Backpropagation Neural Nets Using Sub-Character Primitive Features. In: Proceedings of 10th IEEE International Multitopic Conference on Information Technology (INMIC 2006), Islamabad, Pakistan (2006)
21. The UNIPEN Project, http://unipen.nici.ru.nl/unipen.def

# Architecture Based Reliability and Testing Estimation for Mobile Applications

Vickey Wadhwani[1], Fiaz Memon[2], and M. Muzaffar Hameed[1]

[1] Blekinge Institute of Technology
Department of Systems and Software Engineering, Sweden
vickeywadhwani@gmail.com
[2] University of Sindh, Jamshoro
Institute of Mathematics and Computer Science
fiazmemon@yahoo.com, muzaffar@muzaffar.net

**Abstract.** We propose an architecture-based testing and reliability framework for mobile applications. During our literature study, we explored some of the software testing and reliability techniques available, as well as investigating the techniques that are being used by industry. This paper presents case studies of two companies that are developing mobile applications. These are helpful for finding the strengths and weaknesses of existing software testing and reliability techniques. In light of these strengths and weaknesses, we propose a framework of architecture-based software testing and reliability estimation for mobile applications.

**Keywords:** Software Architecture, Reliability, Software Testing, Architecture-based Estimation, Case Study.

## 1 Introduction

The present age is very competitive and the formula for success is reusability, because reusability helps in saving time and money. Same is true for the software industry. Software firms need good engineering techniques to progress in this competitive era. Good software engineering techniques increase the uses of reusable components. Reusability of software component helps in reducing cost and time of software development. Presently large proportion of development cost is allocated for the software testing phase. This cost can be reduced if implementation of testing can be performed at the early stage of Software Development Life Cycle (SDLC) [1]. At an early stage, we have architecture as an artefact on which we can perform testing. This not only reduces the cost but also helps in increasing the quality of a software system. With the architecture, we can analyze quality attributes such as reliability, modifiability, security and performance at an early stage of SDLC.

"The software architecture of a program or computing system is the structure or structures of the system, which comprise of software elements, the externally visible properties of those elements, and the relationships among them" [2]. Software architecture is described by a software structure at an abstract level consisting of a set of components, connectors and configurations [3]. Software architecture is the main

D.M.A. Hussain et al. (Eds.): IMTIC 2008, CCIS 20, pp. 64–75, 2008.
© Springer-Verlag Berlin Heidelberg 2008

pillar and primary carrier of a software system's quality attributes like performance and reliability [4]. Design of software architecture is an activity at an early stage of SDLC, which creates better software systems by modeling their important aspects in the initial phase of development [5].

"Testing is one of the most expensive activity in the development of complex software systems and represents a mandatory activity to improve the dependability of complex software systems" [6]. Presently, the testing of the software systems is performed by V model like unit, integration and system test, which execute in a bottom-up order to the code, design and specification stages of the development part of the lifecycle [8]. The V model is an extension of the waterfall model. In the V model, each phase of software development life cycle (SDLC) is associated with the corresponding testing phase. Formalized software architecture design languages provide a significant opportunity for testing because they precisely describe how the software is supposed to behave in a high-level view. This allows test engineers to focus on the overall system structure, a form that can be easily manipulated by automated means [5]. Presently, there is a lack of testing techniques for testing at the software architecture level [5].

Software Reliability can be defined as *"the probability of failure-free software operation for a specified period of time in a specified environment"* [4]. The execution sequence of states and individual states determine the reliability of a software system [8]. Software reliability is an important quality attribute. Improving this attribute early in the software life cycle is highly desirable, because it greatly reduces testing and maintenance efforts in the latter stages [9].

Structure of the paper is as follow: Section 2 deals with related work. Section 3 describes architecture-based software testing techniques and architecture-based reliability is explained in section 4. Section 5 deals with qualitative research strategy. Section 6 presents a framework for architecture – base reliability and testing estimation for mobile applications. Section 7 describe conclusion and future work and section 8 list all references used in this paper. Appendix A presents important questions from case study.

## 2 Related Work

Today software architecture is used by many software firms. Many previous studies have shown that software architecture is an important process in SDLC [10], [11]. Architecture provides good understanding of software components that can be re-used in other software products. In the area of architecture based reliability we have found that state-based models, path-based models and additive models are commonly used [12] [13]. A tool for architecture bases reliability is also developed, which follows a white-box approach, following system structures to construct a finite state model. Reliability is then calculated using the state model through a number of matrix computations, such as cofactor, transpose, and determinant [14].

Research is also being conducted in the area of architecture based testing. Different techniques have been proposed in this area. Some of the famous architecture based testing techniques are white box, black box and model-based testing [5], [15]. These techniques can be used for mobile applications.

# 3   Architecture-Based Software Testing Techniques

Testing is an important activity and it is used to improve the quality of software systems but the resources (cost, time and human) consumed by the testing are more than any other activities of the SDLC [6], [16]. This is the reason, why architecture based testing is getting more and more focus from software researchers and practitioners. Architecture based testing is a particular form of specification based testing, which is devoted to check whether the Implementation Under Test (IUT) fulfils the architectural specifications or not [5]. Two general types of testing approaches are used for architecture based testing, i.e. black box testing and white box testing. The decision about when to stop a testing operation is an important consideration for software testing [5], [16]. Most of current testing techniques are either based on implementation or structural information of the system, or based on requirement specification or system design [5], [17].

## 3.1   Issues in Software Architecture-Based Testing

Like all testing techniques, it is very important to know about what to test at the software architecture level; therefore, testing requirements can be defined at this level. Most of the unit testing techniques utilizes program structure to define testing criteria like conditions or data flow, etc. Another issue is about when to stop testing at architecture level [5].

## 3.2   General Properties to be Analyzed and Tested at Architecture-Level

Since we are going to propose our own architecture based testing framework, therefore, first we will list all problems that exist in earlier testing techniques. Following are the requirements for architecture based testing [5].

1. Testing the connectivity and compatibility from a component interface to a connector interface and vice versa [5].
2. Testing the possible data transfer, control transfer and ordering relations among the interfaces within a component and components within the interfaces [5].
3. Testing the connection of two components through a connector [5].
4. Testing the overall architecture structure connectivity. All connections among components and connectors, and all internal connections are to be tested [5].

# 4   Architecture Based Reliability Improvement Approaches

Software reliability is one of the important quality attribute. There are many reasons behind the importance of reliability, such as, company reputation, customer satisfaction, warranty costs, cost analysis, customer requirement and customer advantage [18].

   Many solutions have been proposed for these reasons. During our literature survey we have found many models for the estimation of software reliability and these are classified according to the different classification systems [19]. According to [20], classification of reliability models is based on software life cycle phases, such as, debugging, validation and operational phase. In the past much research has been conducted on modelling the reliability growth during the debugging phase [21], [22]. These are known as black-box model and they just consider the external environment of software without giving any attention to the internal parts of software [23]. These models utilize information from failure data.

Complexity of software is increasing day by day and reliability of complex software cannot be estimated using the black box technique [23]. To be more precise, a technique is needed that can be used to analyze software components and how they fit together [23], [24]. One such technique known as the white box approach estimates reliability from software architecture. This architecture-based approach calculates system reliability from different software components [23], [24]. This information can be used to perform what if analysis, such as: what will happen if a certain component is less reliable? This kind of information is very useful for a company for making decisions. All architecture based reliability models can be categorized into three different approaches, which are state based approach, path based approach and additive approach [25].

In the following sub-section we will explore some of existing approaches for architecture-based reliability, their requirements, assumptions, limitations and their applicability.

## 4.1 Requirements for Architecture-Based Reliability Improvement Approach

Different architecture-based reliability models have been proposed. However, all models have the following common set of requirements [19], [25].

### 4.1.1 Component Identification
It is a logically independent entity that performs well-defined operations. In an architecture-based qualitative analysis, usually analysis on software components is performed.

### 4.1.2 Software Architecture
Software architecture defines the behaviour of software resulting from the interaction of different modules of software system. Data obtained from the interaction of these modules is used as a parameter for the architecture based reliability. In this way, the impact of interaction of different modules on the reliability of software system can be identified.

### 4.1.3 System Malfunction Behaviour
Malfunction might occur when a module is executed or when control is transferred between different modules. A malfunction behaviour determines the impact of a failure on a software system. It also gives information about the severity of faults in a software system.

## 4.2 State Based Approach

This approach can only be used if architecture of the system is represented by a control flow graph [19]. This approach assumes that transfer of control between modules follow the Markov property. According to the Markov property, only the present states give any information about the future states. Past states do not add any new information to the future states [25]. In this approach, the architecture can be modelled as Discrete Time Markov Chain (DTMC), Continuous Time Markov Chain (CTMC) or Semi Markov Process (SMP). State based models can be further categorized into composite or hierarchical categories. In the case of a composite model, architecture is merged with the failure behaviour and then solution obtained. However, for the hierarchical approach, first, the architectural model is solved, and then the failure behaviour is imposed on the solution to predict reliability of the system [26].

### 4.3  Path Based Approach

Parameters and steps of this approach are similar to the state based approach but with one difference. The difference is in the estimation of reliability that is based on the possible execution paths of the program, i.e. experimentally, by testing or algorithmically [26].

### 4.4  Additive Approach

Unlike the state based approach and the path based approach, this approach does not explicitly consider software architecture. The focus of this approach is to estimate overall application reliability based on the components failure data. Component reliability can be estimated by a Non Homogeneous Poisson Process (NHPP) [19] [26]. This approach is named as an additive approach because the system reliability is expressed as the sum of component failure intensities [26].

## 5  Qualitative Research Strategy

When we are exploring any problem, a case study would be an ideal solution to conduct qualitative research [27]. In our case study, we will try to explore the various testing and reliability techniques that are being used in the mobile applications, and problems related to reliability and testing of mobile application such as cost and efficiency. For the case study, two different companies are selected and their selection criterion is based on the usage of the software reliability and testing techniques in mobile applications.

Our qualitative study involves two iterations with two steps in each iteration first step being the formal and the other informal. The two iterations are as follows.

Finding weaknesses or flaws in the traditional software reliability and testing techniques.

Factors affecting traditional software reliability and testing techniques.

As mentioned earlier, two steps were followed, i.e. formal and informal. In the formal approach, we have framed some set of pre-planned questions and presented them to the organizations selected for our case study. The second step is an informal way of interviewing a member or developer of the organization. In this case, questions are not pre-planned; instead, they are framed during the conversation based on the reply given by the member of that organization. The member is also allowed to frame his/her own questions during the conversation and therefore sharing of information takes place. The observations made during the conversation are noted down and stored for future usage. This information sharing helps in finding weaknesses, flaws, strengths and many other factors that affect the reliability and testing techniques based on the different models being used by different organizations.

## 5.1 Data Collection

The data collection procedure of the case study consists of interviews with two selected organizations and observations drawn from the interviews. The interviews are performed by researchers for the purpose of gathering information about the current reliability and testing techniques used by the organizations and the problems related to the traditional techniques. The interviews are recorded using a video camera and the observations documented in text format. The information thus obtained will include conversations that took place during the interview and aspects such as voice, body expressions, and attitude and proper answers can be deduced from this information.

## 5.2 Data Analysis Procedures

The information collected through the recorded video will be articulated in a text document. The text format will facilitate the researchers in analyzing the data and deriving observations. Conclusions are drawn by comparing the results obtained from the two case studies performed on two organizations. The analysis of the interviews and observations highlights the weaknesses, flaws and impact of different factors on the traditional testing and reliability techniques. The data collected by interviews and observations are analyzed by the researchers individually and the final mutually agreed findings are consolidated to find the vulnerabilities in the approaches derive the missing information and prepare a new approach. The data analysis procedure is also useful to draw the necessary conclusions with a proper solution for a better reliability and testing technique.

## 5.3 Validity of Qualitative Study

A threat to the validity of this study is a contradiction of opinions among researchers. This issue is resolved by using the case study in presence of practitioners in real world. The validity of the study is generalized by combining the practitioners' views with the researchers' views, since the researchers' group is small. We have selected two organizations with different environments to mitigate the validity threat.

## 5.4 Expected Outcome

The expected outcome of this study is a greater understanding of weaknesses and problems in different software reliability and testing techniques that are currently being used by the organizations. This study explores some of the factors that influence the reliability and testing techniques. A set of factors and weaknesses that are included in the new approach, as well as observations and conclusions will be derived from the case study. Finally, a new approach based on software architecture is proposed considering the weaknesses and factors affecting the current reliability and testing techniques. Even though the case studies may be influenced by individual scenarios within the organizations, the observations and conclusions will be generalized to some extent by involving practitioners from two different organizations in addition to researchers.

## 5.5   Description of the Two Case Studies

The two organizations have participated in this case study on the condition that their name should not be disclosed. Therefore, they are referred as Organization 1 and Organization 2 in subsequent discussion.

### 5.5.1   Case Study – I

Organization 1 deals in IT consultancy and customized software development. This organization is ISO 9001:2000 certified. They are on the way of achieving CMMI (Capability Maturity Model Integration) level2 certification. They are targeting mobile game applications in the field of mobile software. They are using eclipse, J2ME and C++ for mobile game programming. According to them, quality and cost are the most important factors in software development. They did not follow any standard for software project documentation. According to them, their skilled manpower and quality standards are the strengths of their development process. They are using architecture for the software development process but they are unaware of architecture based testing and reliability. They are reusing software components from old applications. They have three software testers, but they are not highly experienced and the testing team is responsible for measuring the reliability of products. They are using both black box and white box testing. Time and cost are factors affecting the software testing process. They allocate 20-25% of time and budget for software quality assurance activities (SQA). They are planning to build research department in their organization. They have a customer support department but they do not have online helpdesk support for the customers. They have customer feedback repository but it is only accessible to the company management.

### 5.5.2   Case Study – II

Organization 2 provides front and back end IT solutions to several major corporations. They are using Red Hat, Novell, Digium and MySQL, J2ME and C++ for mobile game programming. According to them cost and functionality of applications are the most important factors in the software development. They don't have any mechanism for testing documentation. They are not using architecture but they are using object-oriented designs for the software development process. According to them, their skilled manpower is the strength of their development process. They think that cost is only the issue that lays down the quality standards of their products. They are not reusing software components from old applications. They have five software developers who are also playing the role of software testers and measure reliability of products by software testing. They are using both black box and white box testing. They have no clue about reliability techniques. Time and cost are the factors affecting the software testing process. They allocate 15-18% of time and budget for SQA. They have a customer support department but they do not have online helpdesk support for their customers.

### 5.6  Our Analysis About the Case Studies

We found that both companies are using traditional approaches for testing, which create problems regarding budget and time in the software development process. We observed that both cost and time have huge impact on the quality attributes of software products. According to researchers, architecture based testing may help organizations to cut down their budgets and reduce time for software quality assurance activities. We found that there is no proper use of reliability techniques in both organizations; hence, the quality of software is sacrificed. It is suggested that these organizations should include any kind of reliability measurement approach in their software development process. Authors also understand that these companies cannot directly adopt architecture based reliability approaches in their software development processes. Therefore, initially they should concentrate on traditional approaches of reliability measurement.

## 6  Proposed Framework

### 6.1  Proposed Framework for Architecture Based Testing

Software architecture is an integral part of any component based software system, and software architecture and components are the two sides of the same coin [28]. From figure 1, we can see that components represent the core artefact and software architecture constitutes study of how such components can be integrated to fulfil the desired functional and non-functional requirements. A component-based system is formed by assembling implemented components.

In the case of software architecture, a system is decomposed into units and their collaborations. A unit can be a component, a module, a layer, data storage or a connector. An architectural component can be an abstract (non-concrete) element, a unit, a module or a layer. The architectural design process gets functional and quality attributes as inputs and returns an architecture complying with the requirements and requisite qualities. Figure 1 shows that software architecture is an artefact, i.e. it can be used to describe the architecture of a component-based system and specified using model based specifications and formal specifications. Model based testing employs test cases that are derived as a whole or in part from a model that describes some aspects of the system under test [28]. Model based testing techniques utilize low-level design models and usually focus on behavioural aspects. On the other hand, software architecture models are more abstract and require various models for expressing software architectures.

Specification based testing is a combination of model based testing and formal testing, which is also known as architecture based testing as shown in figure 1. In next step, architecture based test cases are used on the criteria of functional and non-functional requirements. Then architecture level test cases are mapped to code level test cases in order to evaluate the results of architecture based testing as can be seen in figure 1.

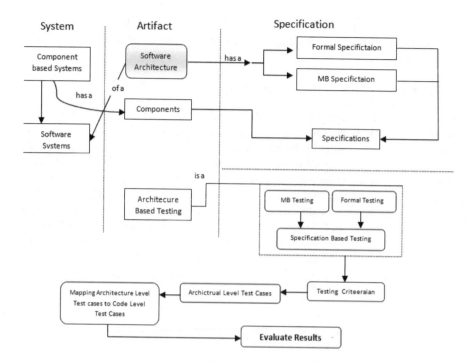

**Fig. 1.** Proposed Architecture-based Testing Framework [28]

## 6.2 Proposed Framework for Architecture Based Reliability

We have used architecture analysis to estimate the reliability of a system. There are different approaches of architecture analysis such as categorization, scenarios, etc. [29]. However, in our approach we have used scenarios for the analysis purpose. The main reason behind the scenario-based analysis is model driven architectural context [29], [30]. Architecture is very subjective in nature; therefore, context is a better choice in this case. Scenarios are used to compare design alternatives [29], [30]. Scenario can be used to measure the quality attributes that are important for the stakeholders [29]. After that, we analyze the architecture under these constraints and measure how well quality attributes (security, portability, reliability, etc.) are supported by the architecture [29].

In the suggested framework as shown in figure 2, the first step is to identify the stakeholders and components from the architecture. During the second step, we will select failure scenarios related to the stakeholders and components, and availability scenarios of the components. After the second step, the selected scenarios are then analyzed and finally we can calculate system's reliability by deducting availability scenarios from failure scenarios because availability scenarios are inversely related to failure scenarios.

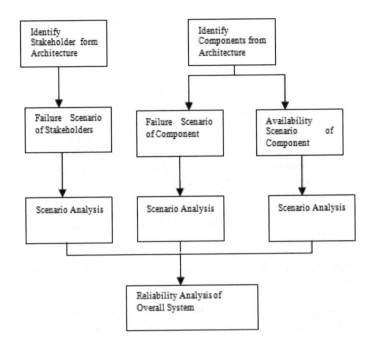

**Fig. 2.** Proposed Architecture-based Reliability Framework [29]

## 7   Conclusion and Future Work

This paper explores traditional techniques of software testing and reliability. The main aim of this study is to find the strengths and weaknesses in the traditional techniques of software testing and reliability with the help of case studies. We have conducted our case studies in two organizations based in Pakistan. Data collected during the case studies helped us find the strengths and weaknesses of the traditional techniques of software testing and reliability, and helped us in proposing a new framework for architecture based estimation of reliability and testing for mobile applications.

We have proposed separate frameworks for architecture based estimation of testing and reliability techniques for mobile applications. However, in future we want to merge both frameworks into a single framework. We will also perform experiments to compare the traditional techniques of testing and reliability with the proposed techniques.

## References

1. Yacoub, S.M., Ammar, H.H.: A methodology for architecture-level reliability risk analysis. Software Engineering. IEEE Transactions on Software Engineering 28(6), 529–547 (2002)
2. Bass, L., Clements, P., Kazman, R.: Software Architecture in Practice, 2nd edn. Addison-Wesley, Reading (2003)

3. Garlan, D., Shaw, M.: An introduction to software architecture: Advances in Software Engineering and Knowledge Engineering, vol. 1. World Scientific Publishing Company, Singapore (1993)
4. Software Engineering Institute cited on September 20 (2007), http://www.sei.cmu.edu/architecture/
5. Jin, Z.: A software architecture-based testing technique. Thesis for Doctor of Philosophy in Information Technology, George Mason University, Fairfax, Virginia (2000)
6. Bertolino, A., Inverardi, P.: Architecture-based software testing. In: Wolf, A.L., Finkelstein, A., Spanoudakis, G., Vidal, L. (eds.) Joint Proceedings of the Second international Software Architecture Workshop (Isaw-2) and international Workshop on Multiple Perspectives in Software Development (Viewpoints 1996) on SIGSOFT 1996 Workshops, San Francisco, California, United States. ACM, New York (1996)
7. Pan, J.: Dependable Embedded Systems. Carnegie Mellon University 18-849b (1999)
8. Wang, W.-L., Wu, Y., Chen, M.-H.: An architecture-based software reliability model. In: Proceedings of 1999 Pacific Rim International Symposium on Dependable Computing, pp. 143–150 (1999)
9. Hu, Q.P., Dai, Y.S., Xie, M., Ng, S.H.: Early Software Reliability Prediction with Extended ANN Model. In: 30th Annual International Computer Software and Applications Conference, vol. 2, pp. 234–239 (2006)
10. Garlan, D.: Software architecture: a roadmap. In Proc of the Conference on the Future of Software Engineering. In: ICSE 2000, Limerick, Ireland, pp. 91–101 (2000)
11. Oquendo, F.: π-Method: a model-driven formal method for architecture-centric software engineering. SIGSOFT Software Engineering Notes 31(3), 1–13 (2006)
12. Goševa-Popstojanova, K., Trivedi, K., Mathur, A.P.: How Different Architecture Based Software Reliability Models Are Related? FastAbstract ISSRE (2000)
13. Dimov, A., Ilieva, S.: Reliability Models in Architecture Description Languages. In: International Conference on Computer Systems and Technologies – Comp. Sys.Tech. (2007)
14. Wang, W.-L., Scannell, D.: An architecture-based software reliability modeling tool and its support for teaching. In Proc. 35th Annual Conference of Frontiers in Education: T4C-15-T4C-20 (2005)
15. Schulz, S., Honkola, J., Huima, A.: Towards Model-Based Testing with Architecture Models. In: 14th Annual IEEE International Conference and Workshops on the Engineering of Computer-Based Systems. ECBS 2007, pp. 495–502 (2007)
16. Dalal, S.R., McIntosh, A.A.: When to stop testing for large software systems with changing code. IEEE Transactions on Software Engineering 20(4), 318–323 (1994)
17. Yu, Y.T., Ng, S.P., Poon, P.-L., Chen, T.Y.: On the testing methods used by beginning software testers. Information and Software Technology 46(5), 329–335 (2004)
18. Relex Software Corporation, http://www.relex.com/resources/overview.asp
19. Goševa-Popstojanova, K., Trivedi, K.S.: Architecture-based approach to reliability assessment of software systems. Performance Evaluation 45(2-3), 179–204 (2001)
20. Ramamoorthy, C.V., Bastani, F.B.: Software Reliability – Status and Perspectives. IEEE Transaction on Software Engineering 8(4), 354–371 (1982)
21. Huang, C.-Y., Kuo, S.-Y., Lyu, M.R.: Effort-index-based software reliability growth models and performance assessment. In: 24th international Computer Software and Applications Conference, pp. 454–459 (2000)
22. Malaiya, Y.K., Denton, J.: What do the software reliability growth model parameters represent? In: Proc. 8th International Symposium On Software Reliability Engineering, pp. 124–135 (1997)

23. Gokhale, S.S., Trivedi, K.S.: Analytical Models for Architecture-Based Software Reliability Prediction: A Unification Framework. IEEE Transactions on Reliability 55(4), 578–590 (2006)
24. Roshandel, R., Medvidovic, N.: Toward architecture-based reliability estimation. In: Proceedings of the Twin Workshops on Architecting Dependable Systems, International Conference on Software Engineering (ICSE 2004), The International Conference on Dependable Systems and Networks (DSN-2004), Edinburgh, UK, Florence, Italy (2004)
25. Ramani, S., Gokhale, S.S., Trivedi, K.S.: Software Reliability Estimation and Prediction Tool. In: Puigjaner, R., Savino, N.N., Serra, B. (eds.) TOOLS 1998. LNCS, vol. 1469, pp. 27–36. Springer, Heidelberg (1998)
26. Goševa-Popstojanova, K., Mathur, A.P., Trivedi, K.S.: Comparison of architecture-based software reliability models. In: Proceedings of 12th International Symposium on Software Reliability Engineering, ISSRE, pp. 22–31 (2001)
27. Creswell, J.: Research Design: Qualitative, Quantitative and Mixed Approaches. Sage Publications Ltd., Thousand Oaks (2002)
28. Muccini, H.: What makes Software Architecture-Based Testing Distinguishable? In: Proceedings of the Working IEEE/IFIP Conference on Software Architecture (WICSA), p. 29 (2007)
29. Tekinerdogan, B., Sozer, H., Aksit, M.: Software Architecture Reliability Analysis Using Failure Scenarios. In: 5th Working IEEE/IFIP Conference on Software Architecture, WICSA, pp. 203–204 (2005)
30. López, C., Astudillo, H.: Use case- and scenario-based approach to represent NFRs and architectural policies. In: Proceedings of 2nd International Workshop on Use Case Modeling (WUsCaM-2005), Use Cases in Model-Driven Software Engineering Held in conjunction with MoDELS 2005, Montego Bay, Jamaica (2005)

# Appendix A

1. Which particular area of application you are targeting on mobile software?
2. Which tools you are using for development purpose?
3. Which are the critical factors in your software development process?
4. What are strengths in your software development process?
5. How many software testers do you have and what experience they have?
6. How many reliability engineers do you have and what experience they have?
7. What are the testing techniques being used by your company?
8. What are the reliability techniques being used by your company?
9. What are common problems you are facing in testing and reliability of software?
10. Which is the critical factor in software testing and reliability for you?
11. What are the strengths of your testing and reliability teams?

# A Framework to Achieve Reverse Channel Multicast in IEEE 802.16 Based Wireless Broadband Networks for Interoperability with IP Networks

Abdul Qadir Ansari[1], Abdul Qadeer Khan Rajput, and Manzoor Hashmani

Department of Computer Systems & Software Engineering,
Mehran University of Engineering and Technology, Jamshoro, Pakistan[2]
qadir.ansari@ptcl.net.pk, aqkrajput@muet.edu.pk,
mhashmani@yahoo.com

**Abstract.** The purpose of this paper is to present a framework design to achieve reverse channel multicast support in IEEE-802.16 networks. The unavailability of reverse channel multicast in IEEE 802.16 networks is one of the issues to be resolved in order to achieve its interoperability with IP networks. The interoperability of IEEE 802.16 networks and IP networks is not yet addressed. Forward channel multicast support is inherent in the IEEE-802.16 standard. Achieving reverse channel multicast support, we can enable IP services requiring native multicast support over IEEE-802.16 networks for seamless interoperation. IP network rudiments are mature enough and not likely to be redefined for any new network design; rather the reverse of this is more likely to be worked on. A framework design to achieve reverse channel multicast support is presented introducing a concentrator layer (Multi Node Access Layer) before Base Station (BS) and MARS (Multicast Address Resolution Server) to handle multicast requests.

**Keywords:** BWN-Broadband Wireless Networks, BWA-Broadband Wireless Access, BS- Base Station, BSC-Base Station Controller, MNAL-Multi Node Access Layer, MARS-Multicast Address Resolution Server.

## 1 Introduction

The IEEE 802.16 standard specifies the air interface for fixed and mobile Broadband Wireless Access (BWA) systems that support multimedia services. The IEEE 802.16-2004 standard, which was also previously called 802.16d or 802.16-REVd, was published for fixed access in October 2004. Understanding of the standard can be found in [1], [2], [3], [4], [5], [6], [7]. The standard has been updated to the 802.16e standard, as of October 2005. IEEE 802.16e amends the base specification with Physical (PHY) and Medium Access Control (MAC) functions for supporting mobile terminals while adopting the same data link principles also for mobile networking systems.

---

[1] Co-Affiliated as Engineer Switching (Fixed/ Wireless) with PTCL, Pakistan.
[2] Research is funded by Higher Education Commission of Pakistan under its program NRGPU.

D.M.A. Hussain et al. (Eds.): IMTIC 2008, CCIS 20, pp. 76–87, 2008.
© Springer-Verlag Berlin Heidelberg 2008

The principle of operation of BWA networks is relatively simple. A BS sends radio signals with typical channel bandwidth allocations 20 or 25 MHz (United Stats) or 28 MHz (Europe) in 10-66 GHz, or various channel bandwidths among 1 to 30 MHz in 2-11 GHz [4] either directly or through other BS to destination node. The IEEE 802.16 standard specifies two modes for sharing the wireless medium: point-to-multipoint (PMP) and mesh (optional-we do not consider mesh mode in framework design).

The 802.16 is a point-to-multipoint network without bi-directional native multicast support [8]. Multicasting is provisioned in forward direction only; as per IEEE 802.16 standard specifications but it is not available on reverse channel. Reason behind this shortcoming is one-to-one communication possibilities over reverse link. Since every communication from a MS is first seen at BS and then relayed to where it is destined. There exists no communication directly between MSs (except in mesh mode of operation) thus limiting the multicasting over reverse channel. IEEE 802.16 is different from existing wireless access technologies such as IEEE 802.11as 802.16 is a point-to-multipoint network and an 802.16 subscriber station is not capable of broadcasting (e.g., for neighbor discovery or dynamic address binding) or direct communication to the other nodes in the network. This lacking of facility for native multicasting for IP packet transfer results in an inadequacy to apply the basic IP operation like IPv4 Address Resolution Protocol or IPv6 Neighbor Discovery Protocol. IEEE 802.16 defines the encapsulation of an IP data-gram in an IEEE 802.16 MAC payload but complete description of IP operation is not present [9], [10].

## 2   IEEE 802.16 MAC -Architecture Overview

Fig. 1 illustrates the architecture of IEEE 802.16. The Convergence Sub-layer (CS) provides any transformation or mapping of external network data that is received through the CS Service Access Point (SAP) and converts them into MAC service data units (MSDUs) received by the MAC layer through the MAC SAP. This sub-layer includes classifying external network SDUs and associating them to the proper MAC Service Flow Identifier (SFID) and connection ID (CID). The MAC Common Part Sub-layer (CPS) provides the core MAC functionality of system access, bandwidth allocation, scheduling, contention mechanism, connection establishment, and connection maintenance. It receives data from various CSs through the MAC SAP, which is classified to particular MAC connections. The IEEE 802.16-2004 standard supports four quality-of-service scheduling types: unsolicited grant service (UGS) for the constant bit rate (CBR) service, real-time polling service (rtPS) for the variable bit rate (VBR) service, non-real-time polling service (nrtPS) for non-real-time VBR, and best effort service (BE) for service with no rate or delay requirements.

The upper-layer protocol data units (PDUs) are inserted into different levels of queues with an assigned CID in the MAC layer after the SFID-CID mapping. These data packets in these queues are treated as MSDUs and then will be fragmented or packed into various sizes according to the MAC scheduling operations. During the initial ranging period, the SS will request to be served in the DL via the particular burst profile by transmitting its choice of DL interval usage code (DIUC) to the BS. Afterwards, the BS will command the SS to use a particular uplink burst profile with

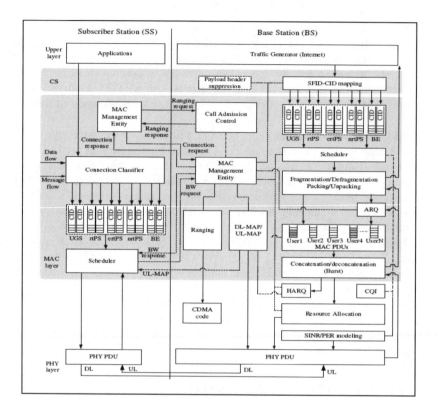

**Fig. 1.** IEEE MAC Architecture Overview [7]

the allocated UL interval usage code (UIUC) with the grant of SS in UL-MAP messages. The DL-MAP and UL-MAP contain the channel ID and the MAP Information Elements (IEs) which describes the PHY specification mapping in the UL and DL respectively. Although IEEE 802.16 defines the connection signaling (connection requests and responses) between SS and BS, it does not define the admission control process. All packets from the application layer are classified by the connection classifier based on the CID and are forwarded to the appropriate queue. At the SS, the scheduler will retrieve the packets from the queues and transmit them to the network in the appropriate time slots as defined by the UL-MAP sent by the BS. The UL-MAP is determined by the scheduler module based on the band width request messages. These messages report the current queue size of each connection in SS [7].

## 3   Internet Protocol Multicast and IEEE 802.16 Standard

The Internet Protocol (IP) is designed for use in interconnected systems of packet-switched computer communication networks. The IP provides for transmitting blocks of data called data-grams from sources to destinations, where sources and destinations are hosts identified by fixed length addresses. The addressing in IP is divided in unicast, multicast, and broadcast in native domain. IP multicast is based on concept of group-cast, thus multicast addressing specifies an arbitrary group of hosts. The members of that group want to receive traffic destined to the group [11]. Multicast-enabled nodes that run the TCP/IP suite of protocols can receive the multicast messages. Internet Group Management Protocol (IGMP) is used to dynamically register individual host in a multicast group. IETF-recommended standard RFCs [12], [13] further provide specifications as contained in Versions 1 and 2. With IP multicast, applications send one copy of the information to a group address. The information reaches all the recipients who are members of that group and want to receive it. Multicast technology addresses packets to a group of receivers rather than to a single receiver; it depends on the network to forward the packets to the only network intends to receive. Existing multicast support for fixed users can be extended to mobile users in wireless environments. However applying such support to wireless networks is difficult for reasons like, bandwidth scarcity, frequent topology changes, hidden wireless node problems, frequent membership changes, variable path profile and thus difficult to maintain over all GoS and reliability.

In order to enable IP Protocols to run over IEEE 802.16 networks we have to realize the uni-cast, broadcast and multicast transport. Uni-cast transport is available in IEEE 802.16; broadcast is not a valid choice for wireless transport for obvious reasons like bandwidth scarcity, network management, limited processing and power issues of mobile stations. Forward channel multicast is provisioned in IEEE 802.16 networks; however reverse channel multicast support is not available. We present a framework design to achieve reverse channel multicast support for IEEE 802.16 networks to enable IP services over such networks. We also know that various IP associated protocols like IPv4 Address Resolution Protocol or IPv6 Neighbor Discovery Protocol needs native multicast support for various advertisement processes and auto reconfigurations with respect to change in network topology.

## 4  Framework Design Issues

In this part we present the issues that are closely related and have lead to this framework design. IEEE 802.16 is a connection oriented access technology for the last mile without native multicast support [14], [15]. IEEE 802.16 only have downlink (forward channel) multicast support and there is no mechanism defined for mobile stations to be able to send multicast packets that can be mapped to uplink (reverse channel) multicast connections [16].

Following observation have built the ground and greatly contributed to design of such a framework.

I.  A particularity of IEEE 802.16 is that it does not include a rigid upper edge MAC service interface. Instead, it provides multiple "Convergence Sub-layers" (CS) with the assumption that the choice and configuration of the upper edge is to be done in accordance with the needs of a specific deployment environment (which might be DSL replacement, mobile access, 802.11 or CDMA backhaul etc.) [10]. Thus, our proposed framework is not to have MAC based multicast transport rather a CS based multicast. From this we also identify that as multiple convergence sub layers are defined on the basis of traffic classification so we present a Multicast Convergence Sub-Layer (Multicast-CS) concept to have centralized access, management and control over multicast transport.

II.   IEEE 802.16 is different from existing wireless access technologies. For example subsequent to network entry, an IEEE 802.16 network subscriber station has no capability whatsoever for data connectivity. Especially, in IP CS case, the criteria by which the Base Station (or other head end elements) sets up the 802.16 MAC connections for data transport are not part of the 802.16 standard, and depend on the type of data services being offered [9], [10]. This further demonstrates the need of a convergence layer for further transport of multicast traffic.

III.  Additionally, as IEEE 802.16 is a point-to-multipoint network and is of connection-oriented nature where connection always ends at BS, an IEEE 802.16 subscriber station is not capable of reverse channel multicasting (e.g. for ND/ARP) and there is no support from 802.16 MAC/PHY (Medium Access Control/ Physical) for direct communication among nodes attached to the same BS within the same subnet (prefix) to discover each other [9]. Thus using a CS layer based multicast traffic concentration mechanism, we shall go ahead with a design of a Multi Node Access Layer (MNAL) where every mobile node is accessible over shared channel. Here we propose a shared link access for mobile nodes to MNAL. Since we are to achieve reverse link multicast for only reason to enable operation of IP Protocols (like NDP, ARP...) so there are no issues of bandwidth conservation as aimed from generic multicast applications. We only need to have small traffic exchange between multicast nodes. Thus shared links access to MNAL a feasible choice.

IV.   Multicast means a group transport. To enable multicast over reverse link in IEEE 802.16 networks a certain group access between multicast-groups and MNAL is proposed. The design mechanism proposed is one used in WLAN ESS (Extended Service Set) with multicast group members as members of BSS (Basic Service Set) [17]. The Multicast-CS provides any transformation or mapping of multicast data that is received through the Multicast-CS Service Access Point (SAP) and converts them into MAC service data units (MSDUs). This sub-layer includes classifying multicast SDUs and associating them to the proper MAC Service Flow Identifier (SFID) and connection ID (CID). MNAL serves as Distribution System (DS) (-defied in WLAN specifications) can be defined to have transport between different multicast groups to gather neighbor information and to achieve inter-node communication. We designate Access Point (AP) (-defined in WLAN specifications) as a head end node that basically provides communication between various multicast group members via MNAL. Thus allocating multicast group address to said AP we can have multicasting services in a similar fashion as seen in IP networks (Fig. 2).

V.    Multicast SDU should identify four addresses (Sender, Receiver, AP-Sender and AP-Receiver) to use routing mechanism of IEEE 802.11 standard. A host address, destination node address, AP address of sender and destination group address i.e. address of destination AP. Basically AP addresses are the group addresses as used in group cast.

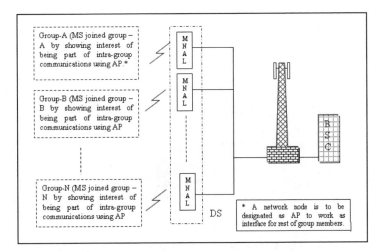

**Fig. 2.** Block diagram showing access scenario to MNAL using ESS design of WLAN

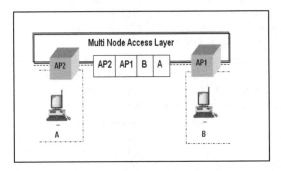

**Fig. 3.** Addressing at MNAL

VI.   Multicast SDU defining "To and from AP" addressing is to be realized for group based multicast transport over shared channel (Fig. 3). All multicast transport is to be converged at Multicast CS (realized through MNAL) first where specific allocations for CID are to take place for multicast transport.

Basically every multicast request generated by a mobile node will be first seen at MNAL as a Unicast transport. Upon identification as a multicast request it is further dealt by MARS. Thus enabling the multicast transport over the reverse channel, as desired by IP protocols when deployed in IEEE 802.16 networks. Therefore we name this as a "Unicast-based- Multicast" approach.

## 5  A Uni-cast Based Multicast Approach

A Unicast approach to realize multicasting is considered in framework design for IP packets originated form IP cloud and destined for clients of BWA network (Fig. 4).

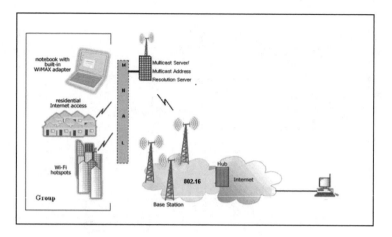

**Fig. 4.** Uni-cast based multicast in IEEE 802.16

Here address resolution requests are sent to Multicast Server that coordinates the address resolution request to MNAL for further transport to APs (group head end nodes). This includes design of Multicast Server having radio access to all MS in the coverage area of a BS through MNAL and APs.

A shared radio channel access method is seen for its feasibility. Carrier Sense Multiple Access/Collision Avoidance (CSMA/CA) scheme with persistence approach is used to resolve issues of contention over shared radio channel. NAV (Network Allocation Vector) is enabled to protect other nodes to transmit when one is transmitting over shared channel to MNAL. It is worth to understand here that CSMA/CD can not be achieved in wireless networks for reasons specially hidden terminal problem. Wireless transports has always encountered variable success rate, fast fading can cause temporal disappearance of node. So if collision detection signal is not reached in time then collision cannot be detected.

*Switched wireless transport at MNAL*: For group based communications we shall replicate the environment of switched Ethernet at MNAL keeping in mind that this will help resolve the multicast access and contention issues for shared access channel to MNAL. In order to minimize the contention issues (realizing the limited processing capabilities) switched wireless access could be realized over IEEE 802.16 by implementing radio switch behavior at MNAL. Breaking up the collision domains will reduce contention and result in better utilization of wireless channel bandwidth.

## 6   Proposed Framework

We present a framework to enable reverse channel multicast and to realize a group based transports of multicast traffic over reverse channel.

Stations want to execute IP services requiring native multicast support can show their interest in joining any such group to enable IP based *services* and to have inter-node communications with in same group. We have incorporated MARS (Multicast

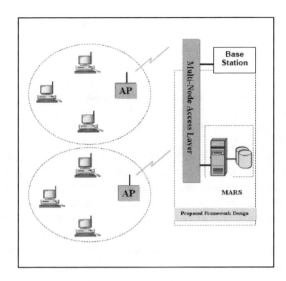

**Fig. 5.** Proposed framework design

Address Resolution Server) as an entity connected with MNAL. All group registration process is carried out at MARS. Fig. 5 depicts the scenario of operation of a multicast request over reverse channel. Here address of an AP is the group address/group ID. Any station member of a group is reached through the group id of AP. A relation table is to be maintained at MARS for membership information. Also any updates may be communicated at some definite interval of time or on dynamic basis (-as any membership changes).

### 6.1 Behavior of Various Components Proposed in Framework Design

a)   AP (Access Point) is like an entity that enables inter node communication with in a group, thus realizing intra-group communications via MNAL. It may be seen similar as one defined in ESS design of IEEE 802.11 WLAN specifications with adaptable radio transport configuration with respect to coverage area. All the multicast services like ND, ARP will be routed to host via AP.

b)   Multi-Node Access Layer: An access layer that could be seen as concentrator. This layer does the job of traffic classification and identifies the multicast traffic and thus allocates a multicast CID (Connection ID) for onward transport to upper layers of protocol. Every multicast request will be given a CID that identifies a multicast specific content.

c)   MARS: Multicast Address Resolution Server is an entity that resolves a Multicast Address attached to a multicast SDUs. MARS is basically consisting of a database server that can resolve multicast addresses. It does maintain the group member ship data and gets updated dynamically when ever membership changes or after a definite time period.

## 6.2  How a Multicast Request will be Executed with Our Design

a)  We assume that a multicast request should be dealt at MNAL and should be seamless at BS.
b)  Every Multicast request should be identified with reference to CS (a multicast convergence sub-layer as proposed in framework design)
c)  Identifying a multicast transport a CID will be allocated to that transport. This CID will be exclusively identified at every intermediate node as Multicast-CID.
d)  The request will be sent to MARS for resolving the multicast address. Identifying the multicast address MARS will execute a lookup query for group memberships so as to further transport to MNAL with necessary transport information in terms of group head-end node (APs).
e)  MARS does maintain Group IDs against each Multicast Group with updated group memberships. With this every neighbor discovery request will be addressed to the group head end node/AP for further transport to all members of that group.

# 7  Evaluation of Proposed Framework

We present an evaluation to our research work based on parameters such as scalability and reliability. We present two network design scenarios. Simple network and Complex network to evaluate the proposed framework for achieving reverse channel multicast support for interoperability with IP networks.

## 7.1  Simple Network Design

We consider a simple network for which we have only one BS and we have to realize MNAL beside it to enable native multicast support for IEEE 802.16 network when IP

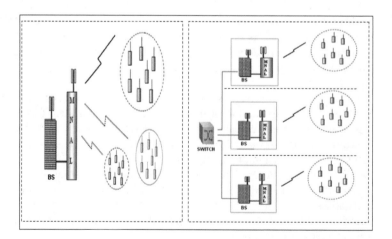

**Fig. 6.** Simple and Complex network designs

protocol is enabled (Fig. 6). We further explain simple network operation in context of native multicast provision with our design. An IP enabled MS will access the MNAL over shared link for execution of IP services like ND, ARP, etc that require native multicast. Here we have proposed CSMA/CA based multiple access technique over shared link to MNAL. We do not see any complications to arise as similar method is used at scaled networks. We are also considering multiple sectors being formed at one BS. We do not see any complications to arise while realizing MNAL per sector.

### 7.2 Complex Network Design

The complex network design is an extension to simple network design. Here we present a network consist of various BS sites. Every BS Site may be sectored for improved access and higher GoS. We see a possibility of inter MNAL group communications from this we mean that group head end nodes- APs can communicate with each other via MNAL-BS-MNAL link. As it does not utilize the shared channel capacity for accessing MNAL, thus there are no issues like bandwidth conservation on shared link. However the group formations such that members of group served by different BSs may be constrained keeping in view that it require massive updating of MARS database, hard to maintain routing and may destroy network hierarchical design parameters that include size, authorized access and network delimiters.

### 7.3 Scalability

IEEE 802.16 based networks use an open system architecture that supports a scalable solution for IP based services over IEEE 802.16 wireless networks [18]. We state that our designed framework supports scalability in a potentially fast expanding network scenario. With our proposed design we have an access layer (MNAL) per BS or per sector over a BS that works as a concentrator for traffic generated from various mobile stations looking for multicast support. The scalability of design could be referenced with scalability of deploying MNALs. We have proposed MNAL per BS or per sector at BS as per network requirement. We do not put any limit to number of MNAL in the network and also realize that MNAL processing and switching capabilities should be such that when deployed per BS/Sector it should be capable to handle all the multicast traffic generated by and for that BS/Sector. Shared channel specifications viz. frequency band, modulation & encoding techniques should be so chosen so as to provide adequate bandwidth for group communications and require limited processing.

### 7.4 Reliability

IEEE 802.16 is a connection oriented access technology with increased GoS for multimedia centric applications. To achieve an acceptable level of reliable transport we have to ensure that Received Signal Strength (RSS) at MNAL should be within threshold to ensure communication. For which we have to minimize any interference caused due to design so as to keep the RSS with in threshold. Strict compliance to group formation regulations as proposed may be assured so that network delimiters be exclusively defined and usefulness of CSMA/CA be achieved. In our design we have proposed persistent strategy to further eliminate the chance of simultaneous transport that may result in collision even in avoidance method and thus improved the efficiency and reliability.

# 8 Conclusion

We conclude our research work with a note that a generic framework design to achieve native multicast in IEEE 802.16 based networks is now available. Adapting the proposed design we can answer the inadequacy of current standard specifications in providing reverse channel multicast support for interoperability of IEEE 802.16 networks with IP networks. There are certain issues relating to hardware design and technology selection at access layer to MNAL which are not included in framework design, realizing the fact that any particularity at generic framework design may affect the utility and benefits of such a design.

# References

[1] Eklund, C., Marks, R.B., Stanwood, K.L., Wang, S.: IEEE Standard 802.16: A Technical Overview of the WirelessMAN™ Air Interface for Broadband Wireless Access. IEEE Communications Magazine, 98–107 (2002)
[2] Ghosh, A., Volter, D.B., Andrews, J.J., Chen, R.: Broadband Wireless Access with Wi-MAX/802.16: Current Performance Bandchmarks and Future Potential. IEEE Communication Magazine, 129–136 (2005)
[3] Yagoobi, H.: Scalable OFDMA Physical Layer in IEEE 802.16 WirelessMAN. Intel Technology Journal 08 (2004)
[4] IEEE Standards: IEEE 802.16-2004, IEEE standard for local and metropolitan area networks, part 16: Air Interface for Fixed Broadband Wireless Access Systems (2004)
[5] IEEE Standards: IEEE 802.16 e, IEEE standard for local and metropolitan area networks, part 16: Air Interface for Fixed Broadband Wireless Access Systems (2005)
[6] IEEE Standards: Std 802.16e™-2005 and IEEE Std 802.16™-2004/Cor1-2005 (Amendment and Corrigendum to IEEE Std 802.16-2004) (2006)
[7] Chen, J., Wang, C.C., Tsai, F.C., Chang, C.W., Liu, S.S., Guo, J., Lien, W.J., Sum, J.H., Hung, C.H.: The Design and Implementation of WiMAX Module for ns-2 Simulator. In: WNS 2006 (2006)
[8] Madanapalli, S.: Analysis of IPv6 Link Models for IEEE 802.16 Based Networks, Internet RFC 4968 (2007)
[9] Jee, J., Madanapalli, S., Mandin, J., Montenegro, G., Park, S., Riegel, M.: IP over 802.16 Problems and Goals, Internet-Draft, Network Working Group (2006)
[10] Jee, J., Madanapalli, S., Mandin, J., Montenegro, G., Park, S., Riegel, S.: IP over 802.16 Problem Statement and Goals (2007)
[11] Internetworking technologies handbook: Chapter 43: Internet Protocol Multicast, pp. 1-16 (2001)
[12] Deering, H.S.: IGMP Ver. 01, Extention for IP Multicasting, Internet RFC 1112 (1989)
[13] Fenner, W.: IGMP Ver. 02, Extention for IP Multicasting, Internet RFC 2236 (1997)
[14] Cicconetti, C., Lenzini, L., Mingozzi, E., Eklund, C.: Quality of Service Support in IEEE 802.16 Networks. IEEE Communication Magazine, 50–55 (2006)

[15] Shetiya, H., Sharma, V.: Algorithm for routing and centralized scheduling to provide QoS in IEEE 802.16 Mesh Networks. ACM 20051-59593-183-X/05/0010, 140–149 (2005)

[16] Patil, B., Behcet, F.X., Sarikaya, S., Madanapalli, S., Choi, J.H.: IETF charter IPv6 Over IPv6 Convergence sublayer in WiMAX Networks, Network Working Group (2006)

[17] Forouzan, B.A.: Wireless LANs-IEEE 802.11, Data Communications and Networking (2004)

[18] Fong, B., Ansari, N., Fong, A.C.M., Hong, G.Y., Rapajic, P.B.: On The Scalability Of Fixed Broadband Wireless Access Network Deployment. IEEE Radio Communications, 12–18 (2004)

# A Dynamical System and Neural Network Perspective of Karachi Stock Exchange

Syed Nasir Danial[1], Syed Raheel Noor[2], Bilal A. Usmani[3], and S. Jamal H. Zaidi[4]

[1] Dadabhoy Institute of Higher Education, SNPA-17/B, Block 3, K.C.H.S.U. Ltd., Karachi
nasirdanial@yahoo.com
[2] Saudi Orger, Kingdom of Saudi Arabia
raheelbinnoor@yahoo.com
[3] NED University of Engineering & Technology, University Road, Karachi
bausmani@gmail.com
[4] Bahria University, Karachi Campus, 13 Stadium Road, Karachi
sjamalzaidi@lycos.com

**Abstract.** This study discusses the evolution of KSE-100 index returns as a dynamical system. We present application of nonlinear time-series analysis. Our results show that estimation of correlation dimension for the case of KSE-100 index returns is not possible. We further go into nonlinear analysis and construct a model of the series based on feedforward neural network with backpropagation training. We construct many neural networks and the one with Levenberg-Marquardt backpropagation is found to give slightly better results compared to ARMA/ARIMA models. Neural networks are found to be applicable in those cases when nonlinear time-series analysis is at failure.

**Keywords:** Neural network, non-linear time series analysis, correlation dimension, KSE-100 index returns.

## 1 Introduction

Over the past few decades the need of forecasting has become a fundamental problem of science. Many approaches to forecasting have been developed from linear [1] to nonlinear type phenomena [2]. Since forecasting is a noisy application as we normally deal with systems which are affected by many inputs, a good forecasting technique should involve all or most of such inputs and should capture the way these inputs have impact on the system. In linear time series analysis, we normally take a few major inputs as the only inputs to the system because the other inputs are taken to act in a complex nonlinear fashion. This, of course, has an adverse effect in a sense that the model behaves in a linear way even if the system does not. Moreover, linear equations can only lead to exponentially growing or periodically oscillating solutions thus all irregular behavior of the system is attributed to some random external input to the system. On the other hand, nonlinear time series methods based on dynamical systems theory argues that random input is not the only source of irregularity in a system's output; a deterministic, nonlinear, chaotic system can also produce irregular output. Now despite that nonlinear analysis is highly data intensive [2] and often very difficult in many practical situations,

D.M.A. Hussain et al. (Eds.): IMTIC 2008, CCIS 20, pp. 88–99, 2008.

detecting nonlinearity in a time-series data is a challenging task. Nevertheless, neural networks (NN) [3] are found to be tools which can model nonlinear phenomena in a relatively more effective way.

The importance of NN is in their successful application across an unusual range of problem domain. They are very useful to solve the problem of prediction and classification in a complex system especially in financial forecasting such as detecting trends and patterns in financial data which are key concepts in a decision making process. Unlike conventional techniques for time series analysis, NNs need little information about the time series data and can be applied to a broad range of problems. However, NN do not shed light on the dynamics of the underlying process and for that reason, we have to employ several linear or nonlinear time-series methods to see what sort of process is responsible for the observed output.

The purpose of this research is three fold: First we attempt to estimate correlation dimension, $d_c$, of Karachi Stock Exchange-100 (KSE-100) index returns using delay-coordinate embedding [4]. To our knowledge, this attempt has never been made on the said problem. Second, we construct a neural network based model using the vectors obtained after delay-coordinate embedding. Finally, we compare the results of our neural network model with those obtained through conventional time-series analysis, i.e., Autoregressive Integrated Moving Average (ARIMA) and Autoregressive Moving Average (ARMA) modeling [1], [5], [6].

Section 2 shows some of the earlier work on KSE-100 index. In section 3 we present important concepts regarding our calculations. Section 4 is our main section which describes the results of $d_c$ estimation, NN construction and the model building using ARMA/ARIMA. Section 5 opens with a discussion and presents our concluding remarks.

## 2  Previous Work

A fine study of the presence of nonlinearities in Asian Stock Markets (including KSE-100) is performed in [7], where the authors use Hinich portmanteau bicorrelation test with windowing technique to detect regions of the time-series when nonlinear behavior is significant. The study, however, reveals that the nonlinearities do not appear to be stable over time and thus modeling of such a process is a difficult task. In [8], a composite model for market liquidity and risk is proposed for KSE-100 index. In another work [9], the authors use annual balance sheet data of the KSE-100 index listed firms from 1981-2000 to identify factors which jointly effect the share prices of KSE-100 index. They explain how the fundamental factors are effecting the price change and argue, on the basis of explanatory power of the fundamental factors, that other factors should also be taken into account for detailed analysis. The Reference [10] uses the day-of- the-week KSE return anomaly and shows that the post 9/11 data set follows a random-walk process. The modeling and forecasting of inflation in Pakistan is performed in [11]. In a recent work [12], multivariate EGARCH and VAR models are used to investigate the impact of Country and Global Factors on stock prices in Pakistan's equity market, i.e., Karachi Stock Market; asymmetric effects of macroeconomic variables are observed on stock returns volatility. Moreover, they show that the stock market is partially integrated. The Reference [13] and [14] presents an NN based forecasting of short-term stock share prices. The authors use Levenberg-Marquardt training algorithm over a feedforward neural network for the daily stock price direction data. They also perform a comparison of forecasting results obtained from NN model with that of those obtained from a linear time-series model and show that NN model's forecast is far better.

## 3  Important Concepts

### 3.1  Phase-Space Reconstruction

Phase Space reconstruction means to transform a time series of scalar measurements $s_1, s_2, \ldots, s_N$ into vectors $\mathbf{s}_N$ (called delay vectors), where $\mathbf{s}_N = (s_{N-(m-1)v}, s_{N-(m-2)v}, \ldots, s_{N-v}, s_N)$, in an appropriate embedded phase space of dimension $m$. The subscript $v$ (or in time units, $\tau = v\Delta t$) is the time difference in number of samples between adjacent components of the delay vectors and is referred to as the delay time or lag time or simply lag. For discrete systems $\Delta t$ is normally set to unity [4]. This method of phase space reconstruction is called the method of delays [15]. This process involves two adjustable parameters, the lag $\tau$ (or $v$) and the dimension $m$. The choice of the values of these parameters is crucial and requires a careful analysis of the time-series using a variety of existing techniques. Once the values of $\tau$ and $m$ are known the whole scalar series can then be converted into state vectors. Usually, an estimate of $\tau$ is obtained by applying one or several techniques together. Some of the most widely used techniques are given here as (a) to take the lag of autocorrelation function (ACF) as $\tau$ when it decays to $1 / e$, (b) to take the first minimum of the time-delayed average mutual information (AMI) [16], (c) the visual inspection of delay vectors with various lags.

According to [2], method (a) is a suitable approach for estimating the delay-time $\tau$ but it only considers the linear correlations whereas for the nonlinear correlations they propose method (b) in certain cases. Although, both of these methods (a) and (b) are used in many applications however the estimated delay times are either similar or noticeably different from one another [17]. In the later case, method (c) may be used as an additional check [18]. Normally, decorrelation of the components of delay vectors are observed visually by analyzing the delay vectors of $m = 2$ or a 2D projection of higher order $m$ for different values of $\tau$ [19]. For the choice of an optimal value of $m$ a well known method, viz., false-nearest neighbors (FNN) is used.

### 3.2  False-Nearest Neighbors Method

The method is proposed by [20] and it gives an optimal value of $m$ sufficient to avoid overlapping or crossing trajectories. The method counts, for a given embedding $k$, points which are nearby in the reconstructed phase space but which correspond to points which are far apart in the original phase-space. The smallest such $k$ for which these points disappear (or are sufficiently small) is the optimal value of $m$. One has to compute

$$R_k(i) = \sqrt{\sum_{p=0}^{k-1}[s(i+p)-s(j+p)]^2}. \tag{1}$$

where $s_j$ be the nearest neighbor of $s_i$, with $i \neq j$ in an embedding space of dimension $k$. If the points $s_i$ and $s_j$ are 'false neighbors' the distance in $(k + 1)$-dimensions, $R_{k+1}(i)$, will normally be greater than $R_k(i)$. The criteria of falseness is given as

$$\left[ \frac{R_{k+1}^2(i) - R_k^2(i)}{R_k^2(i)} \right]^2 = \frac{|s(i+k) - s(j+k)|}{R_k(i)} > R_{tol}. \tag{2}$$

where $R_{tol}$ is some threshold value. Ref. [21] recommends $R_{tol} = 15$, but this value is not critical. The method is performed for each point and the statistic $R_k =$ (number of false-neighbors $/ (N - k +1)$ is calculated. $R_k$ will decrease rapidly with increasing values of $k$ until a minimum embedding dimension is reached. This value of $k$ is the optimal embedding dimension which should be used.

### 3.3 Correlation Dimension

It is introduced in [22] and [23]. The basic formula is given here as under

$$d_c = \lim_{r \to 0} \lim_{M \to \infty} \frac{\partial \ln C(\varepsilon, M)}{\partial \ln \varepsilon}. \tag{3}$$

where $C(\varepsilon)$ is the correlation sum. We use a modified version for estimating the correlation sum as described in [2]. Thus $C(\varepsilon)$ is defined as

$$\frac{2}{(M - n_{min})(M - n_{min} - 1)} \sum_{i=1}^{M} \sum_{j=i+n_{min}}^{M} \Theta(\varepsilon - \|\mathbf{x}_i - \mathbf{x}_j\|). \tag{4}$$

where $M = N - (m - 1)\tau$, $\|...\|$ represents some Euclidean norm, and $\Theta$ is the Heaviside step function such that $\Theta(x) = 0$ if $x \leq 0$ and $\Theta(x) = 1$ otherwise. The sum in equation (4) counts the pairs $(\mathbf{x}_i, \mathbf{x}_j)$ whose distance is smaller than $\varepsilon$ in an $m$-dimensional embedding.

In (4), $n_{min}$ represents the minimum number of points after which temporal correlations disappear. The temporal correlation arises due to the fact that the embedding vectors at successive times are often found to be close in phase space because of continuous time evolution. Such a correlation is considered a serious problem in estimating geometric properties of the attractor and therefore must be removed first. The remedy of temporal correlation is simple [24] that we have to eliminate those pairs of points in the calculation of $C(\varepsilon)$ which are close due to time evolution and not because of the attractor geometry. The problem of finding a reasonable value of $n_{min}$ (also called Theiler window) can easily be solved by drawing the space-time separation plot [25], [26], [27] of the time-series data in some embedding $m$.

In the limits mentioned in (3), the dimension $d_c$ is expected to scale like a power law, $C(\varepsilon) \propto \varepsilon^{d_c}$ and the $d_c$ serves as a lower bound to fractal dimension, $d_f$, of the attractor [28]. Note that $t_{min} = n_{min}\Delta t$, where we set $\Delta t = 1$ and $t_{min}$ is obtained using the space-time separation plot.

### 3.4 Neural Network

Neural networks can be defined as parameterized nonlinear maps, capable of approximating arbitrary continuous functions over compact domains. In [29] and [30] it is proved that any continuous mapping over a compact domain can be approximated as accurately as necessary by a feedforward neural network with one hidden layer. In the context of neural network literature, the term neuron refers to an operator that maps $\Re^n \to \Re$ and is described by the equation

$$y = \Gamma(\sum_{j=1}^{n} w_j u_j + w_0),$$     (5)

where $U^T = [u_1, u_2, \ldots u_n]$ is the input vector, $W^T = [w_1, w_2, \ldots w_n]$ is referred to as the weight vector of the neuron, and $w_0$ is the bias. $\Gamma(.)$ is a monotone continuous function such that $\Gamma(.): \Re \to (-1, 1)$. The function $\Gamma(.)$ is commonly called a 'sigmoidal function'; $\tanh(.)$, and $(1+\exp(-(.)))^{-1}$ are some widely used examples. The neurons are organized in a feedforward layered architecture ($l = 0, 1, \ldots L$), and a neuron at layer $l$ receives its inputs only from neurons in the layer $l - 1$.

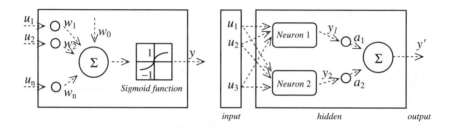

**Fig. 1.** Left: A simple perceptron model. Right: A typical one-hidden layer feedforward NN

A neural network, as defined above, represents a specific family of parameterized maps. If there are $n_0$ input elements and $n_L$ output elements, the network defines a continuous mapping $NN: \Re^{n_0} \to \Re^{n_L}$. A commonly used artificial neuron (Fig. 1) is a simple structure, having just one nonlinear function of a weighted sum of several data inputs $u_1, u_2, \ldots, u_n$; this version, often called a perceptron, computes (5) which is also called a ridge function. However, in order to learn functions more complex than the ridge functions, one must use networks of perceptrons. One such network of perceptrons is the feedforward perceptron network. A typical example of a feedforward perceptron with one hidden layer is also shown in Fig. 1.

## 4   Data Analysis

We collect a time-series, $T$, of daily stock index price of KSE-100 index from May 2, 1990 to August 6, 2007. The data is taken from Bloomberg (www.bloomberg.com). We compute a series of daily stock returns, $R = log(T_{i+1}/T_i)$, where $i \in \{1, 2, \ldots, 3818\}$ represents successive trading days starting with May 2, 1990, for our analysis in the present work. The series $R$ (Fig. 2) shows clear volatility which is common in financial time-series. Power spectral density and ACF (see Fig 2) show that the series obeys weak-stationarity. The ACF quickly drops to $1/e$ (33% of ACF at *lag* 1) for *lags* ≥ 4 with negligible exceptions. The running means are found to be within statistical range, however, the running standard deviations show variability beyond statistical fluctuations. We also compute AMI (Fig. 3) but that does not give any significant value as a minimum as its successive values for *lag* = 0, 1, 2, 3, 4, 5 are 1.428, 0.053, 0.036, 0.039, 0.033, 0.03. This work assumes *lag* = 4 according to ACF criteria as mentioned in section 3.1.

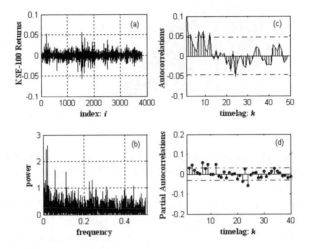

**Fig. 2.** (a) The return series, R, of KSE-100 index from 1990 to 2007. (b) Power Spectral Density of the series R. (c) ACF plot of R: the ACF drops to 1 / e after time lag = 4. (d) The PACF plot.

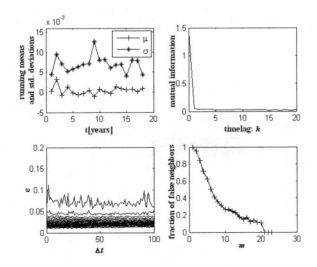

**Fig. 3.** Upper left: running means and standard deviations are depicted. Upper right shows. AMI of series *R*. Lower left: Space time separation plot of *R*; Note the saturation after $\Delta t = 20$; $w = 200 > 20$ is taken for safe-side. Lower right: Fraction of FNN for *R* for *lag* = 1, 2, ...,10.

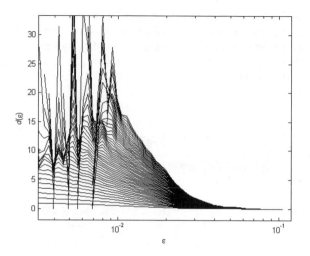

**Fig. 4.** Local slopes of the correlation integral from data of series $R$ for the values of embedding dimension $m \in \{1, 2, ..., 50\}$ counted from bottom to top. No scaling regions are found.

In order to see the effects of temporal correlation and to find the size of theiler window, $w$, we compute space-time separation plot (see Fig. 3). Clearly, $w = 200$ is sufficient for our case. The local slopes of the correlation integral are shown in Fig. 4.

As such, there is no scaling region and we therefore are unable to estimate $d_c$. However, an embedding dimension of $m = 21$ is sufficient for reconstruction of phase space as the fraction of FNN statistic decreases to zero for $m \geq 21$ (see Fig. 3). Failure to find an estimate of $d_c$ is not unusual in financial time series data, where the size of data set is insufficient for use [2]. We obtain similar results while estimating $d_c$ of the residuals of the ARIMA model, however, FNN method results in 0.2 at $m = 14$.

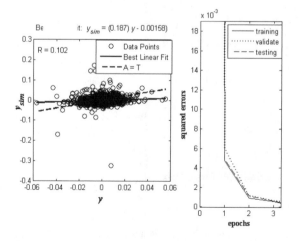

**Fig. 5.** Left: Regression Analysis of Simulation Results of NN-LM7. Right: Squared Errors.

The results we obtain using a feed-forward NN with backpropagation are given here as under: Vectors in the reconstructed phase-space with $m = 21$, $lag = 4$, are divided into three sets – half of them are taken as training set, and one-quarter each for testing and validation sets respectively. We construct several neural networks (see Table 1) with one hidden layer and one output layer, first using Gradient Decent (GD) and then using Levenberg-Marquardt (LM) as training functions. The hidden layer transfer function is chosen to be tangent sigmoidal and on the output layer we use linear function. With GD, even for large epochs, the network could not trace the behavior of the series $R$. A post regression (PR) analysis of all these networks results in low values of the regression coefficient, $r_c$, (as close as zero). On the contrary, LM produces sufficiently well results even for fewer epochs. The NN-LM2 and NN-LM5 with seven and twenty five hidden layer neurons produce mean square error of forecast (MSF) of $4.5545 \times 10^{-5}$ and $4.3149 \times 10^{-5}$ respectively. We also have used a two hidden layer net, NN-LM7, and its MSF gets fairly reduced compared to other NNs. Since testing and validation error (Fig. 5) show similar pattern, it is therefore a sufficient argument against network over-fitting. The PR analysis of NN-LM7 is also

**Table 1.** Various NNs with different arguments. #neuron shows neurons in hidden layer(s). In the column of MSE, the upper row (for each NN model) shows MSE of training and the lower row shows MSE forecast (MSF). * shows a validation stop.

| # | Training Function | NN name | #neurons | Epochs | MSE |
|---|---|---|---|---|---|
| | | NN-GD1 | 5 | 10000 | $1.12 \times 10^{-4}$ |
| | | | | | $5.73 \times 10^{-5}$ |
| | | NN-GD2 | 5 | 50000 | $6.90 \times 10^{-5}$ |
| | | | | | $5.68 \times 10^{-5}$ |
| 1. | Gradient descent back-propagation | NN-GD3 | 5 | 70000 | $6.67 \times 10^{-5}$ |
| | | | | | $5.48 \times 10^{-5}$ |
| | | NN-GD4 | 7 | 10000 | $8.98 \times 10^{-5}$ |
| | | | | | $7.33 \times 10^{-5}$ |
| | | NN-GD5 | 10 | 10000 | $6.60 \times 10^{-5}$ |
| | | | | | $6.11 \times 10^{-5}$ |
| | | NN-LM1 | 5 | 10*/500 | $5.15 \times 10^{-5}$ |
| | | | | | $5.37 \times 10^{-5}$ |
| | | NN-LM2 | 7 | 10*/500 | $5.30 \times 10^{-5}$ |
| | | | | | $4.55 \times 10^{-5}$ |
| | | NN-LM3 | 10 | 10*/500 | $5.26 \times 10^{-5}$ |
| | Levenberg- | | | | $4.91 \times 10^{-5}$ |
| 2. | Marquardt back-propagation | NN-LM4 | 20 | 9*/500 | $5.09 \times 10^{-5}$ |
| | | | | | $4.54 \times 10^{-5}$ |
| | | NN-LM5 | 25 | 9*/500 | $5.03 \times 10^{-5}$ |
| | | | | | $4.31 \times 10^{-5}$ |
| | | NN-LM6 | 30 | 10*/500 | $5.29 \times 10^{-5}$ |
| | | | | | $5.75 \times 10^{-5}$ |
| | | NN-LM7 | (25, 25) | 10*/500 | $3.18 \times 10^{-5}$ |
| | | | | | $3.82 \times 10^{-5}$ |

**Table 2.** Results of Predictions from conventional and NN models

| Steps | Original | Predictions | | | |
|-------|----------|---------|-----------|-----------|--------|
| | | AR(24) | ARMA(22,2) | ARMA(24,2) | NN-LM7 |
| 1 | 0.00420457 | −0.00005 | 0.00005 | −0.00004 | −0.0021057 |
| 2 | −0.0124261 | −0.00057 | −0.00047 | −0.00057 | −0.0055378 |
| 3 | −0.0056032 | −0.00052 | −0.00030 | −0.00053 | 0.0020984 |
| 4 | −0.008853 | 0.00048 | 0.00031 | 0.00046 | −0.0085069 |
| 5 | 0.00125573 | −0.00045 | −0.00053 | −0.00039 | −0.0047921 |
| 6 | −0.0026794 | 0.00006 | 0.00019 | 0.00004 | −0.0070372 |
| 7 | −0.0003627 | −0.00025 | 0.00013 | −0.00032 | 0.0080967 |
| 8 | −0.0001485 | 0.00028 | −0.00015 | 0.00023 | 0.00015748 |
| 9 | −0.0156004 | −0.00024 | 0.00022 | −0.00014 | −0.012605 |
| 10 | 0.00284768 | 0.00010 | 0.00004 | 0.00004 | −0.0071081 |
| MSF | | $5.26 \times 10^{-5}$ | $5.53 \times 10^{-5}$ | $5.28 \times 10^{-5}$ | $3.82 \times 10^{-5}$ |

shown in Fig. 5 which gives $r_c = 0.102$. We have performed similar analysis for each of the other networks shown in Table 1, however, the least value of the $r_c$ is found to be with the case of NN-LM7. We also have checked other NNs (not reported here), with varying number of neurons and layers, and different performance parameters, but we do not found them performing better than those being reported in Table 1.

Finally, we construct conventional models of the series $R$ using ARMA/ARIMA models. Now, based on minimum AICC criterion [1], the AR($p = 24$) model is selected and the Yule-Walker (YW) model equation is formed. The AR(24) passes the diagnostic tests as the Ljung-Box statistic (LBS) is found to be $16.053 < \chi^2_{(0.05)}$ and the order of minimum AICC YW model of the residuals of AR(24) is zero. Moreover, since partial autocorrelations, PACF, (Fig. 2) suggest that AR order $p = 12, 22, 23,$ and 24 should also be considered for model building because of the presence of peaks at 95% confidence interval. We also construct these AR models. The corresponding AICC statistics for AR models with $p = 12, 22, 23, 24$ are computed to be $-26688.58$, $-26679.34$, $-26679.99$ and $-26689.77$ respectively. For peaks below $lag = 12$ in PACF, the models do not pass residual tests, such as, Ljung-Box diagnostic test, and hence are not considered here. Still, two mixed models, ARMA (22, 2) and ARMA(24, 2) are also proposed as they pass diagnostic tests of residuals. The LBS for the former is 28.909; and for the later it is found to be 19.679. We have computed MSF for AR(24), ARMA(22, 2) and ARMA(24, 2) models (see Table 2) and it is of the order of $10^{-5}$.

## 5   Conclusion

KSE-100 index is a difficult place for scientific analysis due to its high variability and socio-economical and political dependence. In this analysis the entire history from the beginning of KSE-100 index is taken into account. Albeit, the original series $T$ is transformed into $R$ to make a possible stationary sample, it is highly likely that so much of noise and irregularities still persist. This is also evident from Hurst Exponent of $R$ which is found to be 0.63; showing long memory and unsuitability of random-walk model. The kind of bearish trend which KSE-100 index often experiences is probably because when economic conditions of the country are good and investors have ample funds to invest money into stock market, they invest their money to gain profits by selling their shares against high rates and then leave the stock market. During this kind of activities a dominant component in KSE-100 returns is deterministic and may best be modeled by even a simple linear regression. But such instabilities can best be understood if so many of them could be recorded in history and the time-series covers an ample period to record such activities. Moreover, if such a time-series is there and a multidimensional phase space is constructed, the global dynamics in this phase space must be nonlinear in order to fulfill the constraints of non-triviality and boundedness. Unfortunately, for the case of financial time-series, such as $R$, such a long time-series record is usually not available.

There are several criteria to check if length, $N$, of a series is ample, e.g., a generally accepted rule is $N \approx 10^d$, where $d$ is the attractor dimension. But various objections to this rule are there, e.g., in [31] the rule: $N = 42^m$ is rather suggested. Alternatively, Ref. [32] claims that only $N = 10^{d/2}$ is sufficient. However, Ref. [33] suggests $N = 10^{2+0.4d}$ or $d = 2.5 \log_{10}N - 5$. Moreover, the minimum $N$ also depends on attractor's type [34]. Besides, if we take FNN result $m = 21$, and use [31] then $42^{21}$ points are needed! Using criterion [33] the maximum $d$ may only be around 4, suggesting $m > 8$. However, a large fraction of false-neighbors, as shown in Fig. 3, from $m > 8$ to 21 suggests the complexity of the dynamics involved. A direct result of all these problems is that $d_c$ of $R$ can not be estimated due to limited points.

The MSF obtained from our NN models are shown in Table 1. Table 2 compares the results of NN-LM7 with conventional models. Amazingly, the MSF in all the models (of Table 2) is $O(10^{-5})$! This shows that even NN models are not the best candidates to capture the dynamics of $R$. Our main conclusions are: First, finding $d_c$ (to develop an insight of dynamics) is still an open question. Second, neural networks perform slightly better than conventional models (Table 2) but the overall precision of forecast is not satisfactory. Our conclusion is supported by [7] where the authors say that regularities in KSE-100 returns are not persistent and they come and go over small uneven periods of time. This makes the system difficult for scientific analysis in terms of creating a global model as is done in the present work. We think partitioning $R$ and developing NN models for each of them is an interesting direction. However, such a methodology is not likely to work for estimating $d_c$.

# References

1. Box, G.E.P., Jenkins, G.M., Reinsel, G.C.: Time Series Analysis. Pearson Education Inc., Singapore (1994)
2. Kantz, H., Schreiber, T.: Nonlinear Time Series Analysis. Cambridge University Press, Cambridge (1997)
3. Fausett, L.: Fundamentals of Neural Networks. Pearson Education Inc., Singapore (2004)
4. Sprott, J.C.: Chaos and Time-Series Analysis. Oxford University Press, Oxford (2003)
5. Davis, R.A., Brockwell, P.J.: Time Series: Theory and Methods. Springer, New York (1991)

6. Pandit, S.M., Wu, S.M.: Time Series and System Analysis with Applications. John-Wiley, New York (1983)
7. Lim, K.P., Hinich, M.J.: Cross-temporal Universality of Nonlinear Dependencies in Asian Stock Markets. Economics Bull 7(1), 1–6 (2005)
8. Kanwer, A.A.K., Ali, A.A.A.: A New Fail Safe Method for Exchange Level Risk Management. In: The 25th International Symposium on Forecasting, San Antonio, TX, June 12-15 (2005)
9. Irfan, C.M., Nishat, M.: Key Fundamental Factors and Long-run Price Changes in an Emerging Market – A Case Study of Karachi Stock Exchange. The Pakistan Development Review 41(4) (2002)
10. Kamal, Y., Nasir, Z.M.: Day of the Week Effect in Stock Return Evidence from Karachi Stock Market. Social Science Electronic Publishing, Inc. (2005), http://papers.ssrn.com/sol3/papers.cfm?abstract_id=829524#PaperDownload
11. Bokhari, S.M.H., Ferdun, M.: Forecasting Inflation through Econometric Models: An Empirical Study on Pakistani Data. Doğuş Üniversitesi Dergisi 7(1), 39–47 (2006)
12. Rizwan, M.F., Khan, S.U.: Stock Return volatility in Emerging Equity Market (KSE): The Relative Effects of Country and Global Factors. Int. Rev. Business Res. Papers 3(2), 362–375 (2007)
13. Burney, S.M.A., Jilani, T.A.: Time Series Forecasting using Artificial Neural Network Methods for Karachi Stock Exchange. Technical Report, Department of Computer Science, University of Karachi (2002)
14. Burney, S.M.A., Jilani, T.A., Ardil, C.: Levenberg-Marquardt Algorithm for Karachi Stock Exchange Share Rates Forecasting. In: Proc. World Acad. Sci. Eng. Tech., vol. 3 (2004)
15. Takens, F.: Detecting Strange Attractors in Turbulence. In: Rand, D.A., Young, L.S. (eds.) Dynamical Systems and Turbulence, Warwick 1980. LNM, vol. 898, pp. 366–381. Springer, Heidelberg (1981)
16. Fraser, A.M., Swinney, H.L.: Independent Coordinates for Strange Attractors from Mutual Information. Phys. Rev. A. 33, 1134–1140 (1986)
17. Frede, V., Mazzega, P.: Detectibility of Deterministic Nonlinear Processes in Earth Rotation Time Series-I: Embedding. Geophys. J. Int. 137, 551–564 (1999a)
18. Konstantinou, K.I.: Deterministic Non-linear Source Processes of Volcanic Tremor Signals accompanying the 1996 Vatnajökull Eruption, central Iceland. Geophys. J. Int. 148, 663–675 (2002)
19. Hegger, R., Kantz, H., Schreiber, T.: Practical Implementation of Nonlinear Time Series Methods: The TISEAN package. Chaos 9, 413–435 (1999)
20. Kennel, M.B., Brown, R., Abarbanel, H.D.I.: Determining Embedding Dimension for Phase-Space Reconstruction using a Geometrical Construction. Phys. Rev. A. 45, 3403–3411 (1992)
21. Abarbanel, H.D.I.: Analysis of Observed Chaotic Data. Springer, New York (1996)
22. Grassberger, P., Procaccia, I.: Measuring the Strangeness of Strange Attractors. Physica D 9, 189–208 (1983)
23. Grassberger, P., Procaccia, I.: Characterization of Strange Attractors. Phys. Rev. Lett. 50, 346–349 (1983a)
24. Theiler, J.: Spurious Dimension from Correlation Algorithms applied to Limited Time Series Data. Phys. Rev. A. 34, 2427–2432 (1986)
25. Smith, L.A.: Comments on the Paper of R. Smith, Estimating Dimension in Noisy Chaotic Time Series. J. Royal Stat. Society Series B-Methodological. 54, 329–352 (1992)

26. Smith, L.A.: The Maintenance of Uncertainty, in Past and Present Variability in the Solar-Terrestrial System: Measurement, Data Analysis and Theoretical Models. In: Castagnoli, G.C., Provenzale, A. (eds.) Enrico Fermi. Proceedings of the International School of Physics, Società Italiana di Fisica, Bologna, Italy, vol. CXXXIII, pp. 177–246 (1997)

27. Provenzale, A., Smith, L.A., Vio, R., Murante, G.: Distinguishing between low-dimensional Dynamics and Randomness in Measured Time-Series. Physica D 58, 31–49 (1992)

28. Leok, M., Tiong, B.: Estimating the Attractor Dimension of the Equatorial Weather System. Acta Phys. Polonica A 85, 27–35 (1994)

29. Cybenko, G.: Approximation by Supervisions of a Sigmoidal Function. Mathematics of Control, Signals, and Sys. 2, 303–314 (1989)

30. Hornik, K., Stinchcombe, M., White, H.: Multilayer Feed-Forward Networks are Universal Approximators. Neural Net. 2, 359–366 (1989)

31. Smith, L.A.: Intrinsic Limits on Dimension Calculations. Phys. Lett. A 133, 283–288 (1988)

32. Ding, M., Grebogi, E., Ott, E., Sauer, T., Yorke, J.A.: Plateau Onset for Correlation Dimension: when does it occur? Phys. Rev. Lett. A 70, 3872–3875 (1993)

33. Tsonis, A.A.: Chaos: From Theory to Applications. Plenum, New York (1992)

34. Raidl, A.: Estimating the Fractal Dimension, K2-entropy and the Predictability of the Atmosphere. Czechoslovak J. Phys. 46, 293–328 (1996)

# A Novel Approach to Non Coherent UWB Reception

Aamir Riaz, Jean-François Naviner, and Van Tam Nguyen

École Nationale Supérieure des Télécommunications
Communications & Electronics Department
46, rue Barrault, 75 634 PARIS CEDEX 13, France
aamir_amr@yahoo.com, naviner@enst.fr, vtnguyen@enst.fr

**Abstract.** Ultra Wide Band (UWB), as allowed by the FCC, covers a spectrum that involves multiples of gigahertz. Therefore, reception of such a signal requires a slight change from conventional ways while using the same phenomena like demodulation, post amplification and detection. This paper details a novel approach towards reception of UWB following the current European standards (3.1 - 4.80 GHz and 6.0 - 8.5 GHz) maintaining cost, size and power consumption as the key design parameters. The whole process of implementation in the receiver is discussed, with a block-by-block description from antenna to detector. The step-by-step results give an understanding of the functioning of each block. Finally, different real life phenomena are tested in order to judge the performance of the receptor in a more realistic manner.

**Keywords:** Non-Coherent reception, UWB OOK based receiver, Two prong receiver.

## 1 Introduction

Term Ultra-wide-band (UWB) may be used to refer to any radio technology having bandwidth exceeding the lesser of 500MHz or a system with a fractional bandwidth greater than 25 percent. Fractional or relative bandwidth is the ratio of total signal bandwidth to center frequency of the signal.

Technically speaking UWB systems can be of two types; *Pulse based* where a small pulse covers the whole spectrum instantaneously or *OFDM/OFDMA based* where each carrier must be of 500 MHz minimum. Although both types may have their unique utilizations but UWB applications are primarily short range and indoor communication systems operating at high speeds. It can also be used as a locating and a tracking device. More specific applications involve sensor networks, wireless printing, wireless PAN etc [1].

As in all communication systems, UWB receiver is of particular importance in the whole chain of communication. Therefore many design proposals are around. Broadly they can be divided into two categories *coherent* and *non-coherent* receivers. Coherent receivers are usually used in systems involving phase detecting demodulations resulting in a complex; large sized and expensive design examples include Transmitted Reference (TR), RAKE etc. Non-Coherent receivers on the other hand generally involve envelope detecting demodulations. Although their performance is slightly lower than that of coherent ones but they are advantageous in their simplicity, cheapness and size [2].

D.M.A. Hussain et al. (Eds.): IMTIC 2008, CCIS 20, pp. 100–109, 2008.
© Springer-Verlag Berlin Heidelberg 2008

Some of the receivers being used for reception of UWB signal have some unique features for example receiver [3] that are based on the *RAKE* infrastructure are efficient receptors but they are extremely complex to implement. Same simplicity and size problems lie with *Transmitted reference* approach where transmitted reference pulses have to be correlated to get the message signal. *Matched Filter* Receiver accomplished correlation in digital domain. But the processing requirements for the correlation in the digital domain are extremely high due to the sampling rate in the order of GHz. Moreover quantization errors can occur corrupting the estimates. As opposed to more complex RAKE receivers, estimation of individual pulse shapes, path amplitudes, and delays at each multipath component is not necessary for *Energy Detectors Receivers* [4]. Moreover, energy detectors are less sensitive against synchronization errors, and are capable of collecting the energy from all the multipath components for energy detectors. But it may not give accurate output as compared to the previously discussed techniques. A less complex approach towards non coherent reception is the *Differential Receiver*. The key point in this type of reception is that one does not require generation of template signal at the receiver end. Receiver uses the delayed component of the receive signal to synchronize the input. But the performance of the receiver is compromised by reducing its complexity especially when a chain of zeros is received. *Partial Rake Receiver* is a reduced complexity receiver as compared to selective Rake as it does not have diversity. P-Rake does not require full channel estimation and its performance is greatly dependent on the richness of the channel as well as the number of taps used.

This paper details a new approach towards reception of UWB signal. The proposed receiver involves a two prong strategy to receive an Amplitude Shift Keying (ASK) modulated signal [5]. The following section gives details of the implementation of the system in Advanced Design System (ADS). The received signal passes through a range of block sets performing various different functions before a threshold detector outputs the detected pulse. To evaluate the performance of the receiver in more realistic environment different tests are performed and their results are presented with some future enhancement recommendations at the end.

## 2  Implementation

The proposed receiver was implemented and simulated in ADS [6] using non-coherent approach. An ASK modulated pulse [7] was preferred with following characteristics;

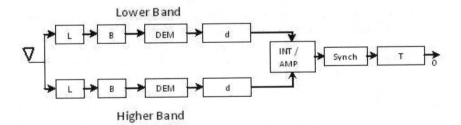

**Fig. 1.** Receiver in Blocks; *L=LNA, B=BPF, D=DEMOD, d=Darlington, IN=integrator, S=Sync. T=detector*

*Period*: 10 nsec, *Rise Time*: 1 fsec, *Fall Time*: 1 fsec, *Period Lower Range*: 590 psec, *Period Upper Range*: 400 psec. Fig. 1 represents a block diagram of the proposed receiver.

The lower and higher end branches explain the two prong strategy used in signal detection. All the blocks are explained in detail next.

## 2.1  Antenna

The design of the antenna was based on research conducted in [8]. The biggest requirement our antenna had to satisfy was that ideally it should give a constant gain in the allowed band of UWB (3.1 - 4.8 GHz and 6.0 - 8.5 GHz) and tries also to block the signal in the unwanted zone (4.8 - 6.0 GHz). For this purpose the antenna devised had the following characteristics and parameters:

a.  *Patch Antenna on FR4 substrate:*
    Thickness = 1.6 mm    Relative Permittivity = 4.4
b.  *Dimensions:*  16.51 mm x 15.01 mm
c.  *Conductor Thickness:*  10 mil

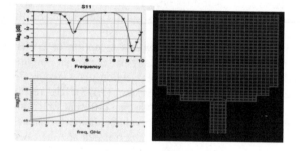

**Fig. 2.** Antenna

The radiation pattern of the antenna at different frequencies with the range 3 - 9 GHz was obtained and that revealed that its pattern is generally Omni directional.

## 2.2  Low Noise Amplifier (LNA)

The signal received by the antenna, in UWB communication, is really low firstly because the transmitted signal was in the orders of −50dBm and secondly due to the phenomena's like attenuation, noise, fading e.tc depending upon the channel in which the system is being operated Therefore it has to be amplified before any further processing could be done. The same LNA is used in both the bands.

A few CMOS designs, proposed in literature, were tried. The proposed approaches focussed mainly on one of the four S-parameters resulting in improvement in one and degradation pf another. Some of them gave good results but the HP's LNA MGA-72543 (ADS part no. sp_hp_MGA-72543_6_19990127), made for adaptive CDMA applications, and provided the perfect solution in terms of all the S parameters.

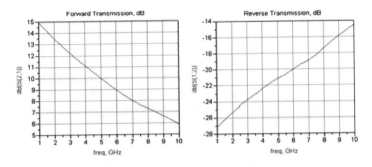

**Fig. 3.** LNA Gain

## 2.3  Band Pass Filters (BPF)

Since we are receiving signals in two separate wide bands, two separate BPFs are used in the lower and upper band branches. The common features between the two filters are:

a. Both of them are passive in order to reduce battery consumption

b. Both are 3rd order filters with maximally flat response in the pass band region.

Although unlike active filter the passive filters do attenuate the received signal to some extent but this could be catered for using the post amplification in the demodulated signal zone.

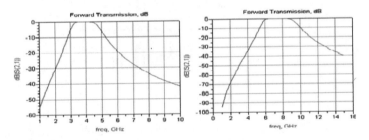

**Fig. 4.** Band Pass Filters

## 2.4  Demodulator and Integrator

A simple envelope detector circuit is used for demodulation. A schottky diode is used in the envelope detector instead of normal diode due to its low voltage drop during forward operation. Darlington circuit using CMOS technology is used in order to ensure that the following circuitry doesn't load the preceding one and the operation is performed using the minimum of battery power.

Integrator is one of the most important blocks in the receptor chain which not only integrates the signal received in two bands but also helps in removing the interference effect up to some extent. Its function can be understood as that of an AND gate which gives high output only when both inputs are high otherwise a low output is the result. This means that in case one of the outputs gets high due to interfering signals (that are going to be common especially in the lower band) then receiver would not generate a false alarm. On the contrary the argument that attenuation might weaken a signal in one band so much so that it is beyond detection cannot be recovered by the receiver is valid. But a receiver operating in an urban indoor environment in for short range communication would face the first situation more than the later. So the compromise in favour of removing interference phenomena is made. CMOS implementation using FETs ensures meeting of the size and power consumption criteria.

## 2.5  Synchronizer

Like in any other receiver synchronizer plays a vital role in reception. In our system its implementation involves the utilization of the first received pulse [9]. When the pulse is detected at the end of the amplifier a source of template signal is triggered. Source produces a signal predefined for a specific baud rate. To change the baud rate simply time period of the pulse train has to be changed. This can be achieved by varying the tuning voltage of the source. Once the template signal is generated, it is correlated with the next incoming pulse. If the pulse received matches in time with the template generated signal, a high pulse is seen on the output as an indication of synchronized output. In cases where the signal received doesn't match in time with the template nothing comes on output. The process of synchronization can be understood by the following flow:

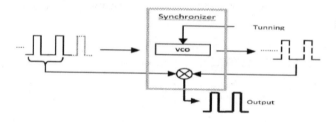

**Fig. 5.** Synchronizer

One thing is to be noted here that the process consumes the first received pulse.

## 2.6  Threshold Detector

One of the challenging issues for the enhancement of UWB receivers is the estimation of the optimal threshold, the other being the synchronization of the incoming signal. Using the approach suggested in [10] we need

- Peak value of the signal when a pulse is transmitted, i.e. a HIGH bit is transmitted. The signal has the effects of interference and noise in it.
- Peak value of the signal when a pulse is not transmitted, i.e. a LOW bit is transmitted. In literal terms, it is the maximum noise value in the whole reception chain. It is the largest amplitude caused by inter-symbol interference when transmit '0'.

It is kept at

*Threshold level = 0.5 x sum (Max Noise Level, Max signal Level)*

# 3   Results

The pulses in two bands used during the simulation had the following frequency domain spectrum:

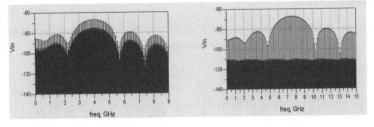

**Fig. 6.** Pulses

Both the spectra have a peak value of about –70dBm which is much lower then allowed transmitting power (–45dBm/MHz) for UWB signals. This is in line with the goal to test the whole receiver for the maximum endurance. The next Figure is the result summary after each block throughout the receptor chain from LNA to detector.

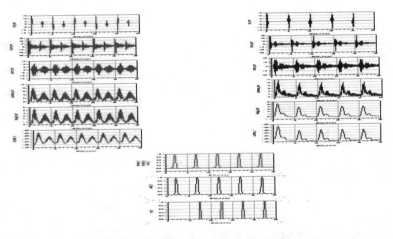

**Fig. 7.** Chain

Results verify the reception of a pulse train using first pulse as a trainer sequence. If you have more than one surname, please make sure that the Volume Editor knows how you are to be listed in the author index.

# 4   Tests

To verify the performance of the receiver, different sets of tests were performed on the receiver. This is done by subjecting the system to more and more severe channel distorted signals. Most of the tests involved the mimicking the same phenomena that the receiver could face during its operation in the real environment in line with the goal to test the whole receiver for the maximum endurance. First these phenomena were introduced one by one and then all at a time.

## 4.1 Multipath Effect

One of the most common of these phenomena's is the Multi-path effect. The following figure will show how our system is subjected to it.

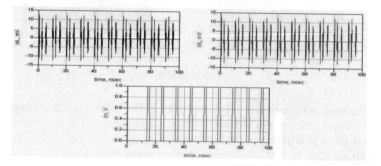

**Fig. 8.** Multi-path Effect

The signal with the highest peak is the one coming from direct path. The second one with delay of 2nsec is the first multi-path components and so on. One thing worth mentioning here is that if the signal from a path other than direct is received with delay less than 2nsec then the integrator will integrate (similar to that in RAKE) it also thereby easing the job of threshold detector.

## 4.2 Noise Plus Multipath

In the next verification noise was added to the same signal (with multi-path). The results were encouraging and are shown below:

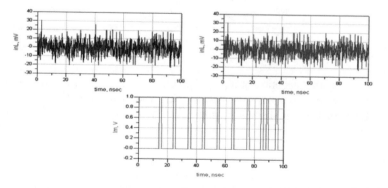

**Fig. 9.** AWGN plus Multi-path

One can clearly observe that the SNR of both the input signals is extremely low and the noise figure is really high. As a result there are some errors in the decision by the threshold. The false alarm can be due to many reasons. For example a multi path signal got a channel that was much better than the one received by the signal coming from direct path. Hence a false alarm occurred.

### 4.3  Narrow Band Interference

This phenomenon is of real importance because of the fact that multiple other systems also work in the same band as that of UWB. Especially many licensed operational systems can be found in the lower portion of the whole band. There can be two scenarios: *Out of band Narrow band Interference* and *In band Narrow band Interference*.

The signals used for Out of band Narrow band Interference had the following spectra:

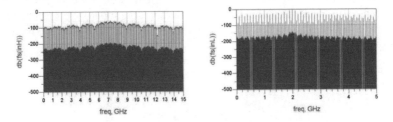

**Fig. 10.** Narrow Band Out of band: Input signal

An interference signal with the following specs was added to the previous multi path noisy signal in lower branch Bandwidth = 200 MHz, Peak power = –5dBm, Centre frequency = 2 GHz. The result of the simulation in time domain is

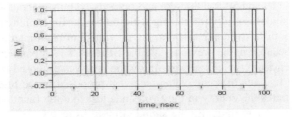

**Fig. 11.** Narrow Band Out of band: Results

At first the system gave a false alarm but as the synchronization is reached the system starts to behave normally without having any effect on its performance due to added interference in the environment.

Multiple simulations were made in this context with different values of power for the narrow band signal. It was notes that system performed satisfactorily when subjected to an interference of –30dBm. As the power is increased errors started to occur. An interfering signal more than –30dBm, means that we are operating our UWB

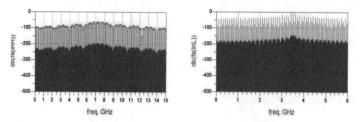

**Fig. 12.** Narrow Band In-band: Input signal

communication system very close to the interference transmitter. The similar graphs, as that of previous case, show the input signal in the frequency domain

Similar to previous case, left hand graph show the old signal in the upper band with noise and multi-paths. The signal on the right graph had following added interference Bandwidth = 200MHz, Peak power = −5dBm, Centre frequency = 3.5 GHz. As the centre frequency is within the band, therefore, the input signal can be regarded as the one affected by a narrow band signal transmitted at high power. The results in time domain are

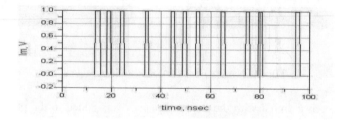

**Fig. 13.** Narrow Band in-band: Results

3 false alarms (19 and 50 and 80nsec) and one missed pulse (about 83nsec) make it evident that the signal is really interfering with the proper functioning of the receiver. The results are poor in the comparison to the previous out band interference which is understandable because in the previous case the BPF blocks out band noise quite well.

As can be seen from the above description the receiver proposed tries to incorporate the positive features of some of the receivers proposed in the literature. For example it accumulates the energy in the two bands just like the functioning of Energy Detector receiver. It can integrate the signal coming from two different paths separated by less than 2msec apart, a concept that is identity of the Rake receiver. Moreover a patch antenna miniaturizes the whole package.

## 5  Conclusion

A unique *Two Branch* implementable model of pulse based UWB non-coherent receiver for European standards has been presented with cost effectiveness, simplicity, size and power consumption as design guidelines. The results of the multiple tests on the receiver to evaluate its performance in the natural environment, with AWGN, multi-paths, fading and interference phenomenon, were satisfactory enough. Although in case of narrow in-band interference the results were not up to the mark but that can be improved by implementing a more sophisticated (more complicated) adaptive threshold detection technique. Moreover the results suggest that if our communication system is at a normal distance from the transmitter of interference the results will be good enough to qualify the receiver.

Further work that can be done in this domain involves testing the receiver in MIMO conditions. This can be approached by using multiple antennas that can be replica of the one suggested. Plus a mechanism allowing a multiple access will increase the utility of the system.

# References

[1] Reed, J.H.: An Introduction to Ultra Wideband Communication System. section 6. Prentice-Hall, Englewood Cliffs (2005)

[2] Kull, B., Zeisberg, S.: UWB Receiver Performance Comparison. In: Conference on Ultrawideband Systems and Technologies (2004)

[3] Durisi, G., Benedetto, S.: Performance of Coherent and Non-coherent Receivers for UWB Communications. In: IEEE conference on Communication, vol. 6, pp. 3429–3433 (2004)

[4] Sahin, M.E., Güvenc, I., Arslan, H.: Optimization of Energy Detector Receivers for UWB Systems. In: IEEE Vehicular Technology Conference, vol. 2, pp. 138–1390 (2005)

[5] Kratzet, S.H.: Elanix, Inc.: Ultra Wide Band (UWB) Transmitter and Receiver Simulation using On/Off Keying (OOK), System View Labs (2002)

[6] Chi, W.S.: Application Notes for ADS, Microwave Laboratory, Department of Electronic Engineering, The Chinese University of Hong Kong (2001)

[7] Win, M.Z., Scholtz, R.A.: Impulse radio: how it works. IEEE Communications Letters 2, 36–38 (1998)

[8] Choi, S.H., Park, J.K., Kim, S.K., Jae, Y., Park, J.Y.: A New Ultra-Wideband Antenna for UWB Applications. Microwave and Optical Technology Letters 40(5) (2004)

[9] He, N., Tepedelenlioglu, C.: Adaptive Synchronization For Non-Coherent UWB Receivers. In: IEEE International Conference on Acoustics, Speech, and Signal Processing (ICASSP), vol. 4, pp. iv-517- iv-520 (2004)

[10] Wang, J., Gong, X., Lin, G., Wang, W., Wang, D.: Design of A Low Complex Impulse Radio Receiver for High Data Rate. In: IEEE Asia-Pacific Microwave Conference Proceedings (2005)

[11] Nathaniel, J.A., Thirugnanam, R., Ha, D.S.: An Adaptive UWB Modulation Scheme for Optimization of Energy, BER, and Data Rate. In: IEEE Conference on Ultra Wideband Systems and Technologies, Kyoto, Japan, pp. 182–186 (2004)

[12] Andrews, J.R.: UWB Signal Sources, Antennas & Propagation. In: Topical Conference on Wireless Communication Technology, Honolulu, Hawaii (2003)

[13] Shen, M., Koivisto, T., Peltonen, T., Zheng, L.R., Tjukanoff, E., Hannu, T.: UWB Radio Module Design for Wireless Sensor Networks. Analog Integrated Circuits and Signal Processing 50(1), 47–57 (2007)

# Load Balancing in EAMC to Integrate Heterogeneous Wireless Systems

Iffat Ahmed[1] and M.A. Ansari[2]

[1] Department of Computer Science, Allama Iqbal Open University, Pakistan
imiffatt@hotmail.com
[2] Dept of Computer Science, Federal Urdu University of Science Arts & Technology, Pakistan
drmaansari@fuuastisb.edu.pk

**Abstract.** In Next Generation (NG) Wireless Networks, end-user is supported with more technologies. Each of the network system has its own characteristic. To utilize best features of each network system, a new architecture *"Enhanced Architecture for ubiquitous Mobile Computing"* has been introduced. Furthermore, load-balancing algorithms have been introduced in this architecture. These algorithms will run after a specified amount of time and all NIAs will broadcast load request to update the load values on each of the NIA in the database. Mathematical performance analysis has been carried out to prove the need for the new architecture and its algorithms.

**Keywords:** Load Balancing, Heterogeneous Networks, Quality of Service, Networks Integration, Hand-off management, Mobility.

## 1 Introduction

There are multiple wireless technologies such as IEEE 802.11; UMTS (Universal Mobile Telecommunication System), Bluetooth and Satellite Networks and all such systems use different radio technologies. Each network system has its own specification and characteristics and these systems are designed for specific purposes. Such networks are required to coordinate with each other. Thus new architecture is required to integrate such wireless systems. Future wireless networks are expected to use features of each other. Therefore, it is the need of today's wireless environment to introduce some architecture for integration of such ubiquitous technologies.

Since each system has its specification and characteristics, therefore each system is designed for providing specific Quality of Service (QoS) parameters. Each of the network system has its own QoS parameters. For example, WLAN may have better signal quality within its domain due to short coverage area, as well as Satellite Network system have larger coverage area. Thus each network system has its own specific Quality of Service parameters.

Each of the network operators are required to have Service Level Agreement (SLA) with one another. In the previous systems, each network had one-to-one SLA with each network, but the architecture proposed by [1] had reduced significant amount of SLAs. In this architecture each network operator is required to have one SLA with AMC.

D.M.A. Hussain et al. (Eds.): IMTIC 2008, CCIS 20, pp. 110–117, 2008.

Thus, if it is desired to utilize the best features of such network systems, they must be integrated. When all such networks are connected then the problems of best network selection and mobility management arise.

Rest of the paper is organized according to the following scheme. Section-II is about related work, Section–III includes the proposed architecture, section–IV describes the proposed solution illustrating the flow diagram and proposed algorithm, Section–V presents mathematical performance analysis to prove the motivation of this research and finally Section – VI concludes the results.

## 2   Related Work

Some work related to integration of different networks has already been done. But still this area requires more advancement. As identified by [1], most of the researchers have tried to integrate pair of wireless systems. Such as [2], the SMART project uses two different network systems to integrate.

### 2.1   Previous AMC Architecture

In the previous AMC architecture proposed by [1], [3] there is only one NIA (Network Inter-operating Agent), which communicates with each of the IG (Inter-working Gateway). Thus the NIA is the single point of communication with all IGs. When the hand-off will take place between two network systems, the IGs of those network systems will communicate with the NIA. Thus NIA is the single point of communication and thus making bottleneck situation.

## 3   Proposed Architecture – Enhanced AMC (EAMC)

### 3.1   EAMC Architecture

Proposed architecture is the enhancement in the already defined AMC architecture. In proposed architecture, there are more than one NIA (that is multiple NIAs), reducing the bottleneck situation. The proposed architecture is shown in Fig. 1.

If we compare the proposed Enhanced AMC architecture to previous one (AMC architecture), we can see that new proposed architecture involves more than one NIA, therefore it is diminishing the bottleneck situation. Thus now when we have more than one NIA, it requires Inter-NIA communication about hand-off. Thus all of NIAs have information about each other. Each NIA has information related to status of each NIA and the load on NIA. These NIAs communicate with other.

The status of NIA is stored in database (described in the next section). The information related to NIA is updated after fixed specified period of time. NIAs also communicate with each other on the basis of load on NIAs. The load on NIA is monitored, if it increases more than the defined threshold value, then it checks the status of other NIAs. Then it compares the loads of each NIA, and the NIA having minimum load is selected for handing-off.

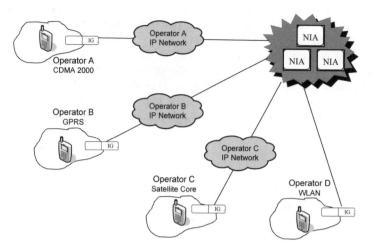

**Fig. 1.** Proposed Architecture: The Enhanced AMC (EAMC)

## 3.2 Components of EAMC

There are two components of previous AMC architecture, which are NIA (Network Inter-operating Agent) and the (Inter-working Gateway) IG. The NIA is the central coordinator, where as IG component reside at each network operator. Whereas, the EAMC incorporates multiple E-NIA and the IG

### 3.2.1 Components of E-NIA & IG

E-NIA and the IG themselves contain sub-components, which are described as follows. The components of IG are shown in Fig. 2 (b).

**Mobility Management Unit**
This component of IG is responsible for implementing Mobile IP (MIP) [4]. To implement the MIP functionalities, it uses Foreign Agent (FA). In [1], to illustrate the authentication process they have used EAP-SIM [5], but there are other techniques like EAP-AKA [6], EAP-SKE [7], EAP-TLS [8], these techniques can be used rather than EAP-SIM. This component also has Seamless Roaming Module, which is basically responsible for seamless inter-system (between different network systems) roaming.

**Traffic Management**
This is another component of IG, which is basically responsible for discarding the upcoming packets from the unauthenticated users

**Authentication & Accounting Unit**
This is responsible for authenticating the user and for billing purposes respectively. If there is only one NIA, it arise bottleneck situation, for this reason the architecture needs more then one NIA, which is our proposed solution.

If there are more than one NIA, then there must be some technique or strategy to handle the inter-NIA communication. For this purpose, all of the NIAs must be aware

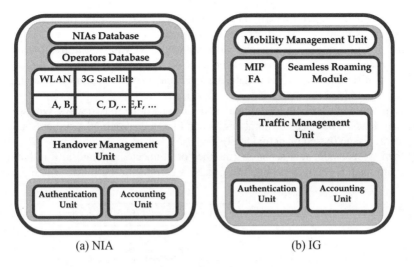

**Fig. 2.** Components of EAMC

of each other, thus it includes additional component in NIA, which will have the information about all the existing NIAs.

The components of E-NIA are shown in Fig. 2 (a).

**Operator Database**
It includes information about all the network operators which have Service Level Agreement (SLA) with NIA. Therefore, those network operators are part of this architecture.

**Handover Management Unit**
This component involves the decision making process about the inter-system handoff (ISHO) should be allowed or not. The ISHO algorithm was proposed by [3], which is responsible for decision-making.

**Authentication & Accounting Unit**
As its name suggest, this component is used to authenticate the users, who are moving between different networks. The Accounting unit is used for billing purposes.

**NIA Database**
Similarly the components of EAMC (proposed solution) also include NIA & IG, but the difference is, there is an additional component of NIA, named as 'NIA Database'. This component is basically the database/repository, in which all the information related to existing NIAs is stored. Also the status of NIA is stored as well as the information related to load on NIA is also present here. The components of EAMC-NIA (Proposed Solution) are depicted in Fig. 4.

## 4   Proposed Solution for EAMC

The flow diagram of 'inter_NIA handoff' is illustrated in Fig. 3. Basically, it continuously monitors the load on NIA. If the load on NIA (denoted by Load) exceeds the threshold value (denoted by Lth), then, it will check / get the status (load values) of all existing NIAs from the 'NIA database' component.

Then it will compare all the load values, and the NIA having minimum load (denoted by Lmin) will be selected, and the inter-NIA handoff initiation will take place. There are different ways to calculate the load and make decision on selecting NIA with minimum load.

The process of checking load on each NIA can be done in two ways. One way is to calculate load on the basis of number of operators under each NIA or based upon the bandwidth usage (based on actual load).

When we consider this scenario (number of connected operators) for finding the minimum load, it entails a problem, if one NIA is transmitting traffic which is mostly

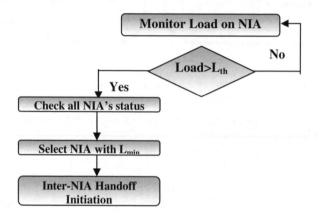

**Fig. 3.** Flow Diagram of Load Management & inter-NIA handoff

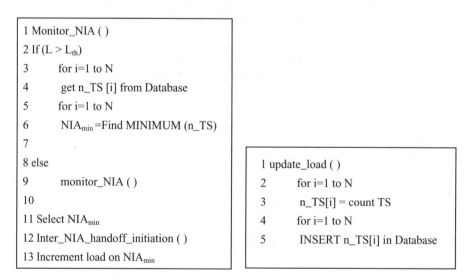

1 Monitor_NIA ( )

2 If (L > Lth)

3       for i=1 to N

4       get n_TS [i] from Database

5       for i=1 to N

6       NIAmin =Find MINIMUM (n_TS)

7

8 else

9       monitor_NIA ( )

10

11 Select NIAmin

12 Inter_NIA_handoff_initiation ( )

13 Increment load on NIAmin

1 update_load ( )

2       for i=1 to N

3       n_TS[i] = count TS

4       for i=1 to N

5       INSERT n_TS[i] in Database

**Fig. 4.** Inter-NIA handoff

**Fig. 5.** DB_Update Algorithm

consisted of real time data and the other NIA is just transmitting very little amount of real time data and more data is non-real time. It does not matter, that how much number of operators is connected with each NIA, rather we can calculate the load based upon the type of traffic it is transmitting and bandwidth usage.

This scenario can be better understood with the help of example. If NIA-1 has 5 number of operators connected to it, and NIA-2 has 8 number connected operators. But suppose NIA-1 is using 30 time slots (TS) and the NIA-2 is utilizing 26 time slots (TS). Thus on the basis of above mentioned scenario, the NIA-1 will be selected as minimum, because it has less number of operators attached to it. If we actually look upon the load on NIAs, then NIA-1 is heavier than NIA-2, because it is occupying most of the bandwidth by utilizing more time slots. This problem can probably be solved by calculating the time slots used by each NIA. The NIA which is using more time slot is the busiest NIA, where as the NIA with minimum time slots usage has the minimum load, because it is occupying less bandwidth. Thus the NIA selection is based upon the number of time slots it is using. This process can be better understood by the algorithm, which is presented in subsequent section (in Fig. 4).

### 4.1  Inter-NIA Handoff Algorithm

The proposed algorithm is based upon the load on NIA. It constantly monitors the load on NIA. Here L denotes the load of the NIA on which algorithm is running at that time and the $L_{th}$ is the fixed threshold load value, that is already predefined. Thus if the load on NIA (L) increases than Lth, then it initiates the process of getting the information about the loads of other NIAs. It gets the values of load on each NIA from the database, which is denoted by n_TS (number of time slots). It compares all the values of n_TS and then selects the minimum n_TS, because the NIA which has minimum load will have the minimum value of n_TS.

Here we are considering the loads based upon the time slot each NIA is using. The NIA acquiring minimum time slots has the minimum load. Therefore, the NIA which has minimum Time Slots is selected. Then after selecting the $NIA_{min}$ that is NIA with minimum load, inter-NIA handoff initiation takes place. When the inter-NIA handoff is initiated, then the load on $NIA_{min}$ is incremented, because the node which was requesting for hand-off is acquiring some of the bandwidth. Thus the time slots used by that node will be added to the load of selected NIA.

### 4.2  Database Load Update Algorithm

There is a need for maintaining the database as well as updating the status of each NIA in the database. Therefore another algorithm (See Fig. 5) is proposed for broadcasting the load request and updating the loads values is proposed.

When this algorithm will run, it will get value of time slots (denoted by TS) used by each NIA. These values are stored in the array named n_TS (Number of time slots). Then each value in an array will be stored/ inserted in the database. Now when the load on NIA will increase than the defined threshold value, then it will get values of NIA's load from the database, in which values of loads are stored by this algorithm. The proposed algorithm will run on each NIA after some fixed specified value of time. This algorithm is necessary for updating the values of load on each NIA.

## 5  Performance Analysis

To measure the load on NIA in AMC, [1] only considers on the ratio for horizontal and vertical handoffs from 3G to WLAN network. But, we must consider handoffs from other networks, like from Satellite to WLAN etc. If we consider one subnet of 3G network, and we denote the number of cells by $R$, then the crossing rate in 3G networks as described by [1] is:

$$R_{sg} = \frac{\rho_g vL_b}{\Pi} \qquad (1)$$

Here $L_b$ is the perimeter of one subnet of 3G network, $v$ is the average user velocity and the $\rho_g$ is the user density of 3G network. Similarly, if we consider the crossing rate between different networks, like from WLAN to 3G or Satellite to 3G:

$$R_{dwg} = N_w \frac{\rho_w vL_w}{\Pi} \qquad (2)$$

$$R_{dgs} = N_g \frac{\rho_g vL_g}{\Pi} \qquad (3)$$

$$R_{dsg} = N_s \frac{\rho_s vL_s}{\Pi} \qquad (4)$$

Here $N_w$, $N_g$, and $N_s$ are the number of networks for WLAN, 3G and Satellite Network Systems respectively. $\rho_w$, $\rho_g$ and $\rho_s$ are the user densities for WLAN, 3G and Satellite Network Systems respectively. Therefore, we have vertical handoffs as follows from equation (2), (3) and (4):

$$H_v = \left( N_g \frac{\frac{\rho_g}{N_g} vL_g}{\Pi} \right) + \left( N_w \frac{\rho_w vL_w}{\Pi} \right) + \left( N_s \frac{\rho_s vL_s}{\Pi} \right) \qquad (5)$$

$$H_v = \frac{N_w \rho_w vL_w + \rho_g vL_g + N_s \rho_s vL_s}{\Pi} \qquad (6)$$

The ratio between the total numbers of vertical handoffs to that of horizontal handoffs is illustrated here.

$$\frac{H_v}{H_h} = \frac{\left( N_w \rho_w vL_w + \rho_g vL_g + N_s \rho_s vL_s / \Pi \right)}{\left( G_s \rho_g vL_b \right) / \Pi} \qquad (7)$$

$$\frac{H_v}{H_h} = \frac{N_w \rho_w vL_w + \rho_g vL_g + N_s \rho_s vL_s}{G_s \rho_g vL_b} \qquad (8)$$

The ratio for vertical handoffs would be much larger than the horizontal handoffs; therefore the load on NIA will be increased. Thus it requires more than one NIA, so that the load can be distributed among all NIAs. Also there is need to control the communication between all the NIAs and load balancing.

# 6  Conclusions

The proposed system is the enhancement in the AMC architecture, named EAMC (Enhanced Architecture for ubiquitous Mobile Computing). The proposed architecture diminishes the bottleneck situation as found in the previous AMC architecture. EAMC also have two components NIA & IG. The NIA of the EAMC has an additional component named "NIA-Database". It stores the status of each NIA and the value of load on each NIA.

Two algorithms are proposed, one for inter-NIA handoff and the other for database update regarding the load on NIAs. First algorithm constantly monitors the load on each NIA, when the load on NIA increases, the threshold value will be checked for the NIA with minimum load, and the handoff will be initiated. Second algorithm is basically introduced to update the load values of each NIA, in the E-NIA component "NIA-Database".

In the performance analysis section it is proved that there is increased load on single NIA, if we compare the ratio of handoffs, therefore, it requires multiple NIAs to diminish the bottleneck situation as well as to make efficient use of resources with the help of load balancing between NIAs.

# References

1. Mohanty, S., Xie, J.: Performance analysis of a novel architecture to integrate heterogeneous wireless systems. Computer Networks 51, 1095–1105 (2007)
2. Havinga, P.J.M., Smit, G.J.M, Wu, G., Vognild, L.: The SMART project: exploiting the heterogeneous mobile world. In: Proceedings of 2nd International Conference on Internet Computing, Las Vegas, NV, USA, pp. 346–352 (2001)
3. Mohanty, S.: A new architecture for 3G and WLAN integration and the inter-system handover management. ACM-Kluwer Wireless Networks (WINET) Journal (in press, 2006)
4. Perkins, C.: IP Mobility Support for IPv4, RFC 3220, IETF (2002)
5. Haverinen, H., Salowey, J.: EAP SIM Authentication, IETF Internet draft, draft-haverinen-pppest-eap-sim-16.txt (2004)
6. Arkko, J., Haverinen, H.: EAP AKA authentication, IETF Internet draft, draft-arkko-pppest-eap-aka-09.txt (2003)
7. Salgarelli, L.: EAP SKE authentication and key exchange protocol, Internet Draft, draft-salgarelli-pppext-eap-ske-03.txt (2003)
8. Aboba, B., Simon, D.: PPP EAP TLS authentication protocol, IETF RFC 2716 (1999)

# A Speech Recognition System for Urdu Language

Azam Beg[1] and S.K. Hasnain[2]

[1] College of Information Technology, UAE University, United Arab Emirates
abeg@uaeu.ac.ae
[2] Pakistan Navy Engineering College, Karachi, Pakistan
hasnain@pnec.edu.pk

**Abstract.** This paper investigates use of a machine learnt model for recognition of individually words spoken in Urdu language. Speech samples from many different speakers were utilized for modeling. Original time-domain samples are normalized and pre-processed by applying discrete Fourier transformation for speech feature extraction. In frequency domain, high degree of correlation was found for the same words spoken by different speakers. This helped produce models with high recognition accuracy. Details of model realization in MAT-LAB are included in this paper. Current work is being extended using linear predictive coding for efficient hardware implementation.

**Keywords:** Urdu speech processing, feature extraction, speaker independent system, machine learning, data pre-processing, modeling.

## 1 Introduction

Speech processing is a diverse, well-researched topic that finds applications in tele-communication, multi-media and other fields. Speech processing in real-time is more challenging than off-line processing. The processing has many facets, for example, distinguishing different utterances, speaker identification, etc.

Many years ago, von Kempelen showed that the speech production system of the human beings could be modeled. He demonstrated this by building a mechanical contrivance that "talked." The paper by Dudley and Tarnocyz [1] relates the history of von Kempelen's speaking machine.

Sounds are mainly categorized into these groups: *voiced* sounds (e.g. vowels and nasals), *unvoiced* sounds (e.g. fricatives), and stop-consonants (e.g. plosives). The speech starts in lungs but is actually formed when the air passes through larynx and vocal tracts [2]. Depending on the status of vocal fold in larynx, the sound can be grouped into: voiced sound that is time-periodic in nature and harmonic in frequency; and the unvoiced sound which is more noise-like [3].

Speech modeling can be divided into two types of coding: *waveform* and *source* [4]. In the beginning, the researchers tried to mimic the sounds as is, and called the technique waveform coding. This method tries to retain the original waveform using quantization and redundancy. An alternative approach makes use of breaking the sound up into individual components that are later modeled separately. This method of utilizing parameters is referred to as source coding.

D.M.A. Hussain et al. (Eds.): IMTIC 2008, CCIS 20, pp. 118–126, 2008.
© Springer-Verlag Berlin Heidelberg 2008

Different characteristics of speech can used to identify the spoken words, the gender of the speaker, and/or the identity of the speaker. Two important features of speech are *pitch* and *formant frequencies* [5]:

(a) Pitch is a significant distinguishing factor among male and female speakers. The frequency of vibration of vocal folds determines the pitch, for example, 300 times per second oscillation of the folds results in 300 Hz pitch. Harmonics (integer multiples of fundamental frequency) are also created while the air passes through the folds. The age also affects the pitch. Just before puberty, the pitch is around 250 Hz. For adult males, the average pitch is 60 to 120 Hz, and for females, it is 120 to 200 Hz.

(b) The vocal tract, consisting of oral cavity, nasal cavity, velum, epiglottis, and tongue, modulates the pitch harmonics created by the pitch generator. The modulations depend on the diameter and length of the cavities. These reverberations are called formant frequencies (or resonances). The harmonics closer to the formant frequencies get amplified, while others are attenuated.

While the humans are speaking, the formants vary depending on the positions of tongue, jaw, velum and other parts of the vocal tract. Two related key factors are: bandwidth of each formant, and formant's membership in a known bandwidth. The vowels for all human beings tend to be similar [5].

Each vowel uttered by a person generates different formants. So we can say that the vocal tract is a variable filter whose inputs are (1) the pitch, and (2) the pitch harmonics. The output of the filter is the gain or the attenuation of the harmonics falling in different formant frequencies. The filter is called *variable filter* model. The transfer function for the filter is determined by the formant frequencies [5].

Rabiner and Schafer's [6] discrete-time model make makes use of linear prediction for producing speech. The vocal tract and lip radiation models use discrete excitation signal. An impulse generator emulates voiced speech excitation; the impulses are passed through a glottal shaping filter. The unvoiced speech is generated by a random noise generator.

Ideally, any features selected for a speech model should be (1) not purposely controlled by the speaker, (2) independent of his/her health condition, and (3) tolerant to any environmental/acquisition noise.

Although the pitch can be easily varied by a speaker, the pitch can be easily filtered for any electrical noise, by using a low-pass filter. Formants on the other hand, are unique for different speakers, and are useful in individual speaker identification. In general, a combination of pitch and formant can be used in a speech recognition system.

Many different schemes have been used in the past for speech feature extraction, for example, *discrete Fourier transform* (DFT), *linear predictive coding* (LPC), *cepstral analysis,* etc. [7]. In this paper, we only use the DFT technique.

Much research has been on done for English (and other major languages) speech processing/recognition, but application of these time-tested techniques has not been investigated for Urdu language. So we consider ours to be the first effort in developing an Urdu language speech recognition system. In this paper, we provide the system (in the form of MATLAB code) so that other researchers can build up on our current efforts.

In Section 2, we include overviews of Fourier transformation and neural networks (NNs). Section 3 covers the speech data acquisition and pre-processing, whereas Section 4 discusses how NNs are used for modeling using MATLAB. In the same section, the results are also included. At the end, we present conclusions and our plan for extending the current work.

## 2  Preliminaries

### 2.1  Overview of Discrete Fourier Transform

DFT is itself a sequence rather than a function of continuous variable and it corresponds to equally-spaced frequency samples of discrete time Fourier transform of a signal. Fourier series representation of the periodic sequence corresponds to discrete Fourier transform of finite length sequence. So we can say that DFT is used for transforming discrete time sequence $x(n)$ of finite length into discrete frequency sequence $X[k]$ of finite length [2].

DFT is a function of complex frequency. Usually the data sequence being transformed is real. A waveform is sampled at regular time intervals T to produce the sample sequence of N sample values; n is the sample number, from $n = 0$ to $N–1$.

$$\{x(nT)\} = x(0), x(T),...,x[(N-1)T]\qquad(1)$$

For length-$N$ input vector $X$, the DFT is a length-$N$ vector $X$ that has elements:

$$X(k) = \sum_{n=1}^{N} x(n).e^{-2\pi j.(k-1).(\frac{n-1}{N})}, \quad k = 1...N.\qquad(2)$$

MATLAB function $fft(X)$ calculates the DFT of vector $X$. Similarly, $fft(X, N)$ calculates $N$-point FFT, but padded with zeros if $X$ has fewer than $N$ points, and truncated if it has more.

### 2.2  Overview of Neural Networks

NNs have proven to be a powerful tool for solving problems of prediction, classification and pattern recognition [8], [20]. The NNs are based on the principle of biological neurons. An NN may have one or more input and output neurons as well as one or more *hidden* layers of neurons interconnecting the input and output neurons. In one of

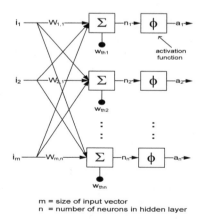

m = size of input vector
n = number of neurons in hidden layer

**Fig. 1.** Structure of a simple feed-forward neural network

the well-known NN types, the outputs of one layer of neurons send data (only) to the next layer (Fig. 1), thus being called *feed-forward NNs*. *Back-propagation* is a common scheme for creating (*training*) the NNs. During the process of NN-creation, internal *weights* ($w_{i,j}$) of the neurons are iteratively adjusted so that the outputs are produced within desired accuracy. The training process requires that the *training set* (known examples/input-output datasets) be chosen carefully. The selected dataset usually needs to be pre-processed prior to being fed to a NN [19], [20].

## 3 Speech Data Acquisition and Processing

The speech recognition system presented in this paper is limited to individual Urdu numerals (0 to 9). The data was acquired by speaking the numerals into a microphone connected to MS-Windows-XP based PC. Fifteen speakers uttered the same number set (0 to 9), specifically, *aik*, *do*, *teen*, *chaar*, *paanch*, *chay*, *saat*, *aat*, and *nau*. All sound samples were curtailed to the same time duration.

The following data processing steps can be generally used for preparing the data for NN training [19], [22]:

(1) *Pre-processing*: This step may involve logarithmic or some transformation. Normalization may also be needed.

(2) *Data length adjustment*: DFT execution time depends on exact number of the samples ($N$) in the data sequence [$xK$]. It is desirable to choose the data length equal to a power of two.

(3) *Endpoint detection*: This step isolates the word to be detected from the following 'silence'.

(4) *Windowing*: A window is applied to select a smaller time than the complete sample. *Hamming window* is one of the widely used windowing algorithms.

(5) *Frame blocking*: Here we make use of the property that the sounds originate from a mechanically slow vocal tract. This assumed stationarity allows overlapping frames of 100 ms or so.

(6) *Fourier transform*: MATLAB's built-in DFT function provides symmetric data from which the lower half can be used for NN training/testing. Frequency spectrum for numeral *aik* (one) is shown in Fig. 2.

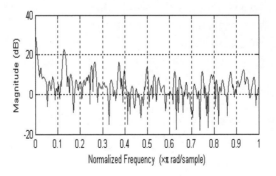

**Fig. 2.** Frequency representation of number *aik* (one)

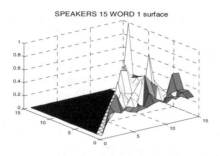

**Fig. 3.** The surface plot of the correlation of the spoken Urdu numbers spoken *aik* (one) by 15 different speakers

We observed that Fourier description of the same number uttered by different speakers had high correlation. Fig. 3 shows the surface plot for correlation of number *aik* (one) by 15 different speakers.

## 4 Neural Network Topology and Implementation in MATLAB

We used multi-layer feed-forward networks in this research. An *N*-layer feed-forward NN is created by *newff* MATLAB command [23]:

```
net = newff(PR,[S1, S2, ..., SN1],{TF1
TF2...TFN1},BTF,BLF,PF)
```

where

| | | |
|---|---|---|
| PR | = | Rx2 matrix of min and max values for R input elements |
| S1 | = | Size of *i*th layer, for N1 layers |
| TF*i* | = | Transfer function of *i*th layer. (Default = 'tansig') |
| BTF | = | Backprop network training function. (Default = 'trainlm') |
| BLF | = | Backprop weight/bias learning function. (Default = 'learngd') |
| PF | = | Performance function (default = 'mse') |

The NN is created and trained with the following set of MATLAB commands:

```
% p[] = array of training inputs
% t[] = array of targets/outputs
% minmax_of_p = array containing mins and maxs of p[]

% network configuration
S1 = 10; % layer-1 neurons
S2 = 10; % layer-2 neurons
net = []; % network reset

% two-layer network declaration
net =
newff(minmax_of_p,[S1,S2],{'tansig','purelin'},'traingd');
```

```
% training and display parameters
net.trainParam.show = 50;
net.trainParam.lr = 0.01; % lrn rate
net.trainParam.goal = 0.01; % trg tol
net.trainParam.epochs = 5000;
% finally the training happens with:
net = train(net,p,t);
```

From the overall dataset of 150 speech samples (10 numbers spoken by 15 different speakers), we used 90% as the *training set*, and set aside the remaining 10% as *validation/testing set*. The learning rates (LRs) of 0.01 and 0.05, and training error tolerance of 0.01 were used. Maximum *epoch* count was limited to 5000.

The NNs used 64 FFT magnitudes as inputs, and produced ten separate predictions (one for each number). This means that we had 64 neurons in the input layer, and 10 neurons in the output layer. We experimented with many different sizes of the hidden layer(s), i.e. 10 to 35 neurons. In single hidden-layer NNs, the best learning accuracy was observed with 10 neurons; additional neurons resulted in accuracy deterioration (Table 1). However, among the 2-hidden layer NNs, the largest network (64-35-35-10) had the best learning accuracy. Table 2 shows the testing/validation accuracies for the trained NNs; the accuracies are reasonably high, i.e. between 94-100%. Unexpectedly, the training iterations (epochs), however, reduced with higher number of hidden layer neurons.

**Table 1.** Training/learning of neural networks with different configurations

| S. No. | Layer configuration | Learning accuracy % |
|---|---|---|
| 1 | 64-10-10 | 81.5% |
| 2 | 64-20-10 | 81.5% |
| 3 | 64-30-10 | 72.6% |
| 4 | 64-40-10 | 83.7% |
| 5 | 64-60-10 | 62.9% |
| 6 | 64-80-10 | 45.9% |
| 7 | 64-10-10-10 | 64.4% |
| 8 | 64-15-15-10 | 69.6% |
| 9 | 64-20-20-10 | 58.5% |
| 10 | 64-25-20-10 | 71.9% |
| 11 | 64-25-25-10 | 71.9% |
| 12 | 64-30-30-10 | 72.6% |
| 13 | 64-35-35-10 | 94.0% |

**Table 2.** Testing of neural network with different speakers

| S. No. | Training speaker | Testing speaker | Prediction accuracy % |
|---|---|---|---|
| 1 | 15 | 5 | 94.3 |
| 2 | 12 | 4 | 95.0 |
| 3 | 10 | 3 | 96.7 |
| 4 | 7 | 2 | 95.1 |
| 5 | 5 | 2 | 100.2 |

The effect of LR on training accuracy is shown in Figs. 4 and 5. In our experiments, a larger value of LR (0.05) took longer to train as compared to the LR of 0.01, although in both cases, the NNs were able to meet the maximum prediction error criterion of 0.01.

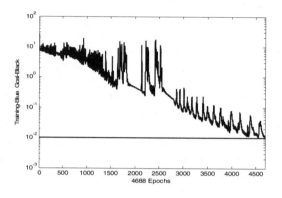

**Fig. 4.** Training epochs and error with LR=0.05

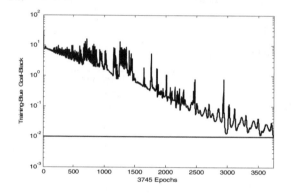

**Fig. 5.** Training epochs and error with LR=0.01

## 5   Conclusions and Future Work

Ours is the first known attempt at using an NN to recognize spoken Urdu language words (numbers, in this case). The DFT of the acquired data was used for training and testing the speech recognition NN. The network made predictions with high accuracy. Currently, we are investigating use of LPC and cepstrum analysis for the same purpose. Our ultimate goal is to implement the Urdu speech recognition in hardware, for example, a DSP chip. The proposed system can be used in applications such as multimedia, telecommunications, and voice-driven customer service, in countries such as Pakistan and India.

# References

1. Dudley, H., Tarnoczy, T.H.: The speaking machine of Wolfgang von Kempelen. J. Acoust. Soc. Am. 22, 151–166 (1950)
2. Parsons, T.: Voice and Speech Processing. McGraw-Hill, New York (1986)
3. Fant, G.C.M.: Acoustic Theory of Speech Production. Mouton, Gravenhage (1960)
4. Deller, J.R., Proakis, J.G., Hansen, J.H.L.: Discrete-Time Processing of Speech Signals. Prentice-Hall, Englewood Cliffs (1993)
5. O'Shaughnessy, D.: Speech Communication: Human and Machine. Addison-Wesley, Reading (1987)
6. Rabiner, L.R., Schafer, R.W.: Digital Processing of Speech Signals. Prentice-Hall, Englewood Cliffs (1978)
7. Varho, S.: New Linear Predictive Systems for Digital Speech Processing. PhD dissertation, Helsinki University of Technology, Finland (2001)
8. Caudill, M.: AI Expert: Neural Network Primer. Miller Freeman Publications, New York (1990)
9. Beg, A., Prasad, P.W.C.: Applicability of Feed-Forward and Recurrent Neural Networks to Boolean Function Complexity Modeling. Expert Systems With Applications 36(1) (in press, 2008)
10. Beg, A., Chu, Y.: Modeling of Trace- and Block-Based Caches. Journal of Circuits, Systems and Computers 16(5), 711–730 (2007)
11. Beg, A., Chu, Y.: Utilizing Block Size Variability to Enhance Instruction Fetch Rate. Journal of Computer Science and Technology 7(2), 155–161 (2006)
12. Singh, A.K., Beg, A., Prasad, W.C.: Modeling the Path Length Delay (LPL) Projection. In: Proc. International Conference for Engineering and ICT (ICEI 2007), Melaka, Malaysia, November 27-28 (2007)
13. Beg, A., Ibrahim, W.: An Online Tool for Teaching Design Trade-offs in Computer Architecture. In: Proc. International Conference on Engineering Education (ICEE 2007), Coimbra, Portugal (2007), http://icee2007.dei.uc.pt/proceedings/papers/523.pdf
14. Beg, A.: Predicting Processor Performance with a Machine Learnt Model. In: IEEE International Midwest Symposium on Circuits and Systems, MWSCAS/NEWCAS 2007, Montreal, Canada, pp. 1098–1101 (2007)
15. Prasad, W.C., Beg, A.: A Methodology for Evaluation (APL) Time Approximation. In: Proc. IEEE International Midwest Symposium on Circuits and Systems (MWSCAS/NEWCAS 2007), Montreal, Canada, pp. 776–778 (2007)
16. Senanayake, N.A., Beg, A., Prasad, W.C.: Learning Monte Carlo Data for Circuit Path Length. In: Proc. International Conference on Computers, Communications & Control Technologies, CCCT 2007, Orlando, Florida (2007)
17. Beg, A., Prasad, W.C., Arshad, M., Hasnain, K.: Using Recurrent Neural Networks for Circuit Complexity Modeling. In: Proc. IEEE INMIC Conference, Islamabad, Pakistan, pp. 194–197 (2006)
18. Prasad, W.C., Singh, A.K., Beg, A., Assi, A.: Modeling the XOR/XNOR Boolean Functions' Complexity using Neural Networks. In: Proc. IEEE International Conference on Electronics, Circuits and Systems (ICECS 2006), Nice, France, pp. 1348–1351 (2006)
19. Beg, A., Prasad, W.C.: Data Processing for Effective Modeling of Circuit Behavior. In: Proc. WSEAS International Conference on Evolutionary Computing (EC 2007), Vancouver, Canada, pp. 312–318 (2007)

20. Prasad, W.C., Beg, A.: Data Processing for Effective Modeling of Circuit Behavior. Expert Systems with Applications 38(4) (in press, 2008)
21. Koolwaaij, J.: Speech Processing, http://www.google.com/
22. Al-Alaoui, M.A., Mouci, R., Mansour, M.M., Ferzli, R.: A Cloning Approach to Classifier Training. IEEE Trans. Systems, Man and Cybernetics – Part A: Systems and Humans 32(6), 746–752 (2002)
23. MATLAB User's Guide. Mathworks Inc. (2006)

# The Future of Satellite Communications in Pakistan

Danish Hussain Memon

Satellite Research & Development Center, Karachi
Pakistan Space and Upper Atmosphere Research Commission (SUPARCO)
P.O. Box 8402, Karachi 75270, Pakistan
danishmemon35@yahoo.com

**Abstract.** The use of satellites in communication systems is a fact of everyday life, as is evidenced by the many homes which are equipped with antennas, or "dishes", used for reception of satellite television (DTH). What may not be so well known is that satellites form an essential part of telecommunications systems worldwide, carrying large amounts of data and telephony traffic, in addition to television signals. Keeping in view the importance of satellite communications as a key to success for any developing country, this paper describes some essential trends in the development of satellite communications and introduces promising future applications for Pakistan.

**Keywords:** VSAT, DTH, Telemedicine, Teleconferencing, SCADA, Air Navigation.

## 1 Introduction

For more than a century, telecommunications was synonymous with telephony, a public utility along with water, electricity, and gas [1]. It is now fully realized that its role in modern society is much more important, and the transmission of information is taking a multitude of forms. New trends are becoming apparent, stemming from the end users placing increased emphasis on particular new aspects of the services they require. There is a growing need for services that imply simultaneous connections between multiple points; these services can be of the conversational type (e.g. video-conferencing), of the distribution type (e.g. TV, electronic mail, information dissemination), or of the retrieval type (e.g. data collection). There is an increased demand of services involving the transmission of video, image, or data material in larger quantities, requiring a wider bandwidth than telephony. The business world is expressing more interest in the setting up of private networks tailored to suit the user's specific needs and offering immediate availability with high reliability. There is an urgent demand for specific telecom services from mobile users traveling in vehicle, aircraft and small boats. More importance is being assigned to communications with remote areas.

Satellites on the other hand, have all the characteristics needed to cope rapidly with these new demands as they emerge.

- They are ideal vehicles for multipoint connections, for distributing and disseminating information to, and collecting data from, geographically dispersed locations.
- They make it possible to establish wide-band links across considerable distances, and the fact that they allow both multiple access at the transmitting end and multiple destinations at the receiving end makes switching centers.

D.M.A. Hussain et al. (Eds.): IMTIC 2008, CCIS 20, pp. 127–132, 2008.

- Present day technology makes it possible for end users to have direct access to satellites by means of small earth stations installed on their own premises, so that private networks can be readily set up without being confronted with difficult interface problems.
- Satellites lend themselves perfectly to the handling of mobile service transmissions for communicating with ships or with land vehicles and aircraft.
- Communications by satellites are insensitive to the distances involved and ignore national boundaries.
- They can serve remote, sparsely populated areas as readily as highly populated centers.

In addition, satellite systems can be built on a time scale that is independent of the location of the users. They can easily be modified to suit the locations to be served and variations in traffic densities over different routes. In the long run, once the traffic pattern is stable, it would be possible for terrestrial networks to take over some to the traffic where this is clearly advantageous.

## 2  Present Scenario of Satellite Communications in Pakistan

In orbit satellite HGS-3 has been acquired by Pakistan from M/s Hughes Global Services on "Full Time Leasing" and relocated to Pakistan's reserved slot at 38 Degree in the geostationary orbit. After a series of orbital maneuvers, the Satellite was stabilized at the final location on December 20, 2002 with 0-degree inclination. The satellite is in position at the Pakistan's-licensed orbital location, 38° east longitude [2].

PAKSAT-1 provides low cost, high quality satellite communications services in both C- and Ku-bands to Pakistan, Africa and the Middle East. PAKSAT-1, 30 C-band transponders and 4 Ku-band transponders provide the total range of satellite communications capabilities. The satellite is in a geostationary orbit and carries high power payloads in both-bands. Currently PAKSAT is providing following services for Public and Private Sector.

- Internet backbone extension
- Business VSAT networks
- Point-to-point data services
- Direct-to-home
- Remote Internet access
- Thin route telephony support
- Broadcast services (video and data)
- Shipboard communications

## 3  Promising Future Applications Using VSAT Networks

### 3.1  Introduction of VSAT Networks

It was in 1965, with the launch of the Early Bird satellite, that a global commercial communication satellite was first used to connect people throughout the world. However, it was in the 1970s that satellites were used for the first time for communication within individual countries, with the US and Canada leading the way. But, at that time, satellite solutions were costly and ground terminals were out of reach of small companies [3].

VSAT technology emerged in the late 1970s as a satellite-based, low-cost approach for connecting multiple locations in private networks for data interchange. VSAT technology in the 1980s was used for corporate communication infrastructure. Finally, developing countries found the use of VSAT networks acceptable only in the 1990s.

During the last five years, satellite communications has emerged as the main technology for broadcast TV channels and high-quality digital audio programming services. Furthermore, Internet over satellite has also become popular. Currently, the request for satellite communications is such that one satellite is going up practically every month over the Asia-Pacific region.

## 3.2  Emerging VSAT Services

Privatization and deregulation were first introduced in developing countries around a decade ago. Before that, telecommunications services were supplied by government-controlled monopoly operators who had their own priorities. In fact, even after privatization and deregulation, the growth of VSAT services remained slow, as rules and regulations remained in favor of monopoly operators. But during the last three to four years, regulations in many countries have been rationalized to a great extent, providing equal opportunities to all service providers. Currently, the economy of Pakistan is once again on the upswing. Therefore, there is great potential for the growth of VSAT services in Pakistan [4].

VSAT services are extensively used to provide voice, data and fax services to various private corporate business houses and service organizations. Most of these organizations, like PSO and others have their own closed user group VSAT services communications network.

During the last five years, the following sectors have been increasingly using VSAT services in the subcontinent:

- Financial sector - including stock exchanges
- Banking sector for online transactions
- Corporate communications for data/voice/video conference
- Rural communications - basic service operators
- Education and training through distance learning
- Security communication by defense, police and intelligence agencies
- Internet
- Health Care services for rural areas

We deal with each of these application areas in turn.

## 3.3  Banking and Financial Sector

At present major financial and banking institutions are extensively using VSAT networks for data and voice transmission. The VSAT networks connect the database centers reliably with the different agencies and branches. Therefore the capacity for reinstallation and configuration is very important for these institutions. A few banking sector applications include

- Credit card approvals
- Automated teller machines
- Teller terminals
- Platform automation
- Cash flow control
- Points of sale transactions

As we know that the banking and financial sectors are having fast and steady growth in Pakistan. VSAT networks can play a vital role in their development and can provide a secure medium for their applications.

### 3.4 Teleconferencing

Teleconferencing allows corporations with branches in different geographic locations to meet without the costs or risks of participant travel. This saves time, which is a key to success of modern companies. In addition to the transmission of images and voices, it is possible to send digitized documents. Keeping in view the reliability and economical concerns, teleconferencing can be used by corporate and public sectors of Pakistan for virtual meetings.

Apart from corporate use, teleconferencing can be used for Education and training in Pakistan for distance learning. Currently a few organizations like virtual university and Allama Iqbal open university are providing distance learning by conventional means. Teleconferencing can be used to make the distance learning easy and simple.

### 3.5 Industries

Data gathered from the SCADA (Supervisory Control & Data Acquisition) system can be transmitted to a database control centre and processes can be controlled from a distance. Various process plants require their monitoring to be made remotely. As is the case of Oil & Gas Exploration Fields, Power Generating Stations, Water Purification Plants, Water Reservoirs Monitoring, National Power Grid Stations, various mines and fields etc. They all require remote data connectivity. As is the case of Pakistan, VSAT networks can play a vital role for these applications.

### 3.6 Commerce Applications

VSAT networks also have applications for the retail environment or points of sale, such as supermarket chains, fast food chains, drugstores and stores that constantly control sales and operations for each branch.

Currently Pakistan State Oil and other companies are using VSAT networks for their control sales and operations applications in Pakistan. There are many more companies those have there setups in different cities of Pakistan. They can use this facility for constantly control sales and operations related tasks.

### 3.7  Thin Route Telephony

Large national telephone companies can use VSAT to extend networks into remote regions. About three to five local access lines are terminated into a centrally-located VSAT terminal. This type of installation, referred to as 'thin-route telephony', can provide suitable and cost-effective communications for remote hilly regions.

### 3.8  Disaster Management

VSAT terminals have been effectively used in developing countries to provide reliable and efficient communication for disaster management. Droughts, floods and coastal cyclones strike South Asian countries almost every year. In addition, Pakistan and other neighboring countries are also prone to earthquakes.

Due to these catastrophes nearly all conventional communication networks totally collapse. But in this situation VSAT network can be installed and made operational in a short time for communications related to disaster management. Subsequently, all district headquarters can be connected through a VSAT network until the main terrestrial public network is restored. The lack of even basic telephony services in most parts of rural and remote areas have made the relief operations difficult. Therefore, it is proposed that VSAT networks can be utilized for disaster management.

### 3.9  Telemedicine

In Pakistan the health care facilities in the rural areas are worst. Providing hospitals with doctors and paramedical staff, to every village in Pakistan is not possible due to shortage of manpower and financial resources. VSAT based Telemedicine network will perform following tasks.

- Patient live interviews and examinations
- Consultation with medical specialists
- Healthcare and educational activates through video conferencing

Pakistan health care system ranges from urban academic institute to small rural clinics through out the country. The legacy of recent decades is an inappropriate distribution of health practitioners and expertise that are concentrated in major urban centers, while people living in rural areas have limited access to basic health care because of geographical isolation. Currently a few Telemedicine centers are established by the government of Pakistan. Keeping in view the scenario of health care facilities in Pakistan it is strongly recommended to install as many as possible VSAT Telemedicine networks in rural areas. The benefits gained from these VSAT based Telemedicine networks are:

- Healthcare facilities accessible to under-served rural populations
- Easy and quick access to specialists
- Cut down cost of travelling and associated costs for patients
- Better-organized and low cost healthcare
- Continuous education and training for rural healthcare professionals
- No need of Land lines (Telephone lines) at remote end
- This service will greatly improve health care services and will reduce medical costs

### 3.10  Airport Infrastructure and Air Navigation

In the field of airport infrastructure and air navigation, the application offers the service of data collection from secondary radar in remote places and transmission to air traffic control centre via satellite. This process will improve air traffic control infrastructure. Keeping in view the benefits of VSAT Networks, it can be implemented in Pakistan for airport infrastructure and air navigation controlling.

Other than the above-mentioned applications of Satellite communications using VSAT Network, satellite communications can be implemented in several other fields.

## 4  Conclusion

In Pakistan currently PAKSAT is working as a key service provider in the field of satellite communications, offering various applications. The development pace required in the field of satellite communications, to cope with the current social and corporate challenges faced in Pakistan is far away.

There are geographical compulsions in developing countries, because general infrastructure is not developed. As a result, the penetration of VSAT networks in developed and developing regions will always be different.

Regulators and government in Pakistan are now more sensitive to the industry needs, and the use of VSATs is expected to pick up fast. Deregulation, rising demand, globalization, technological advances and substantial price decline continue to drive the VSAT market in developing countries.

The major regions where the satellite communications can be implemented using VSAT Networks are Education, Healthcare, disaster management, banking and finance, commerce, and industrial sector. To overcome the challenges of the new era not only government of Pakistan, but private sector is also required to play a vital role.

## References

1. Evans, B.G. (ed.): Satellite communication systems. The Institution of Electrical Engineers (1999)
2. PAKSAT International (Private) Limited (PAKSAT), http://www.paksat.com.pk/
3. Bartholome, P.: The Future of Satellite Communications in Europe. IEEE Journal on selected areas in communications 05(04) (1987)
4. Pakistan Space and Upper Atmosphere Research Commission (SUPARCO), http://www.suparco.gov.pk

# The Small World of Web Network Graphs

Malik Muhammad Saad Missen, Mohand Boughanem, and Bruno Gaume

Institut de Recherche en Informatique de Toulouse (IRIT), 118 Route de Narbonne,
F-31062 TOULOUSE CEDEX 9, France
{Missen,Bougha,Gaume}@irit.fr

**Abstract.** The World Wide Web has taken the form of a social network and if
analyzed carefully, we can extract various communities from this network
based on different parameters such as culture, trust, behavior, relationships, etc.
The Small World Effect is a kind of behavior that has been discovered in entity
networks of many natural phenomena. The set of nodes in a network showing
Small World Effect form a local network within the major network highlighting
the properties of a Small World Network. This work analyzes three different
web networks, i.e. Term-Term similarity network, Document-Document similarity network, and Hyperlinks network, to check whether they show Small
World Network behavior or not. We define a criterion and then compare these
network graphs against that criterion. The network graph which fulfills that criterion is declared to be a Small World Network.

**Keywords:** Information Retrieval, Small World Network, Social Network, Hyperlink Analysis.

## 1 Introduction

Information retrieval systems are designed to help find required information stored in
a certain database. Search engines are the type of information retrieval systems that
not only minimize the time to find particular information but also minimize the
amount of information that must be consulted. The search engines being used to retrieve information from the World Wide Web (WWW) are called *Web Search Engines*. The users express their information need in the form of textual queries to the
Web Search Engine which, on behalf of query terms, searches its indexed collection
of web pages to find such a result set of pages that is likely to satisfy users' information need. This result is then presented to the user in the form of an ordered list. This
ordering is done according to measure of relevance of web pages with the user query,
i.e. the web page with higher measure of relevance with the query is ranked higher
than others in the list. This process is called *Page Ranking* [1].

Classical information retrieval techniques (like *Vector Space Model*) use ranking
algorithms that check the contents of the web pages against the terms in the query, i.e.
they will search the query terms in the content of the web pages and any web page
having most number of query terms will be ranked higher than others in the collection
[1]. But classical information techniques could not do as well as was expected on
web. The reason is that many web page authors want their web pages to be ranked

D.M.A. Hussain et al. (Eds.): IMTIC 2008, CCIS 20, pp. 133–145, 2008.
© Springer-Verlag Berlin Heidelberg 2008

higher so they modify their web pages in many ways. For example, they repeat some important terms many times or add valuable text in invisible font in their pages to get caught by search engines. The process of crafting web pages for the sole purpose of increasing the ranking of these or some affiliated pages, without improving the utility to the viewer, is named as *Web Page Spamming* [2]. Spamming results in wrong ordering of web pages that not only creates problems for the users but also for the web site owners involved in online commercial business. However with the passage of time, a number of techniques with improved ranking algorithms have been developed and *Hyperlink Analysis* is one of the most popular techniques being adopted by web search engines.

Hyperlink Analysis definitely improves the ordering of pages in the result set returned by the search engines. The basic idea behind the concept of Hyperlink Analysis is that link on page A to another page B is in fact the recommendation of page B from page A, i.e. page A and page B both may be on the same topic [3]. However different search engines may implement this idea in different ways [3], [4].

Hyperlinks in fact represent the relationship between individuals (node) of a network (let social network). Given a network of nodes, various patterns of relationship can be observed between different set of nodes that can result in creation of different communities within the network. One of the most popular phenomena discovered in networks is The Small World Phenomena and is explained in next sub-section in detail.

## 1.1 The Small World Networks

*"In mathematics and physics, a Small-World Network is a type of mathematical graph in which most nodes are not neighbors of one another, but most nodes can be reached from every other by a small number of hops or steps or links in other words. A Small-World Network, where nodes represent people and edges connect people that know each other, captures the small world phenomenon of strangers being linked by a mutual acquaintance* [5]".

Examples of Small World Networks are the film actors' network, Paul Erdos' publication network, the organization of board members of big corporation, a grid of electrical power lines in western USA, food web of fish in ocean, etc.

### 1.1.1 Characteristics of the Small World Network

According to Wattz and Strogatz [6], a Small World Network graph has the following properties:

1. The clustering coefficient $C$ is much larger than that of a random graph with the same number of vertices and average number of edges per vertex.
2. The characteristic path length $l$ is almost as small as $l$ for the corresponding random graph.

Also according to [7], the degree distribution graph of a small world network follows a *Power Law of Degree Distribution*, i.e. if the degree distribution graph of a network follows *Power law* then that network can be declared as a Small World Network In short, we have got two ways for a graph to be declared as Small World Network: First, the degree distribution graph of the graph should follow the power law of degree distribution; OR; Second, the clustering coefficient $C$ of the graph should be larger than corresponding random graph and $l$ should be almost as small as $l$ for corresponding random graph.

This research is focused on observing different web networks to check whether they show Small World Network behavior or not. Three web networks graphs being observed in this regard are Term-Term similarity network graph, Document-Document similarity network graph and Hyperlink network graph. These network graphs are compared against the criterion defined above and the network graph which fulfills this criterion is considered as Small World Network.

The organization of the paper is as follows: Section 2 describes the previous work done regarding Small World Networks and in section 3, we define research objectives and research methodology. Results and their interpretations have been given in section 4 and paper summary has been described in section 5 and at the end important references have been given.

## 2    Related Work

The foundation of work regarding Small World Networks was laid by the famous experiments, i.e. small world experiments of Milgram [8]. These experiments were conducted to investigate the Small World Phenomenon [5] by examining the *Average Path Length* for social networks of people in the United States. A certain category of Small World Networks were identified as a class of random graphs by Watts and Strogatz in 1998 [6]. They noted that graphs could be classified according to their clustering coefficient and their mean-shortest path length. The small-world property has been observed in a variety of other real networks, including biological and technological ones [9], [5], and is an obvious mathematical property in some network models, as for instance in random graphs.

At variance with random graphs, the small-world property in real networks is often associated with the presence of clustering, denoted by high values of the clustering coefficient. For this reason, Watts and Strogatz, in their seminal paper, have proposed to define small-world networks as those networks having both a small value of $l$, like random graphs, and a high clustering coefficient $C$, like regular lattices [9].

Adamic [7] showed that the World Wide Web (WWW) is a small world, in the sense that sites are highly clustered yet the path length between them is small. Adamic also demonstrated the advantages of a search engine which made use of the fact that pages corresponding to a particular search query can form small world networks.

## 3    Research Objectives and Methodology

The purpose of this research is to verify the presence of Small World Effects in web networks graphs showing different aspects of relations between their vertices. These networks graphs then are verified against the properties of Small World Networks as defined above.

We use three web network graphs for our work. First network is a network of French language terms where an edge exists between vertices, i.e. terms if two terms are semantically same. The second is the network of documents in which each document is represented by a vertex and two documents are linked if they are similar to some extent. The similarities between the documents are calculated by the Vector Space Model [1] (or *tf.idf* method). The third network, i.e. the network of hyperlinks between documents where the vertices represent the documents and an edge exists between two documents A and B if A has a link destined to document B. First, second and third networks are represented by three matrices as Term-Term Matrix, Document Similarity Matrix and Hyperlink Matrix respectively.

We develop all the programs to calculate the three required properties, i.e. clustering coefficient C, mean of the shortest path *l* and the degree distribution graph using Matlab and Java programming Language.

## 3.1 Clustering Coefficient

The clustering coefficient (C) in fact is the density of triangles in a network graph. We have used the following formula for calculating this coefficient throughout our calculation:

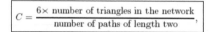

$$C = \frac{6\times \text{ number of triangles in the network}}{\text{number of paths of length two}},$$

**Fig. 1.** Clustering Coefficient [9]

From this equation, it is clear that C is the ratio of number of triangles in the network to the number of paths of length 2. Logically both quantities are related in a way, i.e. if number of triangles in a network are large then understandably the number of paths of length 2 will also be larger. But this situation may change depending upon the graph, i.e. whether it is directed or undirected. Above statement is true when the graph is directed but what happens when the graph is undirected? The impact of being an undirected graph does not affect much the number of paths of length 2 but it affects the number of triangles to great extent.

## 3.2 Mean of the Shortest Distance

Let *l* be the mean of the shortest distance between pair of vertices in a network then

$$\ell = \frac{1}{\frac{1}{2}n(n+1)}\sum_{i\geq j} d_{ij},$$

**Fig. 2.** Small World Effect [9]

where $d_{ij}$ is the shortest distance from vertex $i$ to vertex $j$ and $n$ is the total number of vertices in the graphs. By using simple breadth searching algorithm, we can compute the quantity [9].

### 3.3 Degree Distribution

In graph theory, the degree distribution $P(k)$, is a function describing the total number of vertices in a graph with a given degree (number of connections to other vertices). It gives the probability distribution of degrees in a network [10]. It can be written as:

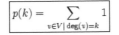

$$p(k) = \sum_{v \in V \,|\, \deg(v)=k} 1$$

**Fig. 3.** Degree Distribution

Where $v$ is a vertex in the set of the graph's vertices $V$, and $deg(v)$ is the degree of vertex $v$.

The Small World Effect can be useful in predicting the spread of news or something else on a network, i.e. how fast news can spread on a network or how many hops needs to be made to reach a certain vertex on the network. These effects can be used to find any kind of social patterns, behavior or attitudes in the social networks. However, Small World Effect can be observed more precisely in networks as, "The networks are said to show the Small-World Effect if the value of $l$ scales logarithmically or slower with network size for fixed mean degree" [9]. The three graphs in our work are built using different relation metrics between their nodes. Therefore, if any of them shows a Small World Phenomenon under the defined criterion, then lots of speculations can be made for behavior of that graph under light of research already done for Small World Graphs. And also it can give us an idea about the kind of relation metrics to be used to build a graph to receive a similar kind of behavior as Small World Networks.

In next section, we analyze the results obtained from our programs for three matrices mentioned above. It is to be noted that all graphs for mean of the shortest distance ($l$) have been redrawn manually just for the sake of better visibility.

## 4   Results and Interpretations

### 4.1  Term-Term Matrix

#### 4.1.1  Preparation of the Matrix
This matrix contained 8430 nodes and was prepared by Gaume [11]. All the terms and the words having semantic similarities with them were linked, i.e. all words (which are French Language Verbs in fact) will be considered as vertices and they will be linked with their synonyms. All words are first indexed with labels and then used while forming the final matrix. It is to be noted here that this matrix is used to evaluate our programs because we already know that this matrix behaves like a Small World Network so if our programs prove it the same, this guarantees that our programs give correct results.

For example, *discuter*, *parler* and *bavarder* almost represent the same semantic so they will form a different row in the final matrix. If the verb *discuter* is labeled as 1, and *parler* and *bavarder* as 2 and 3 respectively as shown in figure below:

| Term | Label |
|------|-------|
| Discuter | 1 |
| Parler | 2 |
| Bavarder | 3 |
| . . . . . | ... |
| . . . . . | ... |

**Fig. 4.** Example of Labeled Indexed Terms

And then in the final matrix these will be represented as:

| First Term Label | Second Term Label | Fixed at "1" |
|------------------|-------------------|--------------|
| 1 | 2 | 1 |
| 1 | 3 | 1 |
| 2 | 1 | 1 |
| 2 | 3 | 1 |
| 3 | 1 | 1 |
| 3 | 2 | 1 |
| ... | ... | ... |
| ... | ... | ... |

**Fig. 5.** Example of Final Term-Term Matrix

## 4.1.2 Clustering Coefficient

The value of $C$ for this graph is turned out to be 0.11298 with (number of paths of length 2 = 2024974 and number of triangles = 38132) while in a random graph with same number of vertices, $C$ is calculated as 0.0012 which is much less than 0.11298.

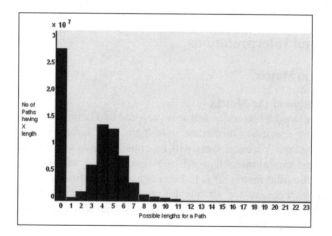

**Fig. 6.** Graph of number of paths of possible lengths

### 4.1.3  Mean of the Shortest Distance

The value of *l* calculated for this graph is 5.9495 which is almost as small as for equivalent random graph with *l* = 6.0123.

Above graph shows the number of paths of different lengths in term-term matrix. Paths of maximum length 21 exist in this graph but because numbers of paths having length greater than 11edges are quite small in number so they are not distinguishable on X-axis. The bar at zero shows the number of vertex pairs not connected at all.

### 4.1.4  Degree Distribution Graph

Degree Distribution Graph has been drawn between *Log (K)* and *Log (P (K))* where *K* is the degree that a certain vertex can have and *P (K)* is equal to the division of total number of vertices having degree *K* and total number of edges in the graph. It is obvious from Figure 7 that this graph follows Power Law of degree distribution.

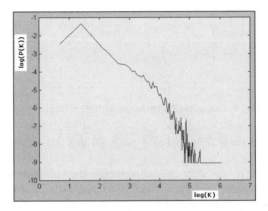

**Fig. 7.** Degree Distribution Graph

### 4.1.5  Conclusion

From above results, it is clear that Term-Term matrix shows the properties of Small World Network under the defined criterion so it can be declared that the graph represented by Term-Term Matrix is a Small World Network.

## 4.2  Document Similarity Matrix

### 4.2.1  Preparation of the Matrix

This matrix contains 1000 nodes and built as a result of the processing of collection of top one thousand documents obtained in respond to a query applied to a popular search engine. All documents were indexed and then each document and indexed terms were labeled by some indices. In Figure 8, the labels of document and terms both can be seen. For example, document www.xyz.com/i.asp has label of 1 and similarly term "Information" has been given label of 1.

This labeling helped us to prepare a two column matrix of the form (as shown below) in which first column of a row was the label of a certain document and second column was the label of the term appearing in that document. In Figure 9 below,

| Document Labels | | Term Labels | |
|---|---|---|---|
| www.xyz.com/i.asp | 1 | Information | 1 |
| www.abc.com/i.htm | 2 | Technology | 2 |
| www.asd,org/i.htm | 3 | IRIT | 3 |
| .......... | ... | .......... | ... |
| .......... | ... | .......... | ... |

**Fig. 8.** Example of document and Term label

| Document Label | Term Label |
|---|---|
| 1 | 2 |
| 1 | 20 |
| 1 | 21 |
| 2 | 21 |
| 2 | 22 |
| ... | ... |
| ... | ... |

**Fig. 9.** Example of documents with their labeled terms

document 1 contain three terms labeled as 2, 20 and 21 and document 2 contains terms 21 and 22.

From this matrix, it was very easy to compute the similarity between two documents using their content, i.e. terms. The Cosine Method of Vector Space Model was used for this purpose and the output was an $A[1000 \times 1000]$ matrix where $A_{ij}$ represents the similarity coefficient between documents $i$ and $j$. According to requirements of our computing algorithms and for better results, this matrix was converted into three final and different matrices each composed of three columns. In fact, to obtain three different matrices, a new threshold value was fixed on $A$ each time and then a different matrix was obtained. All pairs of documents satisfying the threshold value will become first two columns of a row of the new matrix with third column fixed as "1" to show that these two documents are similar.

| Document Label | Document Label | Fixed at "1" |
|---|---|---|
| 2 | 2 | 1 |
| 3 | 2 | 1 |
| 3 | 3 | 1 |
| 3 | 14 | 1 |
| 3 | 51 | 1 |
| ... | ... | ... |
| ... | ... | ... |

**Fig. 10.** Example of final Similarity Matrix

Three threshold values were fixed after reviewing matrix *A* carefully. In Figure 10, documents passing a fixed threshold value have been included in new matrix. For example, document 3 and document 2 can be seen as a row of new matrix.

Three different threshold fixed in this case were 0.0, 0.3 and 0.5. More thresholds were not fixed due to continuation of same results on further threshold values.

### 4.2.2 Clustering Coefficient
The value of *C* is 0.00643 with (number of paths of length 2 = 922582080 and number of triangles = 989459) while for random graph having same number of vertices, it is almost 0.9589.

### 4.2.3 Mean of the Shortest Distance
The value of *l* calculated for this graph is 1.7335 which is less than the *l* calculated for random graph which is 2.6123.

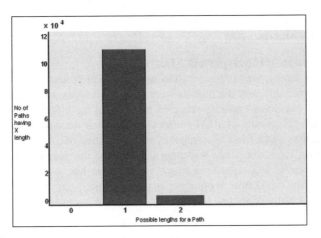

**Fig. 11.** Graph of number of paths of possible lengths

Above graph shows that paths of only two lengths exist in this matrix, i.e. of length 1 and length 2. However, paths of length 1 dominate the paths of length 2 in number. But this chart also gives us indication that vertices in this document similarity graph are tightly bound.

### 4.2.4 Degree Distribution Graph
Degree distribution graphs were drawn on three threshold frequencies (0.0, 0.3 and 0.5) but because of too much resemblance among these graphs, only one with threshold value of 0.5 is shown here.

As clear from the above graph, no way it follows the Power Law of Degree Distribution.

### 4.2.5 Conclusion
The degree distribution graph of this network does not follow power law of degree distribution. The value of clustering coefficient is also very low and also the differences between values of *l* cannot be ignored in this case. Concluding from our results, we can declare that this graph is not a Small World Network.

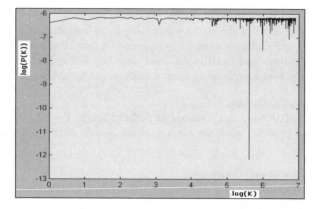

**Fig. 12.** Degree Distribution Graph at threshold 0.5

## 4.3  Hyperlink Matrix

### 4.3.1  Preparation of the Hyperlink Matrix

The hyperlinks matrix was prepared from a TREC [12] collection. From the collection, one thousand different documents were chosen (Results were also calculated on 23545 vertices but due to no difference in results and for coherency with similarity matrix, we kept the results of 1000 vertices). Then each document was processed to extract its outlinks and a file was composed that contained one line for each document and its outlinks. Each line starts with the name of the document followed by all its outlinked documents separated by a space on the same line. In Figure 13, the document www.qwe.com/i2.htm contains two out-links, i.e. one to documents www.bbd.com/1.htm and second to document www.cbf.com/efg.htm.

**Fig. 13.** Example of file representing documents and their out-links

Once this file has been prepared, we come to the phase of labeling all documents present in this file. All the documents were labeled like we did for similarity matrices. Now a final matrix was prepared between the source documents and the targeted documents, i.e. the documents that are pointed out by their out-links. This new matrix contains 3 columns: first for source document itself and second for one of its linked documents (linked by out-links) and third being set for '1'. The idea behind this is the fact that all documents provide link to those documents that have same content, i.e. the probability for the targeted document of having same type of content or related content is much higher. Figure 14 below represents the final matrix where document labels have been mentioned.

| Source Document | Target Document | Fixed at "1" |
|---|---|---|
| 1 | 2 | 1 |
| 1 | 3 | 1 |
| 1 | 10 | 1 |
| 2 | 1 | 1 |
| 2 | 14 | 1 |
| ... | ... | ... |
| ... | ... | ... |

**Fig. 14.** Example of final hyperlink Matrix

### 4.3.2  Clustering Coefficient

The value of $C$ for this graph is 0.5258 (number of paths of length 2 = 192107380 and number of triangles = 16837827) while value of $C$ for equivalent random graph is calculated as 0.7837.

The clustering coefficient tells us the density of triangles in a network and this value may range between 0 and 1. In this case, the value of $C$ is quite high. If compared with the clustering coefficients of previous matrices, then we can see that vertices of this matrix are much more mingled in triangles than other two matrices. The reason behind this is the type of graph. The similarity matrix is an undirected graph and the hyperlink matrix is directed one so it increases the number of triangles three times almost. In other words, the probability that a vertex is connected to the neighbors of its neighbors is quite high in this graph.

### 4.3.3  Mean of the Shortest Distance

Mean of the shortest distance for this graph is 2.6657 which can be verified from figure Fig12 while the value of $l$ for equivalent random graph is 3.1328.

The bar at zero on X-axis shows the number of vertices that are unapproachable from any vertex, i.e. it is the sum of all vertices that cannot be reached from different vertices. It can be observed from the graph that most of the vertex pairs are reachable in maximum 3 edges. There are lot of vertices that can be reached in just one jump and 2 jumps but there are few that need third jump to be reached.

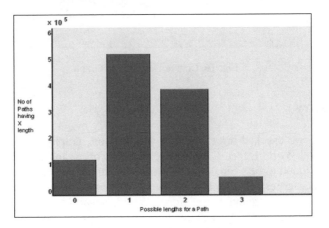

**Fig. 15.** Graph of number of paths of possible lengths

### 4.3.4  Degree Distribution Graph

The degree distribution graph of this matrix in Figure 16 looks very bizarre and of course does not follow the power law of distribution anyway. The lines that are coming down from 0 to the lines at bottom in fact are for those pages that have no degree. In fact as described already that this matrix was taken from a big collection of documents so this contain many documents that are part of the matrix (because other pages contain outlinks destined to these documents) but their degree could not be included in this matrix so they are representing a degree of zero in the Figure 16.

### 4.3.5  Conclusion

It is obvious from degree distribution graph that it is not following the power law of degree distribution so we need to check for values of $C$ and $l$. The value of clustering coefficient $C$ is quite high, i.e. 0.5258 but differences between values of $l$ for this graph and equivalent random graph are too high to be ignored. The reason of having such mixed results is that this matrix was an excerpt of a very large matrix so there are many pages that are included in this matrix but their out-degrees are not part of this matrix. Concluding from these results, we can predict that results can be different if we use the full graph instead of an excerpt of a large graph. However, with the given experimental data, our results show that this graph is not a Small World Network and we will verify it again with larger number of nodes in our future work for final conclusion.

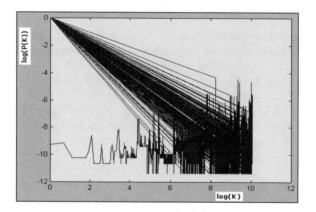

**Fig. 16.** Degree Distribution Graph

## 5  Summary

In this paper, we checked three different web network graphs for their tendency to show the Small World Effect. One of these web network graphs, i.e. Term-Term Matrix has been found to be a Small World Network with given experimental data under already defined criterion. This also evaluated our developed programs. Rest of the two graphs were not found to be Small World Networks. As for as Document-Document similarity matrix is concerned, no one of the three properties showed that it is a Small World Network. However for our third matrix, i.e. Hyperlink matrix, the value of clustering coefficient was quite high (even less than equivalent random graph) and degree distribution graph is quite bizarre also. These two results highlight the possibility of having different results if experimented with larger network, i.e. with full matrix and not with an excerpt of a matrix. We believe that results for this graph can be improved by choosing a larger and consistent web network graph that is part of our future work.

## References

1. Grossman, D.A., Frieder, O.: Information Retrieval: Algorithms and Heuristics 2nd edition, Springer Publishers (2004)
2. Ntoulas, A., Najork, M., Manasse, M., Fetterly, D.: Detecting Spam Web Pages through Content Analysis. In Proceedings of Conference WWW 2006, Edinburgh, Scotland (2006)
3. Kleinberg, J.M.: Authoritative Sources in Hyperlinked Environment. In Proc. Of the Ninth Annual ACM-SIAM Symposium on Discrete Algorithms, pp. 668-667 (1998)
4. Brin, S., Page, L.: The Anatomy of a Large-Scale Hypertextual Web Search Engine. In Proc. of WWW'98, Brisbane, Australia, pp. 107-117 (1998)
5. Small World Experiment, http://en.wikipedia.org/wiki/Small_world_effect
6. Watts, D.J., Strogatz, S.H.: Collective Dynamics of Small-World Networks. Nature, vol. 393, pp. 440-42 (1998)
7. Adamic, L.A.: The Small World Web. In Proc. 3rd European Conf. Research and Advanced Technology for Digital Libraries (ECDL 1999) (1999)
8. Milgram, S.: The Small World Problem, Psychology Today, Vol. 2, 60-67 (1967)
9. Newman, M.E.J.: The Structure and Function of Complex Networks. SIAM Review, Vol. 45 (2003)
10. Erdős, P., Renyi, A.: On Random Graphs. I. Publ. Math. Debrecen, 6, pp. 290-291 (1959)
11. Gaume, B.: Balades Aléatoires dans les Petits Mondes Lexicaux. In I3 Information Interaction Intelligence, CEPADUES edition, pp. 39-96 (2004)
12. TREC Data Collection, http://trec.nist.gov

# Application of Color Segmentation Using Online Learning for Target Tracking

Muhammad Asif Memon[1], Hassan Ahmed[2], and Shahid A. Khan[2]

[1] Pakistan Space & Upper Atmosphere Research Commission (SUPARCO), Gulzar Hijri,
Karachi, Pakistan, 75300
asifmemon@gmail.com
[2] Department of Electrical Engineering, COMSATS Institute of Information Technology,
Islamabad
{h.ahmed,s.khan}@comsats.edu.pk

**Abstract.** Color segmentation techniques find extensive applications in visual tracking as the color information provides a robust reference for identifying a target. Color based tracking systems generally use histograms or static models. However, in the real world the changing surrounding conditions must be taken into account. An online learning method of color segmentation has been adapted to ensure better performance even with changing lighting conditions. A neural network, based on fuzzy Adaptive Resonance Theory (ART), is used to develop the color model that is updated with each frame by the pixels classified within the target. The categories formed by the ART network are restricted to ensure fast processing, and the performance of the system is analyzed at different thresholds for association with the color model. Further, a Kalman filter is added into the loop for predicting the target's position in the next frame and a comparison is made to investigate the improvement in performance.

**Keywords:** Color segmentation, visual tracking, online learning, color model, Kalman filter.

## 1 Introduction

Vision based systems require robust visual tracking especially in case of robotic vision applications. Active vision [1] takes this one step ahead where a moving camera is used for visual tracking. The basic function of tracking camera is to keep the target centered in the image. Robust tracking of object requires algorithms to be efficient enough to perform this function. Mainly, two different techniques are employed to perform visual tracking with active camera: motion based and feature based [2]. Motion based techniques generally employ frame subtraction to identify a moving target. This technique is useful when the background is static, which is not the case in majority of applications. Different improvements have been proposed to overcome this limitation such as background compensation [3] and optical flow [4]. Background compensation manages to overcome problems due to moving scene but still robust detection of object requires some other feature such as shape, color, or texture. Due to

D.M.A. Hussain et al. (Eds.): IMTIC 2008, CCIS 20, pp. 146–155, 2008.
© Springer-Verlag Berlin Heidelberg 2008

the non-homogenous nature of motion at boundaries, optical flow methods are also not precise [4] and require an additional feature for reliable tracking.

Feature based algorithms are based on one of the parameters of the target and recognition of that parameter within successive frames. Edges of an object used as a reference parameter pose difficulty in cluttered scenes and also are computationally expensive [4]. Most of the tracking algorithms in the active robotic vision applications employ color information (also called color cue) for its robust and reliable segmentation ability. The color histogram approach adapted by [2] and color indexing methods employed by [5] and [6] have proven the reliable and computationally efficient performance of color cue. However, these techniques have the drawback of having sensitivity to changing illumination and color change due to object motion and require further color adapting techniques [2, 7, 8, and 9].

Nakamura and Ogasawara [10] built on the work of [11] and [12] to develop a color segmentation technique that is computationally efficient and can adapt to changing surrounding conditions by online learning. Instead of the repetitive principle component analysis (PCA) performed by [12], to find parameters of the color model, fuzzy adaptive resonance theory (ART) [13] is incorporated by [10] to assign each target pixel a category and then matching the successive frames to those categories. The fuzzy ART network is a type of neural network that uses competitive learning and is trained in real-time. This ability of the network makes it a natural choice for color segmentation phase in a visual tracking system. One of the pre-requisites in using fuzzy ART network for classification is to transform the pixels in RGB (Red-Green-Blue) format to YUV color space so that the three pixel coordinates are perpendicular to each other. This transformation causes the classification of similar color pixels to different categories based on a threshold. Further Kalman filter [14] is used for predicting target location in the next frame to place search window at an appropriate point.

In this work, the segmentation technique of [10] is simplified to reduce the computational cost so that the algorithm can be used for fast tracking applications. This also enables the use of another feature in addition with color due to high segmentation speed. In [10], a single object is assigned number of categories and the same process is repeated within search window for each successive frame. This process is reduced by selecting the initial point on the target and using mean of pixels around that point to save a category assigned to that value. All the background objects and color values are not assigned categories as they are not useful. Each pixel in the search region in successive frames is applied to the fuzzy ART network and the category assigned to it is compared against the target category. The mean of all the pixels classified within target category is used to update the target category and also to find the center position of the target or color blob.

## 2 Color-Based Tracking System

The visual tracking system using color segmentation is implemented to work with Pioneer series of Robots to perform real-time tracking. The system is initially implemented to test the performance through different video sequences and then is implemented on the laptop onboard the robot. The PTZ (Pan-Tilt-Zoom) camera mounted on the robot is used to perform tracking and the camera follows the target using its pan and tilt features.

## 2.1  System Architecture

The architecture of visual tracking system is shown in Fig. 1. The system is initialized by the user by clicking on the target at a point with target color. The first step is to transform the color space into HSV format. This format represents each pixel in terms of hue, saturation and brightness, making it possible for the neural network to classify pixels efficiently. The relationship between each component in HSV format is shown in Fig. 2, which illustrates the independence of each component with each other and hence identifying each color with distinguishable value.

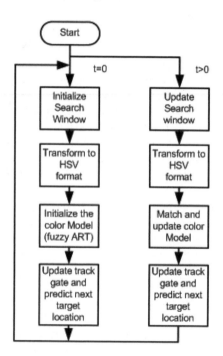

**Fig. 1.** System Architecture and Algorithm Flow

The mean of pixels around the point initialized by the user is fed to the fuzzy ART network to develop the color model. A rectangle is placed around the target point to identify the search region. Each pixel in the search region is compared against the color model represented by the initialized category and is classified as within the target or outside the target The pixels identified as within the target are used to find the center point of the target and the search window is placed on the next center point and the

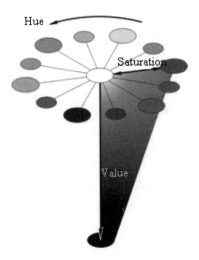

**Fig. 2.** Relationship between each Component for HSV Format

process continues. In the next step, Kalman filter is added in the loop as suggested by [10] to improve the performance of the system and remove jitters in tracking.

## 2.2 Adaptive Color Model

The color model is created and is updated using fuzzy ART network. The architecture of the fuzzy ART network is shown in Fig. 3. It uses fuzzy similarity measures to estimate the category choice. Each ART network includes a field $F_0$ of nodes that represent a current input vector; a field $F_1$ that receives both bottom-up input from $F_0$ and top-down input from a field $F_2$ that represents the active code, or category. Vector $I$ denotes $F_0$ activity; vector $x$ denotes $F_1$ activity; and vector $y$ denotes $F_2$ activity [13].

Each category in field $F_2$ has an associated weight vector representing the input associated with that category. Initially, each category is uncommitted and weights are initialized to '1'. Each input applied to the network is tested against the choice function $T_j$ which is defined by

$$T_j(I) = \frac{|I \wedge w_j|}{\alpha + |w_j|} \qquad (1)$$

Where $w_j$ is the weight vector and the fuzzy intersection $\varpi$ is defined by $(p \varpi q)_i \equiv \min(p_i, q_i)$ and where the norm $|.|$ is defined by

$$\left| p \equiv \sum_{i=1}^{M} |p_i| \right| \qquad (2)$$

The index $j$ denotes the chosen category, where

$$T_J = \max\{T_j : j = 1...N\} \qquad (3)$$

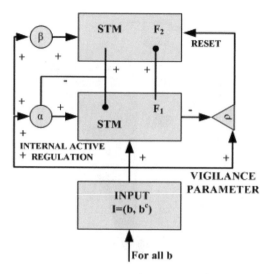

**Fig. 3.** Architecture of fuzzy ART network with three fields

Resonance is said to occur when the match function of the chosen category meets the vigilance criteria or the threshold defined by

$$\frac{|I \wedge w_J|}{|I|} \geq \rho \tag{4}$$

In case of mismatch, the next category is tested for resonance; otherwise the weight vector is updated using Equation 5.

$$w_J^{(new)} = \beta(I \wedge w_J^{old}) + (1-\beta)w_J^{old} \tag{5}$$

The normalization of inputs is carried out by complement coding [13] to avoid category proliferation. As only one category is used, the proliferation is not a problem. The vigilance criterion is the threshold used to assign each input to the categories and its value can be varied to achieve required performance for different scenarios. A number of experiments were performed with different vigilance threshold values.

### 2.3 Search Region and Target Localization

The search region is defined as a square window with '30x30' pixels. However, this region can be increased or decreased depending upon the requirements and the application. This size of this search window must be greater than the size of the target on the image to ensure proper segmentation. Each pixel within the search window is tested against the color model [15] by using Equation 6.

$$\psi_t(x, y) = \begin{cases} 1 & C(x, y) \in CM_t \\ 0 & \text{otherwise} \end{cases} \tag{6}$$

Where $C(x,y)$ and $CM_t$ represent the HSV value of the pixel $(x,y)$ and the color model of the target region respectively.

Based on the membership function in Equation 6, the mean of the spatial locations $(x,y)$ of the pixels classified as within target is used to find the actual location of the target [15] using Equation 7, where R represents the search region.

$$x_t = \frac{\sum_{(x_i,y_i)\in R} x_i \psi_t(x_i,y_i)}{\sum_{(x_i,y_i)\in R} \psi_t(x_i,y_i)}, \tag{7}$$

$$y_t = \frac{\sum_{(x_i,y_i)\in R} y_i \psi_t(x_i,y_i)}{\sum_{(x_i,y_i)\in R} \psi_t(x_i,y_i)}.$$

The mean of all the pixels classified within the target is used to update the weight vector of the target category initialized by the user and the process continues in the next frames. This update for the weights makes it possible for the network to perform reliable segmentation even with changing surrounding conditions.

## 2.4  Kalman Filter for Prediction

The actual center position of the target is used to place the search window in the next frame. However, this makes the tracking system to lag behind one step as the target moves within successive frames. In order to avoid this lag and jitters caused by it, the Kalman filter is used to predict the location of the target in the next frame. The search window is placed on the predicted location and the segmentation is used to find the actual position for that frame. The actual location of the target is used as the ground truth to update the Kalman filter parameters. The characteristic equation for the Kalman filter is the same as used by [11] and [13] and is given in Equation 8.

$$\hat{X}_{t|t} = \hat{X}_{t|t-1} + \hat{G}_t\left(\hat{Y}_t - \hat{X}_{t|t-1}\right) \tag{8}$$

Where, $Y_t$ is the target's mean location and $G_t$ is the Kalman gain matrix.

## 2.5  Camera Calibration for Real-Time Tracking with PTZ Camera

As suggested by Corke and Hutchinson [16] that for robotic vision applications, instead of complex camera calibration techniques, simple pixels to angle transformation can be implemented. Shiao [17] used the focal length of the camera to calculate the pan and tilt angles as given by Equation 9.

$$\theta_r = \tan^{-1}\left(\frac{\varepsilon_r d_r(t + \Delta t)}{f}\right) \tag{9}$$

Where, $r$ represents the axis, $f$ stands for the focal length $\varepsilon$ represents the pixel length and $d_r$ represents the error in pixels for pan and tilt axis.

The field of view of the camera was calculated by taking several points as reference and moving the camera by known angular displacements. The resolution of the camera and its field of view were used to calculate the pixel length and hence to find the angle by which the camera had to be moved in both axis in response to error in pixels calculated by the tracking algorithm.

## 3  Experimental Results

The performance of the tracking algorithm was analyzed at different vigilance parameter values. Three different sequences were used to test the segmentation and tracking performance of the system. It is observed that the value of the threshold for color model can be adjusted to suit a particular condition. A value of 0.85 is a good compromise for a general tracking system to perform reliable tracking under different environments and tracking scenarios. Fig. 4 shows the pixels classification inside the search region with different threshold values for the three sequences.

As the algorithm was first implemented without any prediction, some jitters were present in the tracking performance. However, with addition of Kalman filter in the loop, the jitters were removed and the performance of the system also improved. A list of different performance measures in form of errors calculated for the three sequences is given in Table 1. The error between the actual location of the target and the location predicted by the Kalman filter is calculated and correlation between the two is also provided with error between the actual and predicted value [18].

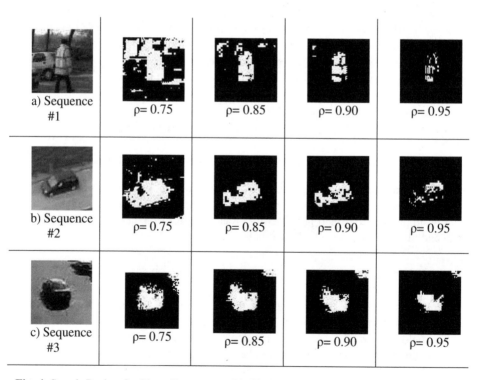

**Fig. 4.** Search Region for Three Sequences with Pixels Classified Within and Outside Target based on Vigilance Criterion ρ

It can be observed from Fig. 4 that as the threshold value is increased, the number of pixels being classified as inside the target is decreased, hence, making the target classification more strict than the case for lower vigilance criterion value. This is particularly useful when the target is surrounded by similar color objects so that the threshold classifies them as different category and thus, outside the target. This is much clear from Sequence #1 where background contains similar color as target and for lower threshold values it is also considered as target, however, for $\rho > 0.90$, only target pixels are classified to reference category.

The algorithm is initially implemented to test with recorded sequences. After testing the performance on different video sequences, a PTZ camera mounted on the Pioneer

**Table 1.** Error Performance for Kalman filter

|  | Mean Square Error | Root Mean Square Error | Correlation |
|---|---|---|---|
| Sequence#1 | 0.60694 | 0.77907 | 0.9996 |
| Sequence#2 | 11.511 | 3.3927 | 0.99754 |
| Sequence#3 | 5.2762 | 2.297 | 0.996 |

**Fig. 5.** Snapshot of the experiment performed on a PTZ camera mounted on a robot. The color cue in second robot is used as tracking reference.

3-DX series of robots is used for active tracking. The Advanced Robot Interface for Application (ARIA) library [19] was used for the robots and camera control. The snapshot of a robot mounted with tracking camera tracking another robot based on its color information as taken from one of the tracked videos [18] is shown in Fig. 5.

The algorithm only uses color information as a reference to track a particular object. When the system was tested with sequences in which the target was occluded by some other object of different color, it was able to maintain the lock. However, when another object with similar color profile as of the target enters the search region the system gets confused and locks the larger blob that is classified as target and hence can loose the actual target of interest. This is an inherent problem as only one cue is being used for tracking and can be solved by addition of some other parameter like motion, edges, etc.

## 4   Discussion and Conclusions

A visual tracking system is implemented using online learning capabilities of fuzzy ART network. The mean of pixels initialized by user is used to initialize a color model and each successive frame is compared within a search window to the color model. The color model is updated based on the mean of pixels classified as target, so the system can adapt to the changing surrounding conditions. The membership threshold for the network can be adjusted according to the requirements of the application. This is further illustrated in Fig. 4 by using three different video sequences. The prediction capability of Kalman filters is used to place the search window at an appropriate location in the next frame and the results show that the predicted value is more than 99% accurate for all the sequences and hence improves the system performance removing jitters in tracking.

The system is able to track the target reliably for all the sequences and is then implemented to track robots using a PTZ camera. The simple approach adapted for color segmentation makes it possible to use the system for real-time applications as experiments show for the PTZ camera tracking. As the system only uses one feature as a reference, it is fast and robust for a single target or even for the condition when the target gets occluded by some other object with different color. The vigilance criterion can be adjusted to distinguish the target from similar color objects, however, if the occluding object is of the same color as that of target in interest, only increasing the threshold will not be sufficient and some other cue like edges, shape or motion must be added to incorporate such conditions. This can be easily achieved as the color segmentation algorithm takes much less time than standard video frame- scan speeds.

The performance of the system can be improved by using some better prediction model instead of simple Kalman filter. Some probabilistic method like particle filtering can be used instead of Kalman filter as they are suitable for non-linear functions. Apart from probabilistic methods, neuro-fuzzy clustering mechanisms can be employed to build a model for target movement and hence improve system performance considerably.

# References

1. Aloimonos, J.Y., Tsakiris, D.P.: On the visual mathematics of tracking. Image and Vision Computing 9(4), 235–251 (1991)
2. Vergés-Llahí, J., Aranda, J., Sanfeliu, A.: Object tracking system using colour histograms. In: Sánchez, J.S., Pla, F. (eds.) Proc. 9th Spanish Symposium on Pattern Recognition and Image Analysis, Universitat Jaume I, pp. 225–230 (2001)
3. Murray, D.W., Basu, A.: Motion Tracking with an active camera. IEEE Trans. on Pattern Analysis and Machine Intelligence 16(5) (1994)
4. Mae, Y., Shirai, Y., Kuno, J.Y.: Object tracking in cluttered background based on optical flow and edges. In: Proc. the 13th Int. Conf. on Pattern Recognition, Vienna, Austria, vol. 1, pp. 196–200 (1996)
5. Swain, M.J., Ballard, D.H.: Color Indexing. IJCV 7(1), 11–32 (1991)
6. Schiele, B., Crowley, J.L.: Object recognition using multidimensional receptive field histograms. In: Buxton, B.F., Cipolla, R. (eds.) ECCV 1996. LNCS, vol. 1064, pp. 610–619. Springer, Heidelberg (1996)
7. Jang, D., Choi, H.I.: Moving object tracking using active models. In: ICIP 1998, vol. 3, pp. 648–652 (1998)
8. McKenna, S.J., Raja, Y., Gong, G.: Tracking Colour Objects Using Adaptive Mixture Models. Image and Vision Computing 17, 225–231 (1999)
9. Schuster, R.: Color object tracking with adaptive modeling. In: Workshop on Visual Behaviors, International Association for Pattern Recognition, pp. 91–96 (1994)
10. Nakamura, T., Ogasawara, T.: On-line Visual Learning Method for Color Image Segmentation and Object Tracking. In: Proc. of IEEE/RSJ Intern. Conf. on Intelligent Robots and Systems, vol. 1, pp. 221–228 (1999)
11. Wren, C., Azerbayejani, A., Darnell, T., Pentland, A.P.: Real-time Tracking of the Human Body. IEEE trans. on PAMI 19(7), 780–785 (1997)
12. Rasmussen, C., Toyama, K., Hager, G.D.: Tracking Objects By Color Alone. Technical Report TR1114, Dept. of Computer Science Yale University (1996)
13. Carpenter, G.A., Grossberg, S., Rosen, D.B.: Fuzzy ART: Fast Stable Learning and Categorization of Analog Patterns by an Adaptive Resonance System. Neural Networks 4, 759–771 (1991)
14. Kalman, R.E.: A New Approach to linear filtering and prediction problem, Transactions of the ASME, Ser. D. Journal of Basic Engineering 82, 34–45 (1960)
15. Memon, M.A., Angelov, P., Ahmed, H.: An Approach to Real-time Color-based Object Tracking. In: 2006 International Symposium on Evolving Fuzzy Systems, UK, pp. 82–87 (2006)
16. Corke, P., Hutchinson, S.: Real-Time Vision, Tracking and Control. In: Proceedings of ICRA 2000, pp. 622–629 (2000)
17. Shiao, Y.: Design and implementation of real-time tracking system based on vision servo control. Tamkang Journal of Science and Engineering 4(1), 45–58 (2001)
18. Memon, M.A.: Implementation of Neural Networks for Real-Time Moving Target Tracking. MSc. Thesis, Dept. of Communication Systems, Lancaster University, UK (2006)
19. Activmedia Robotics Inc., MobileRobots, Pioneer 3-Dx Series, http://www.activrobots.com/ROBOTS/p2dx.html

# Iterative Learning Fuzzy Gain Scheduler

Suhail Ashraf[1] and Ejaz Muhammad[2]

[1] Department of Electrical Engineering, National University of Science and Technology,
Rawalpindi, Pakistan
suhailashraf@yahoo.com
[2] Head of Department of Electrical Engineering, National University of Science and
Technology, Rawalpindi, Pakistan

**Abstract.** Humans learn from experience and from repeating a task again and
again. This paper uses this human capability to iteratively adjust fuzzy membership functions. These learnt membership functions are used to schedule the gain
of a conventional proportional controller. The adjustment of the membership
functions help achieve the design requirements of steady state error and percentage overshoot while the scheduling of the proportional gain gives us an
adaptive controller. Simulation and experimental results demonstrate that this
simple yet robust approach can be applied as an alternative to proportional-
integral-derivative (PID) controllers, which are extensively used for various
applications.

**Keywords:** Fuzzy control, **Learning** control, Gain scheduling, Adaptive control, Iterative learning.

## 1 Introduction

Designing and implementing proportional-integral (PI) and proportional-integral-
derivative (PID) controllers, which are the most used controllers in industry, have
difficulties associated with them [1], namely,

a) They require a mathematical model of the system which is usually not known.
b) Variation of plant parameters can cause unexpected performances.
c) They usually show high performance for one unique action point only.

Extensive efforts have been devoted to develop methods to reduce the time spent
on optimizing the controller parameters like proportional gain, integral gain and derivative gain of these controllers [2]. The PID controllers in the literature can be divided into two main categories. In the first category, the controller parameters are
fixed after they have been tuned or chosen in a certain optimal way. Controller parameters of the second category are adapted based on some parameter estimation
technique, which requires certain knowledge of the process. In most practical cases,
the model of the system is not known and hence conventional PID control schemes
will not deliver satisfactory performance. Also, the dynamics of a system even for a
reduced mathematical model is usually non-linear, making tuning of these controllers
even more difficult [3]. Fuzzy logic [4], [5], [6] has been shown in numerous studies
to tune conventional PI and PID controllers [7], [8], [9].

D.M.A. Hussain et al. (Eds.): IMTIC 2008, CCIS 20, pp. 156–168, 2008.

One effective technique to adjust the parameters of the controllers is gain scheduling. In Conventional Gain Scheduling (CGS), the controller parameters are scheduled according to some monitored conditions. Its main advantage is that controller parameters can be changed quickly. One serious drawback of CGS is that the parameter change may be rather abrupt which can result in unsatisfactory or even unstable performance. Another problem is that accurate models at various operating points may be too difficult or even impossible to obtain. As a solution, fuzzy has been utilized as fuzzy gain scheduling (FGS) to overcome these problems [10].

In fuzzy based scheduling the choices of appropriate membership functions, minimum rule base and suitable fuzzifier and defuzzifier is still a challenging task. Having made these choices, one still needs to tune the fuzzy controller to deliver the desired response. Multiple simultaneous adjustments (rules, membership functions and input/output fuzzy gains) make the optimum tuning even more difficult. Many techniques have been used to overcome these difficulties including technique for rule base formulation [11], membership function definitions for PID gains [12], phase plane technique for rule base design [13], rule modification [14], neural network techniques [1], [10], genetic algorithms [15] and gain phase margin analysis technique [16].

Before formulating any rules, membership functions need to be established. Membership functions have uncertainties associated with them. These uncertainties arise because different people have different perceptions about a concept [17]. One way to tackle such uncertainties is by using Type 2 Fuzzy sets (T2 FS) [18], [9]. With such uncertainties it is difficult to determine the exact Membership Functions (MF) for a Fuzzy System (FS) that would give the desired performance. As an example, while designing a gear shift controller for cars, one of the first steps would be to formulate membership functions. Suppose membership function for speed is denoted by $x$, where $x \in [0,140]$ (km/h). To design membership functions the method that is usually employed is to ask the experts. During one such design, 10 experts were asked to indicate the range for slow speed. Table 1 shows the inputs from the experts.

**Table 1.** Data from experts for fuzzy set 'slow'

| Serial No. | Range for 'Slow' (Km/h) |
|:---:|:---:|
| 1 | 10-40 |
| 2 | 0-40 |
| 3 | 10-50 |
| 4 | 10-30 |
| 5 | 15-40 |
| 6 | 20-40 |
| 7 | 10-40 |
| 8 | 10-30 |
| 9 | 15-40 |
| 10 | 20-40 |

The data shows that even experts have different opinions about the range of slowness. This creates regions of uncertainty about the locations of two end points for the MF, as shown in Fig. 1.

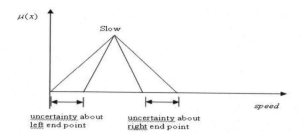

**Fig. 1.** Triangular MFs when end points have uncertainty associated with them

Somewhere in this region of uncertainty are located the lower and upper extremities of our desired MF. There can be $N$ membership functions in this region of uncertainty. We hope to find

$$MF_{desired} = MF_1(x) or MF_2(x)...or MF_N(x)$$

The learning aspect of human mind follows the simple principle of gaining knowledge through repetition of task. This principle has given rise to Iterative Learning Control (ILC). Iterative learning control has shown great success in repetitive tasks [20], [21]. The basic idea behind ILC is that the information learnt from the previous trial is used to improve the control input for the next trial. The control input is adjusted to decrease the difference between the desired and the actual outputs. In the approach presented in this paper, indirect adjustment of control input is made by adjusting the membership functions, using learning laws.

The core of our proposed approach is the Iterative Learning Fuzzy Gain Scheduler (ILFGS). It consists of a fuzzy system to implement the control laws, iterative learning laws to adjust member ship functions and a mathematical formula to calculate controller gains. We begin with an introduction of the 2-D learning [22] process. Then, the approach itself is explained. It is then followed by some simulation results. The results from a motor speed control experiment, using Quanser's DC Motor Control Kit are also presented. Direction of future work is also established.

## 2   Two Dimensional Learning Process

A 2-D learning process is one in which inputs, outputs and system states depend on two independent variables, i.e. its dynamics are propagated along two independent directions. One process, indicated by the variable $k$, reflects the dynamics of the system in terms of time history. The other process, described by the change of variable $j$, reflects the learning iteration and the resultant performance improvement in terms of learning times [23]. For example, $u_j(k)$ expresses the $k^{th}$ item of the input in the $j^{th}$ execution cycle ($j^{th}$ learning process). Here $k = 1...N$ and $j = 1...\infty$.

## 3   Proposed Approach

A typical block diagram of a conventional proportional (P) controller using 2-D representation is shown in Fig. 2.

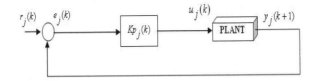

**Fig. 2.** Block diagram of a proportional controller using two-dimensional representations

Here $r_j(k)$ represents the reference signal. The error is represented by $e_j(k)$, the input to the plant is $u_j(k)$ and the next plant output is $y_j(k+1)$. For the proposed scheme $Kp_j(k)$ is adaptively chosen so as to meet design requirements of steady state error (sse) and percentage over shoot (pos). The block diagram of the proposed scheme is presented in Fig. 3.

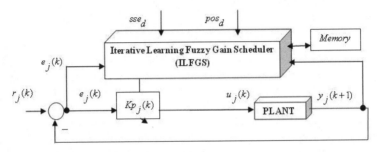

**Fig. 3.** Block diagram of the proposed scheme

Here $sse_d$ is the desired steady state error, $pos_d$ is the desired percentage over shoot, $r_j(k)$, $e_j(k)$, $u_j(k)$, and $y_j(k+1)$ are the reference input, error , input to the plant and next plant output at iteration $j$. Desired steady state error and percentage over shoot are supplied to Iterative Learning Fuzzy Gain Scheduler (ILFGS). The ILFGS adjusts $Kp_j(k)$. It also tunes itself by adaptively adjusting the membership function end points. The aim is to converge with respect to given steady state error and percentage over-shoot. The learnt values of membership function end points are stored in memory to be used in future iterations.

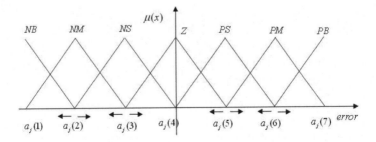

**Fig. 4.** Input Membership functions

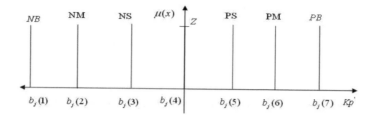

**Fig. 5.** Output Membership functions for Sugeno rule processing

Adapting a Sugeno type rule processing the input (error) and output ($Kp'$) membership functions proposed are of the form given in Fig. 4 and Fig. 5.

The fuzzy system in ILFGS consists of a rule base which contains a set of fuzzy IF-THEN rules. The structure of rules is of the form

$$IF \quad x_1 \quad is \quad A_1^l \quad and...and \quad x_n \quad is \quad A_n^l \quad THEN \quad y^l = c_0^l + c_1^l x_1 + ... + c_n^l x_n \tag{1}$$

where $A_i^l$ are fuzzy sets in $U_i \subset R$, $c_i^l$ are constants, and $l = 1,2,...,M$ where $M$ is the number of rules in the fuzzy rule base and $i = 1,2,...,n$ where $n$ is the number of membership functions. The IF parts of the rules are the same as in the ordinary fuzzy IF-THEN rules, but the THEN parts are linear combinations of the input variables. Here $x = (x_1,...,x_n)^T \in U \subset R^n$ and $y \in V$ are input and output (linguistic) variables of the fuzzy system, respectively. Given an input $x$, the output $f(x) \in V \subset R$ of the TSK fuzzy system with product inference engine, singleton fuzzifier and center average defuzzifier is computed as the weighted average of the $y^l$'s in (1), that is

$$f(x) = \frac{\sum_{l=1}^{M} y^l w^l}{\sum_{l=1}^{M} w^l} \tag{2}$$

where the weights $w^l$ are computed as

$$w^l = \prod_{i=1}^{n} \mu A_i^l(x_i) \tag{3}$$

Equation (2) and (3) gives

$$f(x) = \frac{\sum_{l=1}^{M} y^l \left( \prod_{i=1}^{n} \mu A_i^l(x_i) \right)}{\sum_{l=1}^{M} \left( \prod_{i=1}^{n} \mu A_i^l(x_i) \right)} \tag{4}$$

Suppose that the output range [$Kp'_{min}$, $Kp'_{max}$] can be determined. The values of input and output membership functions can be normalized for convenience. The permissible range of input membership function end points are listed below.

$a_j(7) = [a_j(6), e_{max}]$, $a_j(6) = [a_j(5), a_j(7)]$, $a_j(5) = [a_j(4), a_j(6)]$, $a_j(4) = [a_j(3), a_j(5)]$, $a_j(3) = [a_j(2), a_j(4)]$, $a_j(2) = [a_j(1), a_j(3)]$, and $a_j(1) = [e_{min}, a_j(2)]$

Here $e_{max}$ and $e_{min}$ are the minimum and maximum values of error. The output membership function end points are fixed at

$b_j(7) = Kp'_{max}$, $b_j(6) = 0.6Kp'_{max}$, $b_j(5) = 0.3Kp'_{max}$, $b_j(4) = 0$, $b_j(3) = 0.3Kp'_{min}$, $b_j(2) = 0.6Kp'_{min}$, $b_j(1) = Kp'_{min}$

For the proposed scheme we only move $a_j(5)$ and $a_j(6)$. The end points $a_j(5)$ and $a_j(6)$ can take up any initial starting value within their permissible range. It is recommended, based on experience, to start with values that divide the Universe of Discourse evenly. This leads to values of $a_j(5) = 0.3e_{max}$ and $a_j(6) = 0.6e_{max}$, which divides the range reasonably. The rule base of the fuzzy controller is shown in Table 2.

**Table 2.** Rule base for proposed scheme

| Control | NB | NM | NS | Z | PS | PM | PB |
|---------|-----|--------|--------|--------|--------|--------|--------|
| $e_j(k)$ | $e_{min}$ | $a_j(2)$ | $a_j(3)$ | 0 | $a_j(5)$ | $a_j(6)$ | $e_{max}$ |
| $Kp'$ | $b_j(1)$ | $b_j(2)$ | $b_j(3)$ | $b_j(4)$ | $b_j(5)$ | $b_j(6)$ | $b_j(7)$ |

The following procedure for the up-gradation of $Kp_j(k)$, $a_j(5)$ and $a_j(6)$ is used.

1. The iterative learning mechanism first learns the value of $Kp_j(k)$ such that the system is critically damped i.e.

$$pos \leq \varepsilon \qquad (5)$$

where $\varepsilon$ is some small number. The learning law is

$$Kp_{j+1}(k) = Kp_j(k) \pm \mu_{kp} \left( \|r_j(k)\| - \|y_j(k)\| \right) \qquad (6)$$

Here $Kp_{j+1}$ is the value of gain for the next iteration and $Kp_j$ is the value of gain for the current iteration. The learnt value of $Kp_j$ is stored in memory and is called $K_{pc}$ (critically damped proportional gain).

2. Change the value of $a_j(5)$ such that $sse \leq sse_d$, using the learning law

$$a_{j+1}(5) = a_j(5) \pm \mu_{sse} \left( \|r_j(k)\| - \|y_j(k)\| \right) \qquad (7)$$

Here $\mu_{sse}$ is the step size parameter for correcting steady state error.

3. Change the value of $a_j(6)$ such that $pos \leq pos_d$ using the learning law

$$a_{j+1}(6) = a_j(6) \pm \mu_{pos} \left( \|r_j(k)\| - \|y_j(k)\| \right) \qquad (8)$$

Here $\mu_{pos}$ is the step size parameter for correcting percentage overshoot.

4. The value of $Kp_j(k)$ is set in real time by using equation

$$Kp_j(k) = \left( Kp' \times K_{pc} \right) / e_j(k) \qquad (9)$$

Here $Kp'$ is the output of the defuzzifier against the current error.

## 4   Simulations and Results

PID controllers are the most widely used controllers in the industry. PID controllers can be implemented to meet various design specifications such as steady state error, percentage overshoot and rise time. Despite their wide use, tuning a PID controller can be a very tedious job. Most of the tuning methods require at least some knowledge of the system we want to control. One approach, the Zeigler-Nichols tuning method, which was developed in the 1950's and has stood the test of time is still the most used tuning method. It involves the calculation of critical gain ($K_c$) and critical time constant ($T_c$). Once the values of $K_c$ and $T_c$ are obtained, the PID parameters can be calculated using Table 3.

**Table 3.** Zeigler-Nichols based gain calculation table for P,PI and PID controllers

| Controller | $K_p$ | $K_i$ | $K_d$ |
|------------|-------|-------|-------|
| P | $0.5K_c$ | | |
| PI | $0.45K_c$ | $0.833T_c$ | |
| PID | $0.6K_c$ | $0.5T_c$ | $0.125T_c$ |

The scheme presented in this paper was tested through simulations on a variety of systems. The results obtained by using Zeigler-Nichols P, PI, and PID controllers were also obtained for comparison. In this paper, simulation results of a motor-speed control system are presented. The transfer function of the system is given below.

$$G(s) = \frac{0.05}{0.005s^2 + 0.06s + 0.1025} \tag{10}$$

The aim was to design a controller, which should give less than 5% steady state error (sse) and less than 5% percentage overshoot (pos). The motor was required to run at a speed of 1 rad/sec. For simulation purposes all the ranges were normalized between −1 and 1.

The parameters determined by the Zeigler-Nichols based controllers are listed in Table 4.

**Table 4.** Parameters calculated for P, PI and PID controllers

| System | Zeigler-Nichols P controller | Zeigler-Nichols PI controller | Zeigler-Nichols PID controller |
|--------|------------------------------|-------------------------------|--------------------------------|
| $G(s)$ | $K_p = 122.5$ | $K_p = 110.25,$ $K_i = 0.1083$ | $K_p = 147, K_i = 0.065,$ $K_d = 0.0163$ |

The output obtained by using these designed P, PI and PID controllers as well as the proposed ILFGS based controller are plotted in Figure 6.

As can be seen from the results, the proposed scheme performed much better than the Zeigler-Nichols based controllers. The values of the parameters learnt, for the proposed controller, were $K_{pc} = 11.6857$, $a_j(5) = 0.021$, $a_j(6) = 0.9345$. It took 1 iteration to learn $a_j(5)$ and 6 iterations to learn $a_j(6)$. Comparison of the error between these controllers is exhibited in Figure 7.

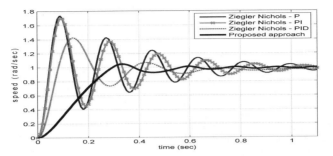

**Fig. 6.** Comparison of responses

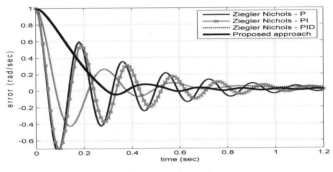

**Fig. 7.** Comparison of errors

Figure 8 below shows the values of $Kp_j(k)$ determined by the proposed controller.

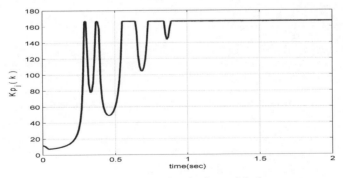

**Fig. 8.** Values of proportional gain used during one run

ILFGS was able to calculate a non-linear range of values for $Kp_j(k)$. The results show that ILFGS satisfactorily controlled the motor speed of the system. In fact it yielded better control performance than the Ziegler-Nichols based controllers. The control surface learnt by the ILFGS is shown in Figure 9.

The over all control surface is non-linear but linear regions can be easily recognized.

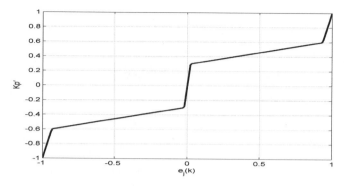

**Fig. 9.** Control surface leant for the system

## 3   Experimental Setup and Results

A QET DC Motor Kit from Quanser [24] was used to test the ILFGS based approach with a practical setup. The kit can be programmed to set proportional, integral and derivative gains. MATLAB was used to communicate with the kit through serial port. ILFGS was implemented in MATLAB. The micro-controller on the kit was used to set proportional gain and speed of the motor. The experimental setup is show in figure 10. The aim was to make the motor run at 100 rad/sec with a steady state error of less than 5% and an over shoot of less than 5%.

**Fig. 10.** Experimental setup with the QET DC Motor Kit from Quanser

The different step size parameters were given values of $\mu_{kp} = 0.01$, $\mu_{sse} = 0.01$ and $\mu_{pos} = 0.01$. The error had a range of $e_j(k) = [-200,200]$ and gain had a range of $Kp' = [-1,1]$. During the learning phase the values of $K_{pc}$, $a_j(5)$ and $a_j(6)$ learnt were $K_{pc} = 0.25$, $a_j(5) = 15.8$ and $a_j(6) = 76.36$. A plot of learning history of parameters $a_j(5)$ and $a_j(6)$ vs. no. of iterations is given in Figure 11.

A 3-D plot of speed, time and number of iterations is presented in Figure 12. The plot was obtained during the process of learning $a_j(6)$.

Figure 12 shows that aj(6) was learnt in 4 iterations. The end point aj(5) was learnt in 6 iterations. The final membership functions after the learning process was completed are given in Figure 13.

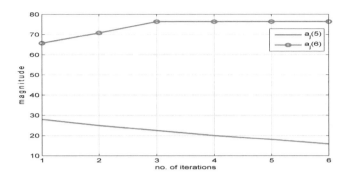

**Fig. 11.** Learning values $a_j(5)$ and $a_j(6)$

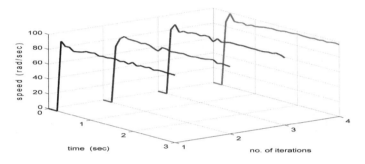

**Fig. 12.** History of system output while learning $a_j(6)$

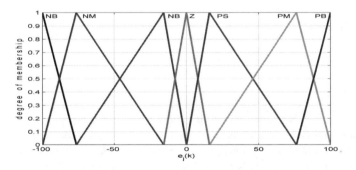

**Fig. 13.** Learnt membership functions

These learnt values were then used in real time to achieve a desired speed of 100 rad /sec. A plot of error between the desired speed and the speed achieved by the motor is exhibited in Figure 14.

During this process the values of proportional gain calculated by ILFGS are shown in Figure 15.

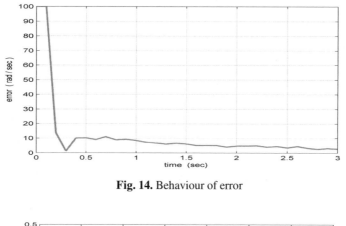

**Fig. 14.** Behaviour of error

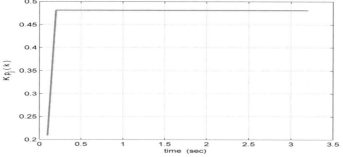

**Fig. 15.** Calculated proportional gain by the ILFGS to control the QET DC Motor

## 6  Conclusions

Fuzzy controllers and for that matter any controller needs to be adaptive in order to compensate for uncertainties, noise, variation in parameters and changes in design requirements. Conventional P, PI, PID etc. controllers also need to be adaptive in order to be more useful in ever increasing complexity of the practical systems, especially when the system parameters are not known. It is now realized that complex real world problems require intelligent systems that combine knowledge, techniques and methodologies from various sources. These intelligent systems are supposed to possess humanlike expertise within a specific domain, adapt themselves and learn to do better in changing environment. The approach presented in this paper combines the conventional control with fuzzy logic and iterative learning to tackle real world problems.

The scheme changes the proportional gain adaptively, through the iterative learning fuzzy gain scheduler (ILFGS). The fuzzy system in ILFGS itself has the capability to adapt because of iterative learning. The learning laws ensure that the steady state error, percentage overshoot etc. requirements are met. The ILFGS adjusts the proportional gain in real time to meet design requirements.

As seen in simulation and experimental results, the proposed scheme is able to learn non-linear control surfaces and is able to perform better than Ziegler-Nichols based controllers. This approach can handle uncertainties in system parameters, system inputs and rules using the learning laws. This attribute makes it a robust approach.

Compared to conventional gain scheduling (CGS) and fuzzy gain scheduling (FGS) alone, the most obvious advantage of ILFGS is that it is able to learn the appropriate parameters adaptively thus avoiding the need of detailed knowledge about the plant dynamics. It is also able to remove uncertainty in formulating membership functions through iterative learning process.

Though the scheme performs very well by adjusting proportional gain alone, the performance effect of integral and derivative gains also need to be researched.

# References

1. Abad, H.B.B., Varjani, A.Y., Asghar, T.: Using fuzzy controller in induction motor speed control with constant flux. Transactions on Engineering, Computing and Technology, Enformatika Society 5, 307–310 (2005)
2. Kuo, B.C., Golnarahgi, F.: Automatic control systems, 8th edn. John Wiley & Sons Inc., Chichester (2002)
3. Cam, E., Kocaarslan, I.: A fuzzy gain scheduling PI controller application for an interconnected electrical power system. Electrical power systems research 73, 267–274 (2005)
4. Zadeh, L.A.: Fuzzy Sets. Information and Control 8, 338–353 (1965)
5. Zadeh, L.A.: Outline of a new approach to the analysis of complex systems and decision processes. IEEE transactions on systems, man and cybernatics 3, 28–44 (1973)
6. Zadeh, L.A.: A fuzzy algorithmic approach to the definition of complex or imprecise concepts. Int. J. man-machine studies 8, 249–291 (1976)
7. Blanchett, T.P., Kember, G.C., Dubay, R.: PID gain scheduling using fuzzy logic. ISA Transaction 39, 317–325 (2000)
8. Lee, J.: On methods for improving performance of PI type fuzzy logic controllers. IEEE Transactions on Fuzzy Systems 1, 298–301 (1993)
9. Rodrigo, M.A., Seco, A., Ferrer, J., Penya-roja, J.L.: Non linear control of an activated sludge aeration process: use of fuzzy techniques for tuning PID controllers. ISA Transactions 38, 231–241 (1999)
10. Tan, S., Hang, C.C., Chai, J.S.: Gain Scheduling: from conventional to neuro-fuzzy. Automatica 33, 411–419 (1997)
11. Wang, L.X.: A Course in Fuzzy Systems and Control. Prentice Hall, NJ (1997)
12. Zhao, Z.Y., Tomizuka, M., Isaka, S.I.: Fuzzy gain scheduling of PID controllers. IEEE transactions on systems, man. And cybernetics 23, 1392–1398 (1993)
13. Li, H.X., Gatland, H.B.: Conventional fuzzy control and its enhancement. IEEE Transactions on Systems, Man and Cybernetics - Part B: Cybernetics 26, 791–797 (1996)
14. Ying, H., Yongquan, Y., Tao, Z.: A new real-time self-adaptive rule modification algorithm based on error convergence in fuzzy control. In: IEEE international conference on Industrial Technology, ICIT, pp. 789–794 (2005)
15. Rafael, A., Jorge, C., Oscar, C., Antonio, G., Francisco, H.: A genetic rule weighting and selection process for fuzzy control of heating, ventilating and air conditioning systems. Elsevier Engineering Applications of Artificial Intelligence 18, 279–296 (2005)
16. Perng, J.W., Wu, B.F., Chin, H.I., Lee, T.T.: Gain phase margin analysis of dynamic fuzzy control systems. IEEE transactions on Systems, Man and Cybernetics - part B 34, 2133–2139 (2004)

17. Zadeh, L.A.: From computing with numbers to computing with words-From manipulation of measurements to manipulation of perceptions. Int. J. Appl. Math. Comput. Sci. 12, 307–324 (2002)
18. Mendel, J.M.: UNCERTAIN rule-based fuzzy logic systems. Prentice Hall, Englewood Cliffs (2001)
19. Mendel, J.M.: Type-2 fuzzy sets and systems: an overview. IEEE Computational Intelligence Magazine 2, 20–29 (2007)
20. Norrlof, M.: An Adaptive Iterative Learning Control Algorithm With Experiments on an Industrial Robot. IEEE Transactions On Robotics and Automation 18, 245–251 (2002)
21. Tayebi, A.: Adaptive Iterative Learning Control for Robot Manipulators. Automatica 40, 1195–1203 (2004)
22. Jerzy, E.K., Marek, B.Z.: Iterative learning control synthesis based on 2-D system theory. IEEE Transactions on Automatic Control 38, 121–124 (1993)
23. Geng, Z., Jamshidi, M., Carroll, R., Kisner, R.: A learning control scheme with gain estimator. In: Proceedings of the 1991 IEEE International Symposium on Intelligent Control, Arlington, USA, pp. 365–370 (1991)
24. Quanser's DC Motor Control Kit, http://www.quanser.com/

# Comparative Analysis of Mobile IP and HAWAII

Aamir Wali, Muhammad Ahmad Ghazali, and Saqib Rehan Ayyubi

National University of Computer and Emerging Sciences (NUCES-FAST),
Block B, Faisal Town, Lahore, Pakistan
{aamir.wali,ahmad.ghazali}@nu.edu.pk, saqibrehan@gmail.com

**Abstract.** Mobile IP is the current standard for IP-based wireless access networks. However, there are some drawbacks of using Mobile IP to support micro-mobility of mobile hosts. Several proposals have been presented to overcome these drawbacks and perhaps the most important of these is Handoff-Aware Wireless Access Internet Infrastructure (HAWAII). In this paper, a comparative analysis of the performance of Mobile IP and HAWAII is performed with respect to the four different issues, namely: Power up, Micro-mobility, Macro-mobility and Paging. The results of this comparison show that HAWAII provides better functionality for Power up, Micro-mobility and Paging, and both Mobile IP and HAWAII provide essentially the same functionality for Macro-mobility.

**Keywords:** Mobile IP, HAWAII, Comparative Analysis, Power up, Micro-mobility, Macro-mobility, Paging.

## 1 Introduction

IP is the dominant internetworking protocol currently in use, and most of the common access networks for different wireless technologies are based on IP [1]. At present, IP based wireless access networks use the Internet standard of Mobile IP. Mobile IP provides a good framework for users to roam outside their home domains without disruption. However, it was not designed specifically for managing micro-mobility and, therefore, results in several limitations when mobile hosts move frequently within the same domain.

There are numerous competing proposals for providing micro-mobility support in wide-area wireless networks e.g. Hierarchical Mobile IP (HMIP), Cellular IP and Handoff-Aware Wireless Access Internet Infrastructure (HAWAII) [2]. To address the shortcomings of the original Mobile IP proposal, these proposals have focused on one or more of the following goals:

- Reducing disruption of data packet flow during handoffs
- Enhancing scalability by reducing updates sent to the Home Agent (HA)
- Providing support for Quality of Service (QOS)
- Allowing efficient usage of network resources

While other proposals have only addressed subsets of the four goals mentioned above, HAWAII fulfills all of the goals. Therefore, we have decided to perform a comparison of Mobile IP and HAWAII. In this paper, we shall not describe the details of the

D.M.A. Hussain et al. (Eds.): IMTIC 2008, CCIS 20, pp. 169–179, 2008.
© Springer-Verlag Berlin Heidelberg 2008

Mobile IP and HAWAII, which can be found in [3] and [4] respectively. Our focus will be on some specific issues related to the efficiency of these two protocols. These issues are: Power up, Micro-mobility, Macro-mobility and Paging. These issues will be discussed in general with respect to Mobile IP and HAWAII and then a comparative analysis of Mobile IP and HAWAII will be performed on the basis of these four issues.

The remaining paper is organized as follows. In the second section, we will be giving a general overview of the four issues mentioned above. In the third and fourth section, these issues will be discussed with respect to Mobile IP and HAWAII, respectively. Comparison of Mobile IP and HAWAII will be done in fifth section. And, finally, conclusions will be presented in the sixth section.

## 2 Evaluation Criteria

This section presents some of the evaluation criteria employed for comparing the Mobile IP and HAWAII protocols.

### 2.1 Power Up

Mobile hosts usually operate on batteries. When they are out of power they are in 'dead state'. This is when they are completely unplugged from the network, non-existent for any base station and hence has nothing to do with receiving and sending messages. Power up is the process when a mobile host is actually 'powered up' or charged to the capacity that it can communicate with base stations and participate in communication of data. Power up also occurs when a mobile host, already in running state, moves into a foreign domain and needs to register with its base station. The power up process can be divided in to two main steps: IP Address Assignment and Registration. Each mobile host just like fixed hosts should have a unique IP address and during power up an IP is assigned to mobile hosts [5]. Normally this address assignment is static but it be made dynamic through the use of Dynamic Host Configuration Protocol (DHCP). This step is ignored in power up process when a running mobile host moves into a foreign domain. Once an IP is assigned, the mobile host associates itself to the nearest base station by sending a registration request. Now it is up to the base station to make necessary arrangements (protocol dependent) to update the network about this 'new host' attached to it. With this update all packets addressed to this 'new host' can be forwarded to the associated base station that in turn forwards it to the mobile host.

### 2.2 Micro-mobility

This term is basically used for defining movement of the mobile host (subscriber) between base stations belonging to its own home domain.

### 2.3 Macro-mobility

Macro-mobility deals with the mobility of hosts at a large scale i.e. mobility between two domains (or wide area wireless access networks). When the mobile host moves from its home domain to another domain, it obtains a care-of-address from the foreign agent. It then informs its home agent about its care-of-address, so that the home agent can forward datagram's destined for the mobile host to the current foreign agent of the mobile host. The foreign agent, in turn, delivers the datagram's to the mobile host. In case of both micro-mobility and macro-mobility, the process of changing the status of the mobile subscriber as it is handed over to another base station from the current base station is called Hand off. Handoff is of two types, i.e. hard and soft hand off.

### 2.4  Paging

Usually fixed hosts remain on-line for most of the time even though mostly they do not communicate. Being "always connected", fixed hosts are reachable with instant access. Mobile subscribers also desire a similar service. In the case of mobile hosts maintaining their location information (for being continuously reachable) would require periodic updates of location, which would consume already limited bandwidth and battery power. As the number of Mobile IP users grow, so will the signaling overhead on the core IP network in support of mobility management. This signaling overhead can be reduced through the introduction of paging [6].

A network that supports paging allows the mobile hosts to operate in two distinct states – an *active state* in which the mobile host is tracked to the finest granularity possible such as its current base station (resulting in no need for paging) and mobile host updates network on every base station change. The other is the *standby state* in which the host is tracked at a much coarser granularity such as a paging area, which consists of a set of base stations. Also in this state the mobile host updates network on every page area change. The network uses paging to locate the mobile host in standby state [7].

Tracking the location of a mobile host is performed through *update* procedures that involve a mobile host informing the network of its location at times triggered by movement, timer expiration, etc. The knowledge about the current location of a mobile host is obtained by a registration procedure through which the mobile hosts announce their presence upon entering a new location area. The new location area will be stored in some database to be used for paging the mobile next time the mobile receives an incoming call [6].

The size of the location area plays an important role in cellular networks because of the tradeoff caused by paging and registration signaling. In networks that support mobility, the precise location of a mobile host must be known before data can be delivered. Therefore, the tradeoff is between how closely the network tracks the current location of a mobile host, versus the time and effort required to locate a mobile host whose location is not precisely known. Locating a mobile host is performed through search procedures that include *paging* the mobile host. Paging typically includes transmitting a request for a mobile host to a set of locations. Idle mobile hosts do not have to respond to the requests if they move within the same paging area.

## 3  Evaluating Mobile IP

### 3.1  Power Up

There are two kinds of registration messages, i.e., registration request and registration reply. Both these messages are sent using User Datagram Protocol (UDP). The data structure of the registration messages is depicted in Fig. 1.

| IP Header | UDP Header | Mobile IP Message Header | Extensions |
|-----------|------------|--------------------------|------------|

Fig. 1. Structure of a Registration Message

The registration process is independent of the way the care-of address is obtained: directly from a foreign agent or from another independent service (DHCP). In the first case, the mobile host first sends the request to the foreign agent, which in turn passes it to the home agent. In the second case the mobile host sends the request to the home agent directly. This is done through the use of care-of address as the source IP address of the request. The structure of registration request comes after the conventional IP and UDP headers. Its details can be found in [3]. Similarly, the details of the structure of registration reply can also be found in [3].

## 3.2  Mobility

All forms of mobility are treated similarly by Mobile IP. It means that a user moving a short distance (e.g. between two base stations within the same domain) has to use the same communication procedure as the user which moves a longer distance (e.g. from one domain to another). The first of these two forms of mobility is called *micro-mobility* and the second one is called *macro-mobility*. The procedure that is used to accomplish these two forms of mobility with reference to Mobile IP is described below:

1. When a mobile host moves into a foreign network, it obtains a care-of-address for the new network (i.e. the foreign network). This care-of-address can be obtained either by listening to the agent advertisement messages or by soliciting such a message from the foreign agent. On the other hand, when a mobile host moves within the domain of its home network it obtains a collocated care-of-address (CCOA), which is a local address associated with one of its interfaces.
2. Each time the mobile host receives a new care-of-address, it informs its home agent about this care-of-address. In case of macro-mobility, this can be done through the foreign agent with whom the node is currently registered.
3. After the home agent has been informed about the care-of-address of the mobile host, the home agent tunnels datagram's that are destined for the mobile host's home address to the foreign agent or send them directly to the mobile host at its collocated care-of address (CCOA).

Although Mobile IP is a good protocol for macro-mobility, it is not an efficient protocol to perform micro-mobility of mobile hosts. Micro-mobility (i.e. movement within a domain) occurs more frequently in the Internet as compared to macro-mobility (i.e. movement across domains). When using Mobile IP, every time a mobile host moves (even within the same domain) it has to signal to the home agent, which results in frequent exchanges being made over the Internet. Therefore, overhead in the form of signaling messages is significantly increased when using Mobile IP to support micro-mobility. Different approaches to overcome this micro-mobility related issue have been proposed and one such approach named HAWAII is discussed in this paper.

### 3.3  Paging

The original Mobile IP proposal [3] does not support paging. Therefore, on every movement the mobile host is required to update the network about its current location. This is an inefficient operation due to the need to send frequent updates to the network, thereby, utilizing precious bandwidth as well as power. Also, without paging, these devices will not be able to perform efficiently because most of the power will be consumed by sending too many updates to the network.

## 4  Evaluating HAWAII

### 4.1  Power Up

The HAWAII network consists of a root domain router, which is a gateway into each domain. Each host has an IP address and home domain. Fig. 2 shows two HAWAII domains each with a root domain router DR1 and DR2. R1, R2, and R3 are intermediate routers. The base stations BS1, BS2 and BS3 lie within one paging area while BS4 and BS5 lie in another. Suppose mobile host MH1 powers up in region close to BS1.

The mobile host first sends a mobile IP registration message to its current base station BS1. The base station checks that the mobile host is powering up according to the power up parameters in the registration message. The base station adds a forwarding entry to forward a packet destined for the mobile host and sends a power up message to the intermediate router R1 that connects to the root domain router. R1 adds a forwarding entry to forward all packets destined for the mobile host to BS1. R1 also sends a registration message to DR1. DR1 adds a forwarding entry for R1 and sends a direct acknowledgement to BS1 that in turn acknowledges the mobile host. This acknowledgement process is defined as registration reply.

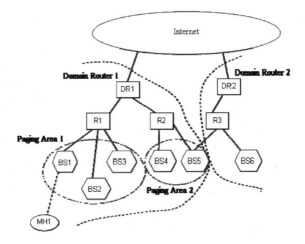

**Fig. 2.** HAWAII Architecture

When a mobile host that is in running state moves into a foreign domain, the same procedure is adopted. However, in this case the procedure is termed as registration. For movement of a mobile host in between different base stations but within the same domain, the HAWAII procedure registers and updates all hosts till the point of cross-over into another domain. For example, referring Fig. 2, if the mobile host moves from BS1 to BS2 within the same domain, then the registration updates are sent as high as R1 in the hierarchy.

## 4.2  Micro-mobility

In case of micro-mobility the mobile host maintains its IP address since its movement is within its home HAWAII domain. During power up, the forward routing table entries have been set up in the domain root router and any intermediate routers to the mobile host forming a tree topology. Crossover router is defined as the router that is at the intersection of two paths, where one path is between the root router and the old base station while the other path is between the new and the old base station.

Four path setup schemes (i.e. Multiple stream forwarding, Single stream forwarding, Unicast non-forwarding, Multi-cast non-forwarding) are used to re-establish path state when a mobile host moves from one base station to another. These four schemes are grouped into two classes based on the way packets are delivered to the mobile host during handoff. In case of the first class, packets are sent from the old base station to the new one before being diverted at the crossover router. This is the class of forwarding path setup schemes. For the second class, packets are diverted at the crossover router, resulting in no forwarding from the old base station. This is the class of non forwarding path setup schemes. Details of these schemes can be found in [4].

## 4.3  Macro-mobility

In case of movement of a mobile host between base stations connected to different Hawaii domains, the domain root router of the home domain acts as a HA for the mobile host which is in a foreign domain. In the new domain the mobile host will acquire a new IP address known as the collocated care of address (CCOA). Macro-mobility will be illustrated using Fig. 3. In the new domain the mobile host will acquire a new IP address of 2.2.2.200 while 1.1.1.100 was its previous home domain IP address.

Mobile host sends a registration message to BS3. BS3 finds out from the message that this is a case of an inter-domain handoff. Therefore, it initiates power up procedure in its domain for establishing host based forwarding entries for 2.2.2.200 CCOA. After power up the base station sends IP registration message (5) to home agent R1. In this figure (Fig. 3) the home domain root router is collocated with the home agent of the mobile host. Thus, home agent forms an entry and will tunnel packets destined for IP 1.1.1.100 to the care of address 2.2.2.200. HA also sends a reply (6) to the IP registration message to the new base station BS3. BS3 then sends an acknowledgement to mobile host (7).

Thus the packets destined for 1.1.1.100 reach the home agent from where they are tunneled to the CCOA address 2.2.2.200. These packets reach R4 based on the subnet portion of the address and then get forwarded through the forwarding table entries of the routers as specified by HAWAII, finally reaching BS3 from where they are sent to the mobile host. Subsequent handoffs within this new network will be dealt locally by HAWAII as explained in the micro-mobility section. Thus HA is updated only when the mobile host crosses a domain boundary resulting in much less handoff latencies.

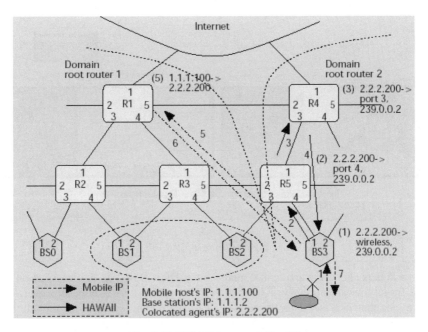

**Fig. 3.** HAWAII Macro-mobility [5]

## 4.4 Paging

Paging operation in HAWAII generally depends on participation from both the mobile host and the network. The mobile host needs to inform the network of its approximate location. Moreover, the network has to pinpoint the exact location of a mobile host when it receives a packet destined for that host. So, in general the paging mechanism can be divided in to two parts depending on role of mobile host or network.

To understand the role of an idle mobile host in paging for HAWAII, consider Fig. 4, which is a close up of the two paging areas depicted in Fig. 2. Note that paging is required only when the mobile host is idle or in standby state. As described previously, PA1 comprises the base stations BS1, BS2, and BS3, whereas, PA2 comprises BS4 and BS5. As a mobile host moves within the paging area PA1 as given in Fig. 4 (a), it is not required to register. However, when it moves into the paging area PA2, it sends an update message for providing its approximate location to the network. The message is first sent to BS4 as given in Fig. 4 (b), which then forwards it to R2.

Moving onto role of the network, assume that the mobile host illustrated in Fig. 2 is idle. The network only has the information that the mobile host is present near one of the base stations covering paging area say PA1 that comprises base stations BS1, BS2 and BS3. In an HAWAII based network, an administratively scoped IP multicast group address is used to identify all the base stations belonging to the same paging area. If IP packets arrive at the root domain router, destined for a mobile host, the network pages BS1, BS2, and BS3 to pinpoint exact location of the mobile host. The procedure employed for delivering the packets to the mobile host is as follows:

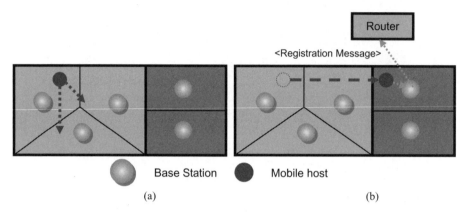

**Fig. 4.** Paging in HAWAII

The data packets first arrive at domain router DR1. Based on the information provided by its forwarding entry, R1 notes that the mobile host is idle. However, DR1 determines that it is not directly connected to the paging area for the mobile host. Therefore, it forwards the data packets downstream to R1. R1 performs similar processing and identifies that it is connected to the paging area of the mobile host. R1 buffers the data packets and initiates a HAWAII page request to the multicast group address. BS1, BS2 and BS3, which belong to the multicast group, receive this message and in turn broadcast a similar message onto their respective wireless channels. The mobile host, which in our case is near BS1, receives the page message and sends a Mobile IP registration message to BS1 [5] and the mentioned registration procedure follows.

The basic idea behind this procedure is to move paging to the leaf nodes, which in our case are the base stations and low-level routers, and away from the domain root router. This provides better scalability [5].

## 5   Comparative Analysis

This section compares Mobile IP and HAWAII based on the evaluation criteria described above in Section 2.

### 5.1   Power Up

As already discussed previously there are two situations when power up occurs. Firstly, when a mobile host actually powers up and registers with the network and secondly when it is already in running state moves into a foreign domain and registers there. In the former situation, the power up procedure is identical for both Mobile IP and HAWAII. In the latter case, however, HAWAII proves to be more efficient. In case of Mobile IP the registration message/update goes all the way up to the home agent. However, for HAWAII the update does not go to the nodes beyond the cross-over router. For example, if the mobile host moves between base stations connected to the same router then registration request message goes upto the associated router and not beyond. This shortcut makes HAWAII more efficient during handoffs.

## 5.2 Micro-mobility

**Mobility Notification.** In case of Mobile IP, as the mobile host moves between different base stations within the same domain, it must notify its home agent (HA). Since, movement within a domain is more common, Mobile IP will entail significant overheads in terms of signaling messages sent to the HA regarding its updated collocated care-of-address.

On the other hand, HAWAII uses a different approach involving updating the forwarding entries in the domain root routers and other intermediate routers. It updates the entries in the domain root router (DRR) only when the router at the next hop of the DRR is changed. Thus, its path setup scheme operates locally in such a way that the higher hierarchical routers mostly remain unaffected by the movement of the mobile host.

We can clearly see from the above discussion that HAWAII provides a much better approach to the notification mechanism as compared to Mobile IP. This not only results in reducing the overhead due to the mobility of hosts within the same domain but also leads to much better scalability.

**Traffic Disruption.** In case of Mobile IP, collocated care-of-address changes during each handoff. This leads to a greater probability of packet loss during handoffs. On the other hand, in case of HAWAII the IP address of the mobile hosts remains the same during intra-domain handoffs. Moreover, various path setup schemes result in less disruption to user traffic by just updating the necessary forwarding entries in the relevant routers.

**Header Overheads.** In case of Mobile IP, the home agent intercepts all the messages destined for a mobile host. It then encapsulates the mobile host's data packets into another IP packet whose destination address is the current collocated care-of-address of the mobile host. This results in overhead in terms of extra header bytes for each data packet. HAWAII uses no such tunneling during micro-mobility, thus it remains free from these overhead bytes.

**Non-optimal Routing.** As described in the previous sub-section, Mobile IP tunnels packets to the collocated care-of-address of the mobile host. It may result in non-optimal route being followed by the data packets. HAWAII uses no such tunneling mechanism; rather it relies on standard intra-domain routing protocols such as RIP and OSPF to detect a router or any link failure. Thus, HAWAII always uses an optimal route to forward data packets to the mobile host.

**Quality of Service (QOS) Support.** IP addresses are used to identify flows. Thus by retaining IP addresses of the mobile hosts during micro-mobility, HAWAII maintains previous QOS information at backbone routers. This ensures simpler QOS support as compared to Mobile IP.

### 5.3  Macro-mobility

**Mobility Notification.** Notification procedures used by Mobile IP and HAWAII are same during macro-mobility of hosts. For both protocols, the mobile host informs the home agent about its macro-mobility, through the foreign agent.

**Traffic Disruption.** HAWAII uses Mobile IP for inter-domain handoffs. Thus, loss of data packets during handoffs remains the same for both protocols.

**Header Overheads.** Both Mobile IP and HAWAII tunnel packets from the home agent to the foreign agent when the mobile host moves between base stations belonging to different domains. Therefore, during macro-mobility the header overheads are the same for Mobile IP and HAWAII.

**Non-optimal Routing.** In the case of macro-mobility, tunneling is done in both Mobile IP and HAWAII. Therefore, both protocols suffer from the non optimal routing that is the consequences of tunneling.

### 5.4  Paging

Apart from HAWAII, paging seems to have received relatively little attention. Its uses administratively scoped IP multicast to distribute paging requests to base stations. This should push paging to the edge of the access network, which assists in *scalability* and *robustness*. Mobile host naturally tracks mobile hosts as they move, through the standard messages to join to/prune from the multicast tree. This obviously has a large amount of location management overhead and moreover at the time when the mobile is idle. This results in not only unnecessary usage of the wireless channel but also needless consumption of power resources of mobile hosts. Thus in terms of paging HAWAII has definitely outclassed mobile IP.

## 6  Conclusion

In conclusion, we would like to state that Mobile IP has some shortcomings that make it ineffective in certain situations. For example, during *power up* a registration message/update goes all the way up to the home agent in case of Mobile IP. However, in HAWAII the power up message does not go to the nodes beyond the crossover router. This provides an efficient way to support *micro-mobility* of hosts. Since, HAWAII uses Mobile IP to support *macro-mobility,* therefore, the performance of both is similar in terms of providing inter-domain mobility. Moreover, HAWAII also provides paging support, which results in efficient utilization of network resources and mobile host's power.

# References

1. Perkins, C.E.: Mobile Networking in the Internet. Mobile Networks and Applications 3(4), 319–334 (1998)
2. Sanmateu, A., Paint, F., Morand, L., Tessier, S., Fouquart, P., Sollund, A., Bustos, E.: Seamless mobility across IP networks using Mobile IP. Computer Networks: The International Journal of Computer and Telecommunications Networking 40(1) (2002)
3. Perkins, C.E.: Mobile IP. IEEE Communications Magazine (May 1997)
4. Ramjee, R., Porta, T.L., Thuel, S., Varadhan, K.: HAWAII: A Domain-based Approach for Supporting Mobility in Wide-area Wireless Networks. IEEE/ACM Transaction on Networking (2002)
5. Ramjee, R., Porta, T.L., SalGarelli, L., Thuel, S., Varadhan, K.: IP-Based Access Network Infrastructure for Next-Generation Wireless Data Networks. IEEE Personal Communications (2000)
6. Campbell, A.T., Gomez, J.: IP Micromobility Protocols. ACM SIGMOBILE Mobile Computer and Communication Review (MC2R) 4(4), 45–54 (2001)
7. Campbell, A.T., Gomez, J., Kim, A., Wan, C.: Comparison of IP Micromobility Protocols. IEEE Wireless Communications (2002)

# Video Transport over Heterogeneous Networks Using SCTP and DCCP

Yousaf Bin Zikria, Shahzad A. Malik, Hassan Ahmed, Sumera Nosheen,
Naeem Z. Azeemi, and Shahid A. Khan

Department of Electrical Engineering, COMSATS Institute of Information Technology,
Islamabad
yusi_2@hotmail.com, smalik@comsats.edu.pk,
h.ahmed@comsats.edu.pk, sumera_nosheen@comsats.edu.pk,
naeemazeemi@comsats.edu.pk, s.khan@comsats.edu.pk

**Abstract.** As the internet continues to grow and mature, transmission of multi-media content is expected to increase and comprise a large portion of overall data traffic. The internet is becoming increasingly heterogeneous with the advent and growth of diverse wireless access networks such as WiFi, 3G Cellular and WiMax. The provision of quality of service (QoS) for multimedia transport such as video traffic over such heterogeneous networks is complex and challenging. The quality of video transport depends on many factors; among the more important are network condition and transport protocol. Traditional transport protocols such as UDP/TCP lack the functional requirements to meet the QoS requirements of today's multimedia applications. Therefore, a number of improved transport protocols are being developed. SCTP and DCCP fall into this category. In this paper, our focus has been on evaluating SCTP and DCCP performance for MPEG4 video transport over heterogeneous (wired cum wireless) networks. The performance metrics used for this evaluation include throughput, delay and jitter. We also evaluated these measures for UDP in order to have a basis for comparison. Extensive simulations have been performed using a network simulator for video downloading and uploading. In this scenario, DCCP achieves higher throughput, with less delay and jitter than SCTP and UDP. Based on the results obtained in this study, we find that DCCP can better meet the QoS requirements for the transport of video streaming traffic.

**Keywords:** Video transport, MPEG4, SCTP, DCCP, heterogeneous networks, ns-2.

## 1 Introduction

Film and television distribution, digitized lectures, and distributed interactive gaming applications have begun to be realized in today's Internet, but are rapidly gaining popularity. Audio and video streaming capabilities will play an ever-increasing role in the multimedia rich Internet of the near future. Real-time streaming has wide applicability beyond the public Internet as well. In military and commercial wireless domains, virtual private networks, and corporate intra-nets audio and video are

D.M.A. Hussain et al. (Eds.): IMTIC 2008, CCIS 20, pp. 180–190, 2008.
© Springer-Verlag Berlin Heidelberg 2008

becoming commonplace supplements to text and still image graphics. The quality of the content delivered by these multimedia applications varies, but they are generally associated with low resolution, small frame size video, and use of transport protocol. Video streams are usually delivered via UDP with no transport layer congestion control. A large-scale increase in the amount of streaming audio/video traffic in the Internet over a framework devoid of end-to-end congestion control will not scale, and could potentially lead to congestion collapse [1].

In today's world the wireless networks are growing rapidly and heterogeneity of networks continues to increase. As most of the transport protocols are designed to tackle the problems related to the wired networks and they simply ignore or assume it will also perform as well in wireless networks. However, in wireless environments, there are a number of issues with traditional transport protocols which have led researchers to develop new protocol and enhance the existing ones to meet the ever growing needs of multimedia applications in heterogeneous wireless networks. SCTP and DCCP belong to this new category of transport protocols. In this work, we focus on evaluating SCTP and DCCP performance for MPEG4 video transport over heterogeneous (wired cum wireless) networks.

This paper is organized as follow. Section II gives a brief description of SCTP and DCCP. Section III provides details of the experimental setup and performance metrics used in this study. In section IV, we discuss simulation scenarios and results. Finally in section V, we present our conclusions and some directions for future work.

## 2  SCTP and DCCP

UDP [2] is the transport protocol mostly used for video streaming platforms mainly because the fully reliable and strict in-order delivery semantics of TCP do not suit the real-time nature of video transmission. Video streams are loss tolerant and delay sensitive. Retransmissions by TCP to ensure reliability introduce latency in the delivery of data to the application, which in turn leads to degradation of video image quality.

Stream control Transmission protocol (SCTP) [3] is a reliable transport protocol. SCTP was originally designed to transport PSTN signaling message over 1P network, but is also capable of serving as a general-purpose transport protocol for text and video transmission. SCTP also allows unordered delivery of data as UDP and provides reliability and flow control like TCP. SCTP also provides attractive features such as multi-streaming and multi-homing that may be helpful in high-mobility environment and additional security against denial of service attack based on SYN flooding [3].

The Datagram Congestion Control Protocol (DCCP) [4] is a transport protocol that implements bidirectional, unicast connections of congestion-controlled, unreliable datagrams. DCCP is intended for applications such as streaming media that can benefit from control over the tradeoffs between delay and reliable in-order delivery. UDP avoids long delays, but UDP applications that implement congestion control must do so on their own. DCCP provides built-in congestion control, including ECN support, for unreliable datagram flows, avoiding the arbitrary delays associated with TCP. It also implements reliable connection setup, teardown, and feature negotiation.

# 3  Simulation and Traffic Models

In this work, we analyze the impact of different transport protocols on the QoS parameters such as jitter, delay and throughput for multimedia video transport over heterogeneous (wireless cum wired) networks. QoS requirements are specified by the type of application used, for example, FTP is not a delay sensitive but demands reliability while MPEG4 video traffic is delay sensitive.

We used Network Simulator, ns-2 [5] for simulations that has a built in support for SCTP and UDP while we had to use a patch [6] for DCCP. ns-2 is object-oriented, discrete event driven network simulator developed at UCB written in C++ and OTcl. ns-2 is widely used for networking research and covers a very large number of applications, protocols, network types, network elements and traffic models.

*Traffic Model*

We model MPEG4 video delivery using SCTP, DCCP and UDP as the transport protocols and evaluate the QoS measures such as throughput, delay and jitter. The traffic model we have used for MPEG4 video applications is developed by A. Matrawy et. al [7]. MPEG4 is the video format that is most suitable for the Internet [8]. It targets low bit rates. It allows real images to co-exist with computer-generated counterparts and also allows their separation and their receiving different treatment due to interaction with the user. The main feature of importance for the network is MPEG4's capability of real-time adaptive encoding. This enhances network utilization and enables MPEG4 senders to be more responsive to changes in network conditions. It generates video in three different frame types (I, P, and B) that serve to encode different portions of the video signal in different levels of quality.

*Performance Metrics*

Performance metrics we have chosen for this analysis are throughput, delay and jitter.

*a. Throughput*
We have calculated the throughput using the relation below

$$Throughput \ = \frac{Number \ \ of \ Received \ \ Bits}{Simulation \ \ Time} \tag{1}$$

*b. Delay*
We have calculated Delay using the relation below

$$Delay = Tr - Ts \tag{2}$$

Where Ts is equal to the time stamp of packet sending and Tr is time stamp of packet receiving. We have taken mean delay for our results and calculated by using Equation 3

$$Mean \ \ Delay \ = \frac{Total \ \ Delay}{N} \tag{3}$$

Where N is the total number of packet sent during simulation time.

*c. Jitter*
We have calculated jitter using the relation below

$$Jitter = Delay\ (j) - Delay\ (i)$$  (4)

Where Delay (j) is a current packet delay and Delay (i) is a previous packet delay. We have calculated mean jitter using Equation 5.

$$Mean\ Jitter = \frac{Total\ Jitter}{N}$$  (5)

Where, N is the total number of packets.

## 4  Simulation Scenarios and Results

*Simulation Scenarios*
For exploring the performance of SCTP, DCCP and UDP, we have studied their behavior in a heterogeneous network. Followings are the three simulation scenarios:

(A) Scenario-I: Downloading (Network to Wireless Node)
(B) Scenario-II: Uploading (Wireless Node to Network)
(C) Scenario-III: Multiple Downloaded Flows

*A. Scenario-I: Downloading (Network to Wireless Node)*
The simulation scenario is depicted in Figure 1. In this scenario, we have used wired node N0 as the sender and wireless node N1 as the receiver. In the above topology R1 act as a Router while R2 is an IEEE 802.11b Access Point. Ethernet node N0 connects with R1 through a 100 Mbps link. The link bandwidth between R1 and R2 is set to 100 Mbps while the wireless link between R2 and N1 operates at a nominal data rate 11 Mbps. We used FIFO DROPTAIL queuing algorithm with a queue size of 100 packets. Simulation is run for 100 seconds.

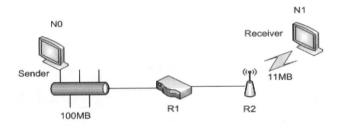

**Fig. 1.** Scenario-I: Downloading (Network to Wireless Node) Simulation Setup

*(i) Throughput*
Figure 2 shows the achieved throughput for the three transport protocols, SCTP, DCCP and UDP as a function of the transmission rate. It can be seen that SCTP, UDP, and DCCP achieve the same throughput for a send rate of 1 Mbps to 3 Mbps. For transmission rates, beyond 3 Mbps and up to 7 Mbps, SCTP show a constant throughput of 3.54 Mbps, whereas UDP achieved 3.9 Mbps to 4.4 Mbps.

*(ii) Delay*

Table 1 shows the delay performance. It can be observed that for all three protocols, higher delays are incurred as transmission rate increases. This is due to the fact that with increasing transmission rates, packets have to wait longer in the queue before being served. As the transmission rate is increased from 1 Mbps to 7 Mbps, delay experienced with DCCP as transport protocol varies from 9 ms to 55 ms; while for SCTP, the delay is in the range of 11 ms to 137 ms and for UDP the delay is in the range of 11 ms to 229 ms. With UDP, higher delay result due to higher number of packets drops, whereas with SCTP increased delay are due the fact that it is a reliable

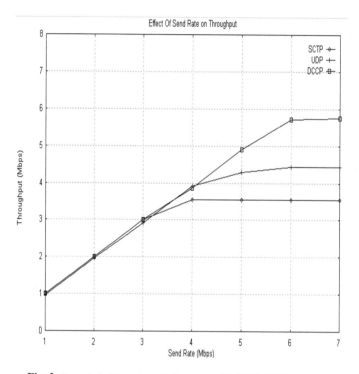

**Fig. 2.** Scenario-I: Throughput Performance with SCTP, UDP and DCCP

**Table 1.** Scenario-I: Delay Performance with SCTP, UDP and DCCP

| Transmission Rate (Mbps) | Delay (ms) SCTP | Delay (ms) UDP | Delay (ms) DCCP |
|---|---|---|---|
| 1 | 11.667744 | 10.805779 | 8.974126 |
| 2 | 22.570612 | 16.953872 | 12.784290 |
| 3 | 40.186953 | 25.319188 | 16.718784 |
| 4 | 137.337322 | 53.100274 | 22.469416 |
| 5 | 137.337322 | 189.204888 | 33.221991 |
| 6 | 137.337322 | 229.317627 | 55.431163 |
| 7 | 137.337322 | 235.946116 | 148.019961 |

protocol. The congestion control mechanism in DCCP efficiently reduces the number of packet drops leading to lower delay values for DCCP as compared to both SCTP and UDP.

*(iii) Jitter*

The jitter performance for each of the three transport protocols is depicted in Table 2. It can be noticed that both SCTP and DCCP exhibit better performance in comparison with UDP. As transmission rate increases, SCTP exhibits a constant jitter past a transmission rate of 3 Mbps, similar to its throughput and delay performance.. For DCCP, the jitter values increase with increasing transmission but range values is significantly smaller than that for UDP.

**Table 2.** Scenario-I: Jitter Performance with SCTP, UDP and DCCP

| Transmission Rate (Mbps) | Jitter (ms) SCTP | Jitter (ms) UDP | Jitter (ms) DCCP |
|---|---|---|---|
| 1 | 0.021150 | 0.041712 | 0.002537 |
| 2 | 0.033607 | 0.054861 | 0.003109 |
| 3 | 0.037686 | 0.055635 | 0.003472 |
| 4 | 0.003971 | 0.065599 | 0.005629 |
| 5 | 0.003971 | 0.407819 | 0.007398 |
| 6 | 0.003971 | 0.745116 | 0.008540 |
| 7 | 0.003971 | 1.135312 | 0.013755 |

*B. Scenario-II: Uploading (Wireless Node to Network)*

Figure 3 shows the network topology for the Uploading scenario. In this scenario, wireless node N1 is the traffic source and wired node N0 is the receiver or traffic sink. The node R1 acts as a Router while R2 is an IEEE 802.11b Access Point The link bandwidth between R1 and R2 is set to 100 Mbps while the wireless link between R2 and N1 operates at a nominal data rate 11 Mbps. We used FIFO DROPTAIL queuing algorithm with a queue size of 100 packets. Simulation is run for 100 seconds.

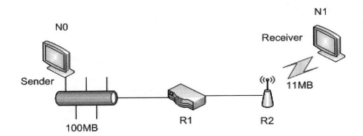

**Fig. 3.** Scenario-II: Uploading (Network to Wireless Node) Simulation Setup

*(i) Throughput*

Figure 4 shows achieved throughput as a function of transmission rate for the three transport protocols in Uploading scenario. It can be seen that up to transmission rate of 4 Mbps, same throughput is achieved with both DCCP and UDP while a smaller

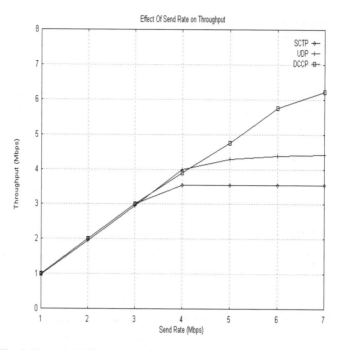

**Fig. 4.** Scenario-II: Throughput Performance with SCTP, UDP and DCCP

throughput of 3.54 Mbps is achieved with SCTP. It is also noticed that with SCTP, a constant throughput of 3.54 Mbps is attained beyond the transmission rate of 4 Mbps. The higher throughput achieved with DCCP results from its efficient congestion control mechanism.

*(ii) Delay*
The delay performance in the Uploading scenario is shown in Table 3. It can be seen that the delay experienced with DCCP is smaller than that with SCTP. For example, the delay values with DCCP range from 9 ms to 105 ms as the transmission rate increases from 1 to 7 Mbps, whereas with SCTP the range of values is from 11 ms to

**Table 3.** Scenario-II: Delay Performance with SCTP, UDP and DCCP

| Transmission Rate (Mbps) | Delay (ms) SCTP | Delay (ms) UDP | Delay (ms) DCCP |
|---|---|---|---|
| 1 | 11.082638 | 10.809174 | 8.998525 |
| 2 | 21.112519 | 17.001962 | 12.741109 |
| 3 | 37.488860 | 25.237682 | 16.533839 |
| 4 | 136.944275 | 53.882709 | 24.251241 |
| 5 | 136.944275 | 203.382689 | 34.796687 |
| 6 | 136.944275 | 222.935928 | 48.422309 |
| 7 | 136.944275 | 239.357967 | 105.738827 |

137 ms. The higher delays with increasing transmission rate result due to larger number of packets being queued as the wireless link bandwidth is limiting factor..

*(iii) Jitter*
Table 4 shows the jitter performance in the Uploading scenario. It is observed that for DCCP, jitter values range from 0.002424 ms to 0.041729 ms as the transmission rate varies from 1 to 7 Mbps. For SCTP, the jitter values lie in the range from 0.023745 ms to 0.004439 ms, with constant jitter exhibited for transmission rates beyond 4 Mbps. The range of jitter values for UDP is from 0.035837 ms to 1.254496 ms. It is to be noted is that the inter-packet delays severely effect the jitter performance. The jitter behavior demonstrated for DCCP is significantly better than that for UDP and SCTP.

**Table 4.** Scenario-II: Jitter Performance with SCTP, UDP and DCCP

| Transmission Rate (Mbps) | Jitter (ms) SCTP | Jitter (ms) UDP | Jitter (ms) DCCP |
|---|---|---|---|
| 1 | 0.023745 | 0.035837 | 0.002424 |
| 2 | 0.035983 | 0.043651 | 0.003951 |
| 3 | 0.040701 | 0.053212 | 0.004703 |
| 4 | 0.004439 | 0.053534 | 0.004716 |
| 5 | 0.004439 | 0.447111 | 0.005244 |
| 6 | 0.004439 | 0.874738 | 0.011779 |
| 7 | 0.004439 | 1.254496 | 0.041729 |

*C. Scenario-III: Multiple Download flows*
Figure 5 shows the network topology used to study performance of three transport protocols, namely SCTP, DCCP and UDP for the case of multiple downloaded flows. The simulation scenario consists of a server and four workstations. Each workstation has established a 3 Mbps connection with the server for downloading an MPEG4 video file. This constitutes four simultaneous MPEG4 flows. The performance measures such as delay, throughput and jitter are evaluated for SCTP, DCCP and UDP.

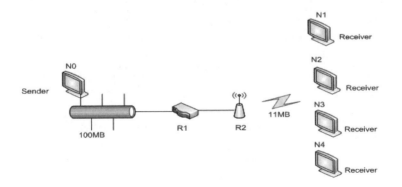

**Fig. 5.** Scenario-III: Multiple Downloaded Flows Simulation Setup

The throughput achieved for each flow can also help in determining whether the protocol is fair to all the flows.

*(i) Throughput*

Table 5 shows the throughput achieved by each of the four flows. It can be seen that with UDP, flow 1 achieves a throughput of 2.642 Mbps which is highest among the four, whereas in case of SCTP, highest throughput achieved is 0.902 Mbps and for DCCP is 2.294. However, DCCP provides the highest cumulative throughput (about 9 Mbps) than that for SCTP and UDP. Further, bandwidth distribution among the four flows is reasonably fair with both SCTP and DCCP whereas UDP is strongly biased towards flows 1 and 2, which start transmissions earlier than flows 3 and 4.

**Table 5.** Scenario-III: Throughput Performance with SCTP, UDP and DCCP

| No. of Flows | Transmission Rate (3 Mbps) | | |
| --- | --- | --- | --- |
| | SCTP | UDP | DCCP |
| | Throughput (Mbps) | Throughput (Mbps) | Throughput (Mbps) |
| 1 | 0.902 | 2.642 | 2.294 |
| 2 | 0.789 | 1.349 | 2.218 |
| 3 | 0.847 | 0.250 | 1.801 |
| 4 | 0.794 | 0.131 | 2.043 |

**Table 6.** Scenario-III: Delay Performance with SCTP, UDP and DCCP

| No. of Flows | Transmission Rate (3 Mbps) | | |
| --- | --- | --- | --- |
| | SCTP | UDP | DCCP |
| | Delay (ms) | Delay (ms) | Delay (ms) |
| 1 | 278.450951 | 245.593325 | 951.344096 |
| 2 | 280.474673 | 257.016502 | 952.220539 |
| 3 | 279.536945 | 263.499839 | 948.377405 |
| 4 | 281.401053 | 266.506301 | 949.284595 |

**Table 7.** Scenario-III: Jitter Performance with SCTP, UDP and DCCP

| No. of Flows | Transmission rate (3 mbps) | | |
| --- | --- | --- | --- |
| | SCTP | UDP | DCCP |
| | Jitter (ms) | Jitter (ms) | Jitter (ms) |
| 1 | 0.180276 | 2.150841 | 0.024230 |
| 2 | 0.142350 | 1.879810 | 0.510569 |
| 3 | 0.195371 | 0.645456 | 0.630514 |
| 4 | 0.203872 | 0.033591 | 0.580231 |

*(ii) Delay*

The delay performance for this scenario is shown in Table 6. It is observed that the delay experience by /incurred for each of the four flows is in the range of 245 ms to 266 ms with UDP. Whereas, the delay values for these flows are around 280 ms and 950 ms with SCTP and DCCP respectively. Here we see that delay values are much higher in case of DCCP as compared to the other protocols. This is primarily due to the maximum achievable throughput on the 801.11b wireless link that is about 7 Mbps for nominal data rate of 11 Mbps. With SCTP, one gets much lower throughput, i.e. in the range of 800-900 kbps but the delay is approximately three times smaller than that with DCCP.

*(iii) Jitter*

Table 7 shows the jitter performance. With DCCP, jitter values for the four flows range from 0.024230 ms to 0.580231 ms while with SCTP, these values range from 0.180276 ms to 0.203872 ms and from 2.150841 ms to 0.033591 ms with UDP. These values, when looked at with throughput and delay performance exhibited, indicate that DCCP provides highest throughput with high latency and moderate jitter. As for SCTP, although the throughput is lower than DCCP, however SCTP is characterized by low latency, low jitter and a fair bandwidth distribution. UDP, on the other hand, is highly unfair and results into very high jitter.

## 5   Conclusions and Future Work

We have evaluated the performance of three transport protocols DCCP, SCTP and UDP for the transfer of MPEG4 video over heterogeneous (wired cum wireless) networks in order to determine which transport protocol can better meet the quality of service (QoS) requirements of video applications. Various performance measures such as throughput, delay and jitter are evaluated for a number of simulation scenarios involving both downloading and uploading of video traffic. UDP performance serves as a basis for comparison.

In the case of downloading (network to wireless node) with a single flow, the results show that DCCP achieves a throughput of up to 38% higher than that of SCTP and 23% higher than that of UDP as transmission rate increases from 1 Mbps to 7 Mbps. For the delay performance, DCCP experiences 60% smaller delay than SCTP and 75% smaller than UDP. Similarly, DCCP show better performance in terms of jitter for video rates of 4 Mbps and beyond.

For uploading scenario with a single flow (wireless node to network), the results show that DCCP has 43% better throughput than SCTP and 29% better than that of UDP. In terms of delay DCCP shows 23% better performance than SCTP and 56% better than UDP. DCCP experiences 90% less jitter than that in SCTP and beyond 3 Mbps SCTP has up to 89% better performance than DCCP while DCCP is up to 95% better than UDP.

In the scenario of multiple-flows where each of four receivers is simultaneously downloading video at a data rate of 3 Mbps from the server, results show that throughput achieved with DCCP is 60% higher than that with SCTP and 39% - 94%

higher than that with UDP. In terms of delay DCCP experiences 70% more delay than that of SCTP and 74% larger than that of UDP. It can be noted that with DCCP, per flow throughput achieved is around 2 Mbps, mean delay of about 950 ms and jitter in the range 0.5-0.6. While with SCTP, per flow throughput achieved is around 0.8 Mbps, mean delay of about 280 ms and jitter in the range 0.18-0.20. With UDP, an unequal bandwidth distribution results into few flows getting much greater throughput and the remaining flows with very low throughputs.

The results presented above clearly indicate that both DCCP and SCTP exhibit better performance than UDP. Both SCTP and DCCP are reasonably fair in distributing bandwidth among the multiple simultaneous flows as compared to UDP. DCCP displays higher throughput performance while SCTP demonstrates better delay and jitter performance in term of all the three protocols. Based on the results gathered in this study, DCCP seems better suited to video streaming application whereas SCTP may be employed for real-time video delivery due to lower delay and jitter in spite of its relatively lower throughput Thus DCCP seems a better choice for fulfillment of QoS requirements for the transport of streaming video traffic over heterogeneous wireless networks. In future, we plan to extend this work for a wider range of applications (VOIP, MPEG4, and Real Audio) and newer transport protocols, enabling a better understanding of the interactions among applications and transport protocols and the underlying access network mechanisms.

# References

1. Balk, A., Gerla, M., Sanadidi, M.: Adaptive Video Streaming: Pre-encoded MPEG-4 with Bandwidth Scaling. Computer Networks on QoS in Multiservice IP Networks 44(4), 415–439 (2004)
2. RFC 768
3. RFC 2960
4. RFC 4340
5. http://www.isi.edu/nsnam/ns/
6. http://lifc.univ-fcome.fr/~dedu/ns2/
7. Matrawy, A., Lambadaris, I., Huang, C.: MPEG4 Traffic Modeling Using Transform Expand Sample Methodology. In: Proc. of 4th IEEE International Workshop on Networked Appliances (2002)
8. Aloimonos, D.P.: On the visual mathematics of tracking. Image and Vision Computing 9(4), 235–251 (1991)

# Implementing Constructivist Pedagogical Model in Dynamic Distance Learning Framework

Shakeel A. Khoja[1], Faisal Sana[2], Abid Karim[2], and Arif Ali Rehman[2]

[1] Learning Societies Lab, University of Southampton, Southampton, U.K.
sk07v@ecs.soton.ac.uk
[2] Department of Computer Science and Engineering, Bahria University, Pakistan
fasialsana@hotmail.com, abid@bimcs.edu.pk, rehmanaf@bimcs.edu.pk

**Abstract.** The objective of this paper is to develop an educational framework, using data mining technologies, with the help of dynamic web technologies that will be used by teachers to organize the course contents on the web according to existing infrastructure, experience, needs, reorganizing it later on if necessary, depending upon the performance of students. The approach to organizing the lecture contents is based on adaptive learning theory, incorporating a Problem Based Learning (PBL) strategy. Presently, course syllabus and handouts on web sites provided to the student are static in nature. Once distributed, these documents cannot be changed or modified, and lack depth. When course materials are placed on the web, students can select a topic in the course outline and look at the description of a topic, and required reading assignments. Instructors can easily change schedules in these on-line documents and inform the students via e-mail. Students can also submit assignments, projects and take-home exams electronically. A course home page comprises a syllabus, assignments, projects and exams, readings and references, class presentation charts and student handouts. Students on a course are mostly assessed based on questions such as Why, How, What, etc. In this way, a student can be graded and ranked, which in turn provides the feedback to the student for future improvement and challenges. Most such web sites are implemented on the theory of constructivism. Constructivists propose that the construction for new knowledge starts from one's observations of events through previous experiences. Hence, learning is the integration of new knowledge and behaviors into a framework and subsequently recalling relevant events in the appropriate situation. This theory is also applied in our educational framework.

**Keywords:** E-learning Systems, Problem Based Learning, Pedagogies, Constructivist theory.

## 1  Introduction

Education is undergoing a major paradigm shift towards learning rather than teaching. Learning is no longer considered as a process transferring and distributing knowledge to the students. It is now viewed as a transformational process whereby students acquire facts, principles, and theories as conceptual tools for reasoning and problem solving in meaningful contexts. From the technological perspective, internet is

D.M.A. Hussain et al. (Eds.): IMTIC 2008, CCIS 20, pp. 191–201, 2008.

playing an important role for such student based learning. The objective of this paper is to develop an e-learning framework that will be used by teachers and knowledge providers to organize the course contents for distance learning through existing infrastructure, experience, needs, and later on reorganizing it if necessary, depending upon the performance of students, by adapting evaluation techniques. The approach to organize the lecture contents is based on the adaptive learning theory. Finally, emphasis is given to evaluation procedures that the tutor can adopt to evaluate the students, in order to provide these students course contents, based on their learning curve.

Learning mechanisms and learning styles are changing at a rapid pace. Learners get more and rapid learning through new technologies such as electronic media. The trend of education has largely shifted towards the web based learning environment with the ever-increasing popularity of the Internet. More and more web based learning sites, VLEs (Virtual Learning Environments) and LMSs (Learning Management Systems) are becoming common, providing various mediums of interaction amongst tutors and learners. The most commonly used mediums are passive posting of study material, Internet chat and email. A number of software and learning systems have been developed using these technologies, such as Blackboard, WebCT, etc.; however the usage of Web2.0 technologies (RSS, social bookmarking, blogs, etc.) is still in its early stage [1]. The major problem among the above-mentioned mediums is the lack of face-to-face communication between the tutor and the learner.

The delivery of information and learning through a web based instructional systems could be done by using text, graphics, sounds, video, etc. To provide true replica of real world learning, these systems should also provide reference sections of the topics that are being offered on that site, which could be in the form of hypermedia. But still it lacks the learning through behaviors, i.e. the learning with the help of gestures and face-to-face question and answer sessions. To overcome this, web based systems use 3-D or video chat sessions (Microsoft Net Meeting, CUseeMe, etc.) and email group discussions. Sites constructed like this are said to be implemented on the theory of constructivism [2]. Once the material distribution issue is resolved the issue of assessment becomes active. In real classroom environment a teacher takes quizzes, hands over course works and assignments. Again this could be done using chat and email capabilities of the system. However to keep track of the efforts put in by a student to achieve the learning and ability outcomes of the subject is not yet standardized. Furthermore, most sites are static in nature as for as the organization of the course contents are concerned.

In this paper, we will try to address the issues of adaptive learning through student feedback. In the beginning the theory of constructivism is explained then a discussion on learning through constructivism, teaching, teaching/learning on the web and assessment is provided and finally we propose our solution to the assessment and reorganization of the course contents, if deemed necessary by the tutor, after assessing student's performance.

## 2  Constructivism

Constructivist theory is highly promising for web-based learning, from which many new strategies are emerging. Constructivists propose that the construction for new knowledge starts from one's observations of events through previous experiences one already has. Learning something new involves changing the meaning of one's previous experiences [2]. It can further be debated that under certain conditions, absolutely new concepts might have been developed instead of altering the old ones.

## 2.1  Learning: Individual and Social Process

Students decide what they need to learn by setting personal learning goals. Students construct for themselves meaningful knowledge as a result of their own activities and interaction with others [3]. Learning strategies include library research, problem and case-based learning, solving assignments and projects, group work, discussions, and fieldwork. On the contrary, classroom teaching is a stimulus to the student's real learning that mostly takes place outside formal classes. Further unstructured (Constructivism) classes with individualized activities, much discussion and optional attendance will provide more chances of learning then traditional class room methods. Students engage actively with the subject matter and transform new information into a form that makes personal sense to them and connects with prior knowledge. Students are placed immediately into a realistic context with specific coaching provided as needed [4].

The constructivist model is learner-centered. The student must control the pace of learning. The teacher acts as a moderator who facilitates the process of learning. Students learn better when they are allowed to discover things by themselves rather than being told what to learn. Variations of the constructionist theory include the collaborative theory of learning, and the cognitive information-processing model of learning. Constructionist-based theories are very popular and have triggered a paradigm shift in the education process towards a student-centered learning approach. This means that there should be some contact between the teacher and the student for the achievement of learning and ability outcomes of a subject. A leaner builds his/her knowledge by building up a knowledge base according to a sequence that suits his/her learning process. Hence every learner has a different experience of learning. A learner can acquire knowledge from social interaction (social-constructivism) with others [2].

According to different theories of learning [4], it could be summarized that the ability to learn is the result of some sort of input and then constructing the meaning from it. Learning consists both of constructing meaning and constructing systems of meaning [5]. There could be many factors involved the process of learning of learner e.g. language, environment and the medium of communications.

## 2.2  Modern Style of Learning

Learning is the integration of new knowledge/behaviors into a framework and subsequently recalling what is relevant in the appropriate situation. To understand learning we must consider how new information is received and the stage through which new information is processed as it progresses from immediate sensory experience to long-term storage. It is also important to understand how novices and experts organize, analyze, or encode, and then retrieve necessary information. Teaching consists of organizing, planning, delivering and evaluating the content of the subject area. Figure 1 shows the basic elements and their relation in an adaptive learning environment.

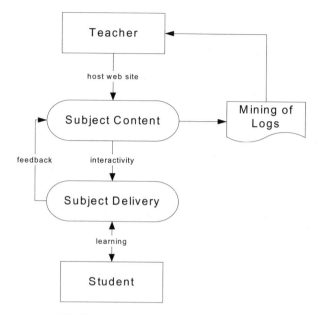

**Fig. 1.** Adaptive Learning methodology

Learning is contingent upon perception. It improves concept formation- the essential element of human thought. Learning is benefited by memory and is achieved through thoughtfulness and critical thinking [5].

## 2.3  Modern Style of Teaching

Teaching may best be defined as the organization of learning. So the problem of successful teaching is to organize learning for authentic results. Teaching may be thought of as the establishment of a situation in which it is hoped and believed that effective learning will take place. This situation is complicated and made up of many parts, such as the requirement of a learner, facilities (stated place, time of meeting, books, aids, etc.), orderly and understood procedures, rules for grading and finally an organizer who brings these parts to a complete form – in other words a teacher. Teaching is the organization of learning. Thus it follows that a teacher is essentially an organizer. The task of any organizer is to enable a group and the individuals in it to function effectively together for the achievement of a common purpose [6].

## 2.4  Teaching/Learning on the Web

Presently, the course syllabus and handouts provided to the student are static in nature. Once published, these documents cannot be changed or modified, and also lack depth. The topics to be covered may be listed in the course outline, but including descriptions makes the document lengthy and thus is rarely done. This makes the course devoid of the requisite flexibility. The costs of duplicating handouts, syllabi, and assignments have proven to be a burden. Sometimes logistical factors prevent a student from receiving an assignment in time.

When course materials are placed on the web, students can select a topic in the course outline and look at the description of a topic, and required reading assignments. Students can select the exam schedule and look at the topics included for that exam. Assignments and projects can be made available on-line on the WWW. Students can access class materials at anytime and from any place (via a computer), save or print handouts, assignments, etc.

Instructors can easily change schedules in these on-line documents and inform the students for example via e-mail. Students can also submit assignments, projects and take-home exams electronically. Alternatively, all the handouts and the hypermedia software can be provided to the students on disk and they can browse the course materials off-line at their convenience.

Traditional course materials were rendered unusable when the software package in which they were created became obsolete. For example, a handout prepared using an old business graphics software cannot be used anymore because it is no longer available in the office or classroom. Instructors also face problems when the software package that they use at home is not available in the office.

Web course materials are software and hardware independent. Instructors can reuse handouts and presentation materials that were created with old or incompatible software, by converting them from the original format straight to a format that can be used on the web, for example GIF files. Browser software can be used across computer platforms.

### 2.5 Assessment of Students

The Web-based informal assessment represents a cognitive behavior modification technique designed to help students develop goal-setting behavior, planning and self-monitoring and provides the opportunity for students to master the concepts. For example, students can regulate and monitor their own learning throughout the course in a sequential and constructive fashion as they respond to the questions and receive ongoing feedback [7]. Students in a course are mostly assessed based on questions such as Why, How, What, etc. In this way, a student can be graded and ranked, which gives in turn student a feedback for future improvements and challenges. There are various modes in which a student can be assessed, such as written exam and quizzes, viva voce, projects, lab reports, thesis, dissertation, self-assessment, peer assessment and many more.

## 3 Proposed Framework for E-Learning Using Adaptive Techniques

The above sections provided details of the different aspect of education. I this section, we are presenting a framework for E-learning which will aid in developing a prototype and online sites incorporating multimedia tools such as audio, video and graphics, along with the above mentioned aspects of education. Figure 2 depicts the basic elements or our model. This is a student centered model and the teacher's job along with providing education, is to tune certain parameters, through student's feedback, to describe what next should be provided to the student.

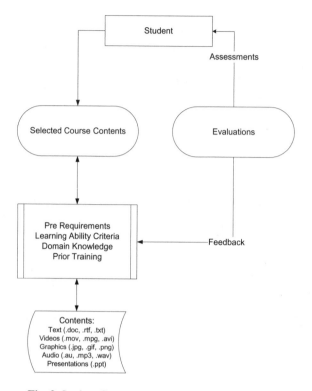

**Fig. 2.** Student Centered Model for Adaptive Learning

PBL utilizes student groups, but each group member is also responsible for independent research. Further, instructor scaffolding is considerably less direct in problem-based learning than in other constructivist models such as anchored instruction. Students are allowed to struggle and induct their own mental model of course concepts with only occasional "life-lines" from the instructor when concept processing falls off-track. Problem-based learning is most similar to case-based instruction, but in its purest form, PBL is more open-ended [8].

Constructivist teaching methods place responsibility on students for managing their own, learning while the teacher guides them and manages their learning environment. Students are given tasks and opportunities, information resources and support, and encouraged to construct their own knowledge structure which is guided through feedback and revision. Learning strategies include library research, problem and case-based learning, doing assignments and projects, group work, discussions, and fieldwork. Classroom teaching is a stimulus to the student's real learning that mostly takes place outside formal classes [9, 10].

## 4 Implementing Framework

The proposed framework would be developed based on both the constructivist theory and PBL. For the case study, Object Oriented Programming courses are inspected.

The course contents are developed on a wider assessment difficulty index of easy, moderate and difficult (to be decided by the tutor). For example, topics in which the syntax and the basics of a language are discussed could be placed in an easy difficulty index category. Topics that discuss the use of utilities already made and supporting files, could be placed in the moderate difficulty index. Finally, the difficulty index could encompass topics which are dependent on some other subjects e.g. networks, videos, etc.

The difficulty index would be different for each course according to targeted audience and would be decided by tutor by allocating less points for topics that lie in the easy level, average points for moderate level and maximum points for topics difficult index level (a linear mechanism). Each topic taught would increase learner's knowledge points by a predefined increment. If a learner's knowledge points match or come closer the tutor defined outcome points of the course, a student would be considered to have good problem solving skills. If no one or only few come closer to outcome points, the tutor can reorganize the course contents in the difficulty index to achieve the expected outcomes.

### 4.1 Web Based Implementation

The site is developed using the server side technology (.net). Tutors (authors of course) and learners are required to register themselves in a course. A profile is stored for every user. Along with, data fields are provided for storing pre-requirements, learning abilities, domain knowledge, prior training and feedback received. Options for Live Video sessions and streaming the stored video clips are also provided. For arranging Live Video sessions, we use standard video conferencing equipment with the capacity to connect up to 200 users. The equipment uses a built-in web server to stream videos. At the client site, Media Player is used to view live videos. For interactive chats, we have developed customized software, where the registered students can login in particular classrooms. In the profile, student thumbnail pictures of the students are also provided. In order to create interactivity with the videos, it is envisaged that flash video format and its player will be used in the system.

In this system, interaction between lecturer and students is done through questions, in form of text messages question. Students ask questions regarding current lecture is centrally maintained in a list and displayed on each students' and teacher's client screen. Teacher or his/her assistant can pick a question from the list and give answer either through text reply or can comment during his/her lecture. To check the involvement of student in a lecture, teacher or his/her assistant sends few questions for students at random time during the lecture. Students will be required to answer these periodically prompting multiple choice questions during the lecture hour. System will verify these answers and makes result on the base of their responses. The main architecture has three major parts: primary server, teacher client and student client, as shown in figure 3.

### 4.2 Primary Server

Primary server is a combination of remote server, streaming server, content management server and database server. Remote server handles communication between

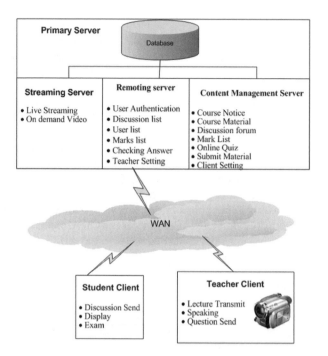

**Fig. 3.** Main System Architecture

Primary and other parts of main system by using .Net Remoting [12]. Streaming sever is used for live transmitting and on demand video. Content management server deals with asynchronous learning. Following are the main features of primary server.

**Discussion list:** This list contains the entire questions that have been asked by students during lecture hours. Student client has ability to add new questions in the list and also teacher can add answers in special case. This list is display on both teacher and students client screen.

**Live Streaming:** This uses media server [13] for live lecture video transmission. This maintains a transmission point for each teacher.

**Online Quiz list:** Teacher client will be able to invoke this method by using their teacher id and course id. It returns all online quiz marking schemes and their respective questions to teacher client.

**Auto Checking Answer Module:** Receive each student multiple choice question response and then make the result based on their answer. This method is invoked by student client.

**Customized Marks List:** This allows each teacher to customize design marks list accordance with their subject requirement. Each marks column has further online quiz option, allowing teacher to add a question list for online quiz.

**Auto Online Quiz:** This function is used for taking remote student examination. This part takes input from customized marks list function that provides quiz question list. After processing each student response, this module update marks list.

**Course Notice:** This contains all notices related to a course which also include university announcements and other student related tasks.

**Course Material:** This provides a way to teacher to post their course materials. This maintains separate material repository for each teacher course. Only students enrolled in the course may view them.

**Discussion forum:** This provides a good means of public communication, allowing instructors and students to post and reply to messages. Discussion forums can be used for assignments, FAQ's, collaborative work, peer critique of work, etc.

**Submit Material:** This function is use to handle all submission record of remote student and also handles the submission update issues.

### 4.3  Instructor Module

Instructor Module part is made for teacher facilitation. This consists of following main features.

**Lecture Delivery Module:** This uses encoder [14] that transmits the live lecture video to primary server. Then primary server part, live streaming further transmits it to student clients.

**Speaking Agent:** We have use agent [15] that will speak up question of discussion list in a user control manner.

**Quiz Module:** It will contain the results of primary server's mark list function. It allows the teacher to view each marking scheme questions and then allow sending these questions to student client for checking involvement.

### 4.4  Student Module

This part is made for students. Following are the main features of student client.

**Discussion Module:** This is used for adding new question to discussion list of primary server.

**Display:** Media player [11] is used to view lectures.

**Checking Module:** This is used to examine the students performs with multiple choice question, sent by teacher, which will appear in this area for short interval during the lecture, and student's response will automatically confirm and store in a database table by using primary server checking answer function.

Instructor is responsible for organizing subject materials according to the introduction of the topic and the sub-topics in a particular lecture are cross-referenced using hyperlinks on to other web pages. Those sub-topics fall in the categories of explanations, examples, questions and references. It is the responsibility of the tutor to develop and maintain the formula by using the above-mentioned entities, to decide whether a student is allowed to proceed further. The development and study of this formula is beyond the scope of this paper. Currently in our framework, we have adopted linear incremental method for modifying difficulty index.

Each course on the site is created based on course title. Followed by a topic, and sub-topics in a particular index of difficulty, learners can navigate a topic using links on a page for its sub-topic. Each link clicked increases a student's points for directed unsupervised activity. If a page is clicked from its sub-topic it will not constitute to the points of a student. Once the learner has finished reading through the topic, he/she can check his/her skills by appearing in a quiz. The scored marks in the quiz are added to already scored points through navigation of pages. This way, students' performance is checked by comparing their old scores with the current ones.

At the end of the course a tutor can calculate the average of students' performance, which in turn will be matched with his expected points for each difficulty index. If there is any difference found then the tutor can change the criteria of difficulty index.

## 5  Conclusion and Future Aspects

The next step in this direction is to analyze the learning pattern of a student over a period of time by mining his/her performance throughout. This will result in an improvement in the customized delivery of course material to students. Currently many researchers are putting their efforts towards adaptive learning environment. So the results from our E-learning framework will help to dynamically organize notes according to student's learning abilities. This research is not limited to text only but will also incorporate extensive research in voice and video for semantically organized content-based retrieval. This will aid in conducting online lectures to simulate a real-world classroom environment.

The paper has considered theoretical and research issues associated with design and use of an E-learning framework. Overall this study has found evidence that an E-learning framework incorporating adaptive learning techniques, its models and architectures for WWW, provide more help to students for interactive learning as compared to traditional systems.

## References

1. Ebner, M.: E-learning2.0 = e-Learning 1.0 + Web 2.0? In: Second International Conference on Availability, Reliability and Security (ARES 2007), pp. 1235–1239 (2007)
2. Liu, E.Z.-F., Lin, S.S.J., Chiu, C.-H., Yuan, S.-M.: Web Based Peer Review: The learner as both Adapter and Reviewer. IEEE Transactions of Education 44(3), 246–251 (2001)
3. Montgomery, S.M.: Addressing Diverse Learning Styles Through the Use of Multimedia. In: ASEE/IEEE Frontiers in Education Conference vol. 1, pp. 3a2.13–3a2.21(1995)
4. Pelz, B.: My three principles of effective online pedagogy. Journal of Asynchronous Learning Networks (JALN) 8(3), 33 (2004)

5. Fardouly, N.: Learner-Centered Teaching Strategies. In: Principles of Instructional Design and Adult Learning Series, The University of New South Wales, Australia (1998), http://www.fbe.unsw.edu.au/learning/instructionaldesign/strategies.htm

6. Hein, G.E.: Constructivist Learning Theory. In: International Committee of Museum Educators Conference (CECA), Jerusalem, Israel, pp. 15–22 (1991)

7. Buriak, P., McNurlen, B., Harper, J.: Systems Model for Learning. In: Proc. ASEE/IEEE Frontiers in Education Conference, Atlanta, USA, vol. 1, pp. 2a3.1–2a3.7 (1995)

8. Ashcroft, K., Foreman-Peck, L.: Managing Teaching and Learning in Further and Higher Education. The Falmer Press, Routledge (1994)

9. Hazari, S., Schno, D.: Leveraging Student Feedback to Improve Teaching in Web-based Courses, Feature Article. THE Journal 26 (1999)

10. Sana, F., Khoja, S.A.: A low cost interactive distance learning solution. In: Proc. 6th Annual IEEE/ACIS International conference on Computer and Information Science (ICIS2007), Melbourne, Australia (2007)

11. McLean, S., Naftel, J., Williams, K.: Microsoft. NET Remoting. Microsoft Press, Washington (2002)

12. Microsoft Technical Manual, Window Media Server, http://www.microsoft.com/windowsserver2003/technologies/winmedia/default.mspx

13. Microsoft Technical Manual, Window Media Encoder, http://www.microsoft.com/windows/windowsmedia/forpros/encoder/default.mspx

14. Deitel, H.M., Deitel, P.J., Deitel, T.R.: Visual Basic. Net: How to Program. Prentice Hall, New Jersey (2002)

15. Microsoft Technical Manual, Window Media Player, http://www.microsoft.com/windows/windowsmedia/player/9series/default.aspx

# Effect of Schedule Compression on Project Effort in COCOMO II Model for Highly Compressed Schedule Ratings

Sharraf Hussain[1], Shakeel A. Khoja[2], Nazish Hassan[1], and Parkash Lohana[3]

[1] Bahria University (Karachi Campus), Pakistan
[2] University of Southampton, UK
[3] Usman Institue of Technology, Karachi
sharraf@bimcs.edu.pk, sk07v@ecs.soton.ac.uk, plohana@uit.edu

**Abstract.** This paper presents the effect of 'schedule compression' on software project management effort using COCOMO II (Constructive Cost Model II), considering projects which require more than 25 percent of compression in their schedule. At present, COCOMO II provides a cost driver for applying the effect of schedule compression or expansion on project effort. Its maximum allowed compression is 25 percent due to its exponential effect on effort. This research study is based on 15 industry projects and consists of two parts. In first part, the Compression Ratio (CR) is calculated using actual and estimated project schedules. CR is the schedule compression percentage that was applied in actual which is compared with rated schedule compression percentage to find schedule estimation accuracy. In the second part, a new rating level is derived to cover projects which provide schedule compression higher than 25 percent.

**Keywords:** COCOMO II, Project Schedule Compression, Compression Ratio, Schedule Estimation Accuracy, Rating Level.

## 1 Introduction

COCOMO II is a model that allows one to estimate the cost, effort, and schedule when planning a new software development activity. It consists of three sub-models [1], each one offering increased fidelity the further along one is in the project planning and design process. COCOMO II is the only model in which project scheduling has its own effect on the overall cost. Among its seventeen cost drivers [2], one is used for scheduling, which is named as SCED (Schedule Cost Driver). This driver has five rating levels (Table 1) depending on the project schedule compression, expansion or nominal schedule. The ratings according to COCOMO II research are based on study of 161 industry projects and ranges from 25 percent compression to 60% expansion of schedule [3].

It has been studied that the range of compression rating levels in COCOMO II is from very low (75% of nominal) to very high (160% of nominal). Nominal schedule is the schedule without any compression or stretch-out [4]. A project with schedule of less than 100% will fall in the area of compression and a project with greater than 100% of schedule will fall in the area of stretch-out.

D.M.A. Hussain et al. (Eds.): IMTIC 2008, CCIS 20, pp. 202–214, 2008.
© Springer-Verlag Berlin Heidelberg 2008

In COCOMO II, an increase in compression, of more than 25% will approximately increase project's cost to 50%. It has been analyzed that increasing the compression rate increases project cost exponentially. Due to this reason, a maximum compression of 25% has been included. Above these compression ratings, the project is considered in impossible region where either its schedule cannot be compressed anymore, or the cost overruns take place.

**Table 1.** COCOMO II SCED Cost Driver Rating Scale [3]

| SCED Descriptors | 75% of nominal | 85% of nominal | 100% of nominal | 130% of nominal | 160% of nominal |
|---|---|---|---|---|---|
| **Rating Level** | Very Low | Low | Nominal | High | Very High |
| **Effort Multiplier** | 1.43 | 1.14 | 1.00 | 1.00 | 1.00 |

It has been studied that the range of compression rating levels in COCOMO II is from very low (75% of nominal) to very high (160% of nominal). Nominal schedule is the schedule without any compression or stretch-out. A project with schedule of less than 100 percent will fall in the area of compression and a project with greater than 100 percent of schedule will fall in the area of stretch-out [3].

In COCOMO II, an increase in compression, of more than 25% will approximately increase project's cost to 50%. It has been analyzed that increasing the compression rate increases project cost exponentially. Due to this reason, a maximum compression of 25% has been included. Above these compression ratings, the project is considered in impossible region where either its schedule cannot be compressed anymore, or the cost overruns take place.

## 2 Experimental Investigation

A study of two experiments, extracted from Boehm's *et al.* [4] research, is included in this study. The first experiment is about checking whether in estimating project effort, the SCED cost driver is rated accurately or not, and the second experiment is calculating Ideal Effort Multiplier (IEM) of SCED for compressed schedules of more than 25%. This IEM value is then applied on projects to check its accuracy level. In order to carry out these experiments, 15 industrial projects of leading software houses of Karachi, Pakistan have been assessed. The experiments are described as follows:

### 2.1 Experiment I: SCED Rating Quality

This experiment is performed on COCOMO II datasets of 15 industry projects to determine the rating of SCED quality. Since it is recognized that the SCED rating in every data point comes from a subjective judgment, the authors have tried to logically derive a more accurate SCED rating by analyzing the data. To calculate the Derived

SCED, estimated effort without Rated SCED using Equation1 are computed and its results are used to calculate the estimated schedule Total time to develop TDEV_est by using Equation2. Further Equation 3 is used to calculate the schedule compression ratios CR to determine the derived SCED.

$$Estimated\_effort = A*(KSLOC)^{(B+0.01*(\sum_{i=1} SF_i))} *(\prod_{j=1}^{16} EM\_But\_SCED_j) \qquad (1)$$

$$TDEV\_est = C*(PM_{est})^{(D+0.2*(E-B))} \qquad (2)$$

$$CR = TDEV_{actual} / TDEV_{est} \qquad (3)$$

Where,

    i.    A, B are model constants, calibrated for each different version of CO-COMO model.

    ii.    C is schedule coefficient that can be calibrated

    iii.    D is scaling base-exponent for schedule that can be calibrated

    iv.    E is the scaling exponent for the effort equation

    v.    SF are five scale factors including PMAT, PREC, TEAM, FLEX, and RESL

    vi.    EM_But_SCED are effort multipliers except SCED, including RELY, DATA, CPLX, RUSE, DOCU, TIME, STOR, PVOL, ACAP, PCAP, PCON, APEX, PLEX, LTEX, TOOL, and SITE

    vii.    A nominal Schedule is under no pressure, which means no schedule compression or expansion; initially set to 1.0.

SCED rating quality can be obtained for each project, by comparing the Derived SCED and the Rated SCED. The five steps being performed in this experiment are shown in Figure1 and are defined as:

**Step 1:** Compute estimated effort assuming that schedule is nominal. Formula in Equation2 shows estimated effort assuming nominal schedule (SCED is equal to 1).

**Step 2:** Compute estimated schedule TDEV_est. Formula in Equation 2 shows estimated schedule using estimated effort, computed in step1. TDEV_est is estimated time to development.

**Step 3:** Compute Actual Schedule Compression/Stretch-out Ratio (SCR). Every data point comes with an actual schedule. For example, in COCOMO II, it is named *TDEV*$_{actual}$ (time to development).

Actual Schedule Compression/Stretch-out Ratio (SCR) can be easily derived through the following equation:

$$SCR = Actual\ Schedule\ /\ Derived\ Schedule \qquad (4)$$

For example, if a project's TDEV is 6 months, and the estimated nominal schedule TDEV_est is about 12 months, then we consider the actual schedule compression as 50% (= 6/12).

**Step 4:** Obtain "derived" SCED rating. COCOMO II SCED Driver Definition Rating [4] (Table 3) has defined rating ranges. Using Equation 6 (discussed in 2.2), compute the actual schedule compression/stretch-out ratio, look up in the SCED driver definition table and check for the closest matched SCED rating. Then a new set of

SCED ratings is produced which reflects the project's schedule compression level more accurately.

**Step 5:** Compare "derived" and "rated" SCED to analyze SCED Rating Quality. The comparison of derived SCED and rated SCED will be done. The above steps will result in a matrix table showing a comparison of derived SCED and rated SCED rating levels which will give clear picture of SCE rating quality observed after performing experiment I.

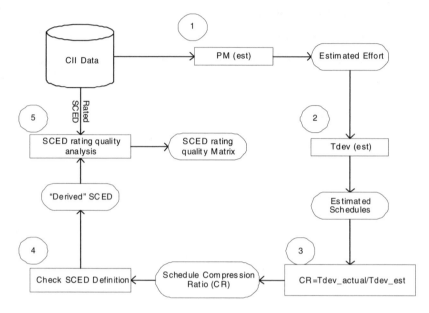

**Fig. 1.** SCED Rating Quality Study Steps

## 2.2 Experiment II: Ideal Effort Multiplier (IEM) Analysis on SCED

SCED cost driver is one of the important cost drivers in COCOMO II. Methods have been established to normalize out contaminating effects of individual cost driver attributes, in order to get a clear picture of the contribution of that driver (in this case, the SCED) on development productivity [5]. It has been slightly modified the original definition to give it a meaning of working definition:

For the given project P, compute the estimated development effort using the CO-COMO estimation procedure, with one exception: do not include the effort multiplier for the cost driver attribute (CDA) being analyzed. Call this estimate PM(P, CDA). Then the ideal effort multiplier, IEM(P, CDA), for this project/cost-driver combination is defined as the multiplier which, if used in COCOMO, would make the estimated development effort for the project equal to its actual development effort PM(P, actual). i.e.,

$$IEM(P, SCED) = PM(P, actual) / PM (P, SCED) \qquad (5)$$

### 2.2.1  Steps for IEM-SCED Analysis

The following steps (Figure 2) were performed to complete the IEM-SCED analysis on the COCOMO II database.

**Step 1:** Compute the PM(P, CDA), using the following formula

$$PM(P,CDA) = A * (KSLOC)^{(B+0.01*(\sum_{i=1}^{5} SF_i))} * (\prod_{j=1}^{16} EM\_But\_SCED_j) \qquad (6)$$

**Step 2:** Compute the IEM(P, CDA) using Equation (6)

**Step 3:** Group IEM(P, CDA) by the same SCED rating (i.e. VL, L, N, H, VH)

**Step 4:** Compute the median value for each group as IEM-SCED value for that rating. This step involves the computation of the median value of IEM-SCED for each rating level. This will give the new rating scale for extra-low level of SCED.

**Step 5:** Comparison of IEM results and COCOMO II

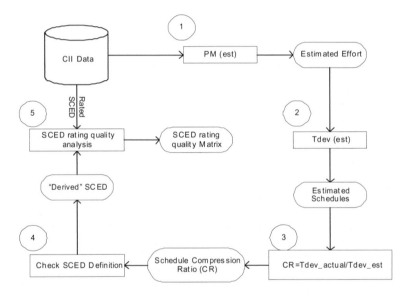

**Fig. 2.** IEM analysis Steps

## 3  Implementing Model

The above described experiments I and II with detailed steps, are being applied on dataset of 15 industry projects. These projects were estimated using COCOMO II Model, belonging to the leading software houses of Karachi, Pakistan.

The projects were developed using COCOMO II Model Estimation, which took place after the requirements and preliminary design was completed. Due to that reason COCOMO II's Post-architecture Model was used. Based on the datasets, Size of the 15 projects was calculated using Function Point Method as given in the following Table 2.

After calculating the size of projects, Effort estimation in Person Month (PM) calculated by using COCOMO II Post-Architecture Model equation:

$$PM = A * (SIZE)^B * (\prod EM_j)$$
(7)

We are taking SCED nominal; therefore total time for developing project can be calculated by following formula

$$TDEV = \left[3.67 * (\overline{PM})^{(0.28+0.2*(B-0.91))}\right] * (\frac{SCED\%}{100})$$
(8)

**Table 2.** Basic Information of the Projects considered, their derived, rated SCED and compression percentages

| Project | Project Name | Organization | Size (KLOC) | Derived SCED | Rated SCED | Derived % of compression w.r.t. Nominal |
|---------|--------------|--------------|-------------|--------------|------------|------------------------------------------|
| Project 1 | Prepaid Card Sales System | GO CDMA | 48.86 | VL | N | 51% |
| Project 2 | SITE Construction System | GO CDMA | 37.31 | VL | N | 57% |
| Project 3 | HR | Supernet Ltd | 28.67 | VL | L | 68% |
| Project 4 | MIS User Admin | Supernet Ltd | 11.024 | VL | N | 60% |
| Project 5 | Franchisee online | GO CDMA | 23.95 | VL | N | 74% |
| Project 6 | BTS Inventory | GO CDMA | 33.97 | VL | N | 62.7% |
| Project 7 | WNO | GO CDMA | 7.473 | VL | N | 67.56% |
| Project 8 | SME | GO CDMA | 12.93 | VL | N | 62.42% |
| Project 9 | SOP | Telecard Ltd. | 8.162 | VL | VL | 65.83% |
| Project 10 | LDI Installation System | GO CDMA | 31.694 | VL | N | 68.69% |
| Project 11 | Telco System | GO CDMA | 24.115 | VL | N | 71.5% |
| Project 12 | Complaints Management System | GO CDMA | 41.49 | VL | N | 74.55% |
| Project 13 | Promotional material management System | Telecard Ltd. | 18.974 | VL | VL | 74.62% |
| Project 14 | Corporate Stock Inventory mgmt System | Telecard Ltd. | 23.53 | VL | L | 73.3% |
| Project 15 | Customer Services IS | GO CDMA | 35.404 | VL | N | 72.42% |

**Table 3.** SCED Driver definition Ratings table

| Rating | Range | Median Value | Range |
|--------|-------|--------------|-------|
| VL | <0.77 | 0.77 | (0, 0.77) |
| VL-L | >=0.77 <0.82 | 0.80 | (0.77, 0.82) |
| L | >=0.82 <0.90 | 0.87 | (0.82, 0.90) |
| L- | >=0.90 <0.95 | 0.93 | (0.90, 0.95) |
| N | >=0.95 <1.10 | 1.03 | (0.95, 1.10) |
| N-H | >=1.10 <1.22 | 1.16 | (1.10, 1.22) |
| H | >=1.22 <1.37 | 1.30 | (1.22, 1.37) |
| H-VH | >=1.37 <1.52 | 1.45 | (1.37, 1.52) |
| VH | >=1.52 | 1.53 | [1.52, >1.52) |

Here TEDV is estimated total development time and it can be represented as TDEV(estimated), however total development time in actual can be represented as TDEV(actual). After having the TDEV(estimated) and TDEV(actual) values, schedule compression ratio of 15 projects was calculated by the following formula:

$$CR = \frac{TDEV(actual)}{TDEV(estimated)}$$

Table 3 is a standard index of rating levels provided by [4], [9], used here in order to know the rating of CR calculated above.

Calculations have been carried out to compute the actual CR of all the 15 projects. As the projects are of almost same working environment, therefore the SCALE FACTORS and COST DRIVERS rating values taken are the same for all the projects.

Table 4 shows a comparison of derived SCED and rated SCED. The rated SCED is obtained from the subjective judgment of development team at the time of effort estimation, while derived SCED is obtained from calculation of CR calculated from actual person months of project.

Table 2 shows a big difference between derived SCED and the rated SCED. The table further shows that the result of subjective judgment was very optimistic but was incorrect. In general, a project team does not consider SCED as important and use its default rating, i.e. Nominal rating and estimate project's cost and effort. But this should be considered seriously at the time of estimation, because a slight change of SCED level results in a huge change in cost and effort. As this is known fact that compression of schedule increases the project effort exponentially so this SCED cost driver has great importance. The SCED can be rated easily by dividing effort by total development time. A rating analysis has been performed in the form of matrix, counting each rating level's number of occurrences for derived and rated SCED both.

A rating analysis has been performed in the form of matrix, counting each rating level's number of occurrences for derived and rated SCED both.

Table 4 is a matrix representation of results in which, rows show derived SCED rating levels and columns show rated SCED rating levels. The intersection of each row and column is the number of occurrences counted from Table5. The circled value shows 11 occurrences of N ratings, as rated by subjective judgment but is derived to be Very Low ratings from Experiment1. This matrix is the final result of Experiment 1, i.e. "SCED accuracy analysis".

After analysis of the matrix of Table 4, it has been proved that SCED is not rated accurately in estimating effort. Keeping in mind its impact on effort it should be rated correctly to get accurate results

### 3.1 Ideal Effort Multiplier (IEM)

This experiment is carried out to propose a new SCED rating level Extra-low and its respective effort multiplier.

Formula for calculating Ideal Effort Multiplier is as follows:

$$IEM\,(P, SCED) = \frac{PM\,(P, actual)}{PM\,(P, SCED)} \qquad \text{(see Equation 5)}$$

**Table 4.** SCED Rating Quality Analysis in COCOMO II database

| SCR | (0, 0.77) | (0.77, 0.82) | (0.82, 0.90) | (0.90, 095) | (0.95, 1.10) | (1.10, 1.22) | (1.22, 1.37) | (1.37, 1.52) | (1.52, +) |
|---|---|---|---|---|---|---|---|---|---|
|  | VL | VL-L | L | L-N | N | N-H | H | H-VH | VH |
| VL | 2 | X | X | X | X | X | X | X | X |
| VL-L | X | X | X | X | X | X | X | X | X |
| L | 2 | X | X | X | X | X | X | X | X |
| L-N | X | X | X | X | X | X | X | X | X |
| N | (11) | X | X | X | X | X | X | X | X |
| N-H | X | X | X | X | X | X | X | X | X |
| H | X | X | X | X | X | X | X | X | X |
| H-VH | X | X | X | X | X | X | X | X | X |
| VH | X | X | X | X | X | X | X | X | X |

SCED: Reported In Data

SCED: Derived from the experiment

The results of each project IEM are shown in the last column of Table 6. This multiplier is the perfect SCED multiplier for that particular project. If this is applied in the formula the estimated effort will become equal to the actual effort.

Table 5 shows that percentages of derived SCED are less than 75% of nominal. Here we can suggest a new rating level named Extra-Low which will cover the projects having compressions of more than 25%. The group of Extra-Low level rating is shown in Table 6a.

Given that extreme values (outliers) exist in our databases. Those outliers could give great impact to the mean values. To avoid that, the median value is used since it is not as sensitive to outliers.

To calculate the median of the group data of IEMs, first we have to sort them in ascending order as shown in Table 6b. The mid-term will be the median in case of odd

**Table 5.** Results of IEM calculation

| Project | Derived % | TDEV (actual) | Staff | PM (actual) | PM (est) wout SCED | PM (est) | IEM |
|---------|-----------|---------------|-------|-------------|--------------------|----------|-----|
| P1 | 51% | 6 mths | 6 | **36** | **23.8** | 23.8 | **1.51** |
| P2 | 57% | 5 mths | 6 | **30** | **18.16** | 18.16 | **1.65** |
| P3 | 68% | 4.5 mths | 4 | **18** | **11.65** | 13.28 | **1.54** |
| P4 | 60% | 3.75 mths | 2 | **7.5** | **5.647** | 5.647 | **1.32** |
| P5 | 74% | 4 mths | 2 | **8** | **3.63** | 3.63 | **2.20** |
| P6 | 62.7% | 4 mths | 2 | **8** | **6.289** | 6.289 | **1.27** |
| P7 | 67.56% | 3.75 mths | 2 | **7.5** | **3.968** | 3.968 | **1.89** |
| P8 | 62.42% | 5 mths | 3 | **15** | **13.959** | 13.959 | **1.07** |
| P9 | 65.83% | 3 mths | 3 | **9** | **5.36** | 7.66 | **1.67** |
| P10 | 68.69% | 4.3mths | 5 | **21.5** | **15.43** | 22.06 | **1.39** |
| P11 | 71.5% | 5.5 mths | 3 | **16.5** | **11.73** | 11.73 | **1.40** |
| P12 | 74.55% | 6.75 mths | 5 | **33.75** | **20.20** | 20.20 | **1.67** |
| P13 | 74.62% | 4 mths | 3 | **12** | **9.23** | 13.20 | **1.30** |
| P14 | 73.3% | 4.25 mths | 4 | **17** | **11.45** | 13.057 | **1.48** |
| P15 | 72.4% | 6.25 mths | 5 | **31.25** | **17.24** | 17.24 | **1.81** |

**Table 6a.** IEM of projects

| P1 | P2 | P3 | P4 | P5 | P6 | P7 | P8 | P9 | P10 | P11 | P12 | P13 | P14 | P15 |
|----|----|----|----|----|----|----|----|----|-----|-----|-----|-----|-----|-----|
| **1.51** | 1.65 | 1.54 | 1.32 | 2.20 | 1.27 | 1.89 | 1.07 | 1.67 | 1.39 | 1.40 | 1.67 | 1.30 | 1.48 | 1.81 |

**Table 6b.** Sorted IEM of projects and its median

| P8 | P6 | P13 | P4 | P10 | P11 | P14 | P1 | P3 | P2 | P9 | P12 | P15 | P7 | P5 |
|----|----|-----|----|-----|-----|-----|----|----|----|----|-----|-----|----|----|
| **1.07** | 1.27 | 1.30 | 1.32 | 1.39 | 1.40 | 1.48 | 1.51 | 1.54 | 1.65 | 1.67 | 1.67 | 1.81 | 1.89 | 2.20 |

**Table 7.** COCOMO II SCED New rating scale

| SCED Descriptors | <75% of Nominal | 75% of Nominal | 85% of Nominal | 100% of Nominal | 130% of Nominal | 160% of Nominal |
|---|---|---|---|---|---|---|
| **Rating Level** | Extra Low | Very Low | Low | Nominal | High | Very High |
| **Effort Multiplier** | **1.5*** | 1.43 | 1.14 | 1.00 | 1.00 | 1.00 |

* *derived through experiment.*

number of data records. In case of even data, two of the mid terms are taken and their mean value is calculated.

IEM value at 8th term is the median which is found to be 1.51. Hence the value 1.51 is the rating value for Extra Low range of SCED cost driver, as shown in Table 7.

## 4  Applying IEM (SCED) Rating in Effort Estimation

To check the accuracy of IEM(SCED), new rating has been applied on the same projects and re-calculated effort with a change of SCED driver value equals to 1.51. The resulted effort is named IEM-PM(est) as shown in the last column of Table 8.

From Table 8 it is observed that estimated effort using new SCED rating is much closer to the actual effort than the previous estimation, and now on the basis of these results model accuracy will be calculated.

**Table 8.** Calculation results of IEM-PM(est) using IEM(SCED) = 1.51

| Project | Schedule % | PM(actual) | PM(est) | IEM-PM(est) |
|---|---|---|---|---|
| Project 1 | 51% | **36** | 23.8 | **36** |
| Project 2 | 57% | **30** | 18.16 | **27.43** |
| Project 3 | 68% | **18** | 13.28 | **17.59** |
| Project 4 | 60% | **7.5** | 5.647 | **8.52** |
| Project 5 | 74% | **4** | 3.63 | **5.48** |
| Project 6 | 62.7% | **8** | 6.289 | **9.5** |
| Project 7 | 67.56% | **7.5** | 3.968 | **6** |
| Project 8 | 62.42% | **15** | 13.959 | **21** |
| Project 9 | 65.83% | **9** | 7.66 | **8** |
| Project 10 | 68.69% | **21.5** | 22.06 | **23.3** |
| Project 11 | 71.5% | **16.5** | 11.73 | **17.71** |
| Project 12 | 74.55% | **33.75** | 20.20 | **30.502** |
| Project 13 | 74.62% | **12** | 13.20 | **13.94** |
| Project 14 | 73.3% | **17** | 13.057 | **17.29** |
| Project 15 | 72.4% | **31.25** | 17.24 | **26.03** |

## 4.1  Calculating Model Accuracy with Magnitude of Relative Error (MRE)

The MRE [6], [7] as a percentage of the actual effort for a project is defined as:

$$MRE = \left| \frac{Effort_{ACTUAL} - Effort_{ESTIMATED}}{Effort_{ACTUAL}} \right| \qquad (9)$$

In addition, we have used the measure prediction level Pred. This measure is often used in research studies [10], [11] and is a proportion of a given level of accuracy:

$$pred\ (l) = \frac{k}{N} \qquad (10)$$

A common value for $l$ is 0.25 [2], [8], which is used for this study as well. The Pred(0.25) gives the percentage of projects that were predicted with an MRE equal or less than 0.25. Conte *et al.* [2] suggests an acceptable threshold value for the mean MRE to be less than 0.25 and for Pred(0.25) greater or than 0.75. In general, the accuracy of an estimation technique is proportional to the Pred(0.25) and inversely proportional to the MRE and the mean MRE.

In Table 9, MRE is calculated using Equation 9. Actual and estimated effort is listed in the table. These two efforts are used to calculate MRE and the absolute value of the answer has to been taken. Table 9 shows two MREs, first one is without IEM(SCED), calculated using effort estimated using old SCED rating. Second one is with IEM(SCED) rating, this is calculated using effort estimated with IEM(SCED) rating value.

**Table 9.** Calculation results of MRE of PM(est) with and without IEM(SCED)

| Project | PM (actual) | PM(est)without Ex-Low rating | PM(est) with Ex-Low rating | MRE without IEM(SCED) | MRE with IEM(SCED) |
|---------|-------------|------------------------------|----------------------------|-----------------------|---------------------|
| Project 1 | 36 | 23.8 | 36 | 0.33 | 0 |
| Project 2 | 30 | 18.16 | 27.43 | 0.39 | 0.085 |
| Project 3 | 18 | 13.28 | 17.59 | 0.26 | 0.022 |
| Project 4 | 7.5 | 5.647 | 8.52 | 0.24 | 0.136 |
| Project 5 | 4 | 3.63 | 5.48 | 0.0925 | 0.37 |
| Project 6 | 8 | 6.289 | 9.5 | 0.21 | 0.187 |
| Project 7 | 7.5 | 3.968 | 6 | 0.47 | 0.2 |
| Project 8 | 15 | 13.959 | 21 | 0.069 | 0.4 |
| Project 9 | 9 | 7.66 | 8 | 0.14 | 0.11 |
| Project 10 | 21.5 | 22.06 | 23.3 | 0.026 | 0.083 |
| Project 11 | 16.5 | 11.73 | 17.71 | 0.28 | 0.073 |
| Project 12 | 33.75 | 20.20 | 30.502 | 0.401 | 0.096 |
| Project 13 | 12 | 13.20 | 13.94 | 0.1 | 0.161 |
| Project 14 | 17 | 13.057 | 17.29 | 0.231 | 0.017 |
| **Project 15** | 31.25 | 17.24 | 26.03 | 0.448 | 0.167 |

The median value for MREs, i.e. without IEM(SCED) sorted is calculated as 0.24, and with IEM(SCED) sorted is calculated as 0.11. Prediction level is calculated to find out the proportion of a given level of accuracy.

## 4.2  Measure Prediction Level Pred(*l*) for Level of Accuracy *l*

The prediction level has been calculated on three standard percentages 20, 25 and 30 using Equation 10. In current situation for l = 0.20, k is the number of observations with MRE <= 0.20 and N is the total number of MRE observations. The calculations are shown in Table 10.

**Table 10.** Pred(l) calculation for MRE without and with IEM(SCED)

| *MRE Without IEM(SCED)* | *MRE with IEM(SCED)* |
|---|---|
| Pred(0.20)=5/15=0.33 | Pred(0.20)=13/15=0.86 |
| Pred(0.25)=8/15=0.53 | Pred(0.25)=13/15=0.86 |
| **Pred(0.30)=10/15=0.66** | Pred(0.30)=13/15=0.86 |

The derived IEM(SCED) values from Table 10 have been applied into the well-calibrated COCOMOII database and improvement has been observed in the accuracy of the model. This increase in accuracy is shown in Table11.

**Table 11.** Accuracy Analysis results of COCOMO11

| *Database* | | *Pred(20)* | *Pred(25)* | *Pred(30)* |
|---|---|---|---|---|
| **COCOMO II** | **Without IEM** | 33% | 53% | 66% |
| | **With IEM** | 86% | 86% | 86% |

The table shows that by applying the IEM(SCED) values into COCOMOII, all three accuracy levels - Pred(20), Pred(25), and Prec(30) - increase by 53%, 33%, and 20%.

## 5  Conclusions

The two experiments are performed, one for SCED accuracy analysis and other for deriving new rating level for projects with schedule compression of more than 25%. The result of experiments may lead to following conclusions.

Data reporters often carry out inaccurate subjective judgments for compression level of project schedule, resulting in under estimation of project effort. So it is recommended to choose the exact level of schedule compression level.

The new derived rating level is named extra-low. This level will address projects having compression levels between 25% and 50%. The effort multiplier for this SCED rating is equals to 1.51. This derived rating is applied on the same projects and their effort is re-estimated. The results show improvements in COCOMO II model accuracies, i.e. by 53% for Pred(20), 33% for Pred(25), and 20% for Pred(30).

# References

1. Boehm, B., Abts, C., Brown, A.W., Chulani, S., Clark, B.K., Horowitz, E., Madachy, R., Refier, D., Steece, B.: Software Cost Estimation with CocomoII. Prentice Hall, Englewood Cliffs (2000)
2. Selby, R.C.: Software Engineering: Barry W. In: Boehm's lifetime contributions to software development, management and research (practitioners). Wiley-IEEE Computer Society, Chichester (2007)
3. Baik, J., Chulani, S., Horowitz, S.: Software Effort And Schedule Estimation Using The Constructive Cost Model: COCOMO II, Center for Systems and Software Engineering, University of Southern California, USA, http://sunset.usc.edu/csse/research/COCOMOII/cocomo_main.html
4. Yang, Y., Chen, Z., Valerdi, R., Barry, B.: Effect of Schedule Compression on Project Effort. In: Proc. 27th Conference of the International Society of Parametric Analysis, Denver, CO, USA (2005)
5. Bradford, C.: Quantifying the Effects of Process Improvement on Effort, Software Metrics Inc., pp. 65–70. IEEE Computer Society Press, Los Alamitos (2000)
6. Kemerer, C.F.: An Empirical Validation of Software Cost Estimation Models. Communications of the ACM 30(5), 416–429 (1987)
7. Chulani, S.: Bayesian Analysis of Software Cost and Quality Models, IBM Research (1997)
8. Chulani, S., Clark, B., Barry, B.: Calibrating the COCOMO II Post-Architecture Model. In: Proc. International Conference on Software Engineering, Kyoto, Japan, pp. 477–480 (1998)
9. Chiang, R., Vijay, S.: Improving Software Team Productivity: A Quantitative Process Design Approach. Communications of the ACM 47(5) (2004)
10. Simmons, D.B., Ellis, N.C., Fujihara, H., Kuo, W.: Software Measurement – A Visualization Toolkit for Project Control & Process Improvement. Prentice Hall, Englewood Cliffs (1998)
11. Osama, A.: Pakistan's Software Industry Best Practices & Strategic Challenges, Pakistan Software Export Board (PSEB), http://www.pseb.org.pk/UserFiles/documents/Best_Practices_Study.pdf

# The Concept of Bio-fitness in Bio-inspired Computational Metaphor Abstraction

Anjum Iqbal[1], Mohd Aizaini Maarof[2], and Safaai Deris[2]

[1] Informatics Complex, P.O. Box 2191, Islamabad, Pakistan
anjum.pk@gmail.com
[2] Faculty of Computer Science and Information Systems
Universiti Teknologi Malaysia, 81310 UTM Skudai, Johor, Malaysia

**Abstract.** A computational metaphor, which is based on some biological phenomena or structures, is known as Bio-Inspired Computational Metaphor
(BICM). Though computation and biology are distinct fields, there should exist
some criteria for evaluating the strength of a BICM. The criteria would be
based on the biological phenomena or structures that have inspired the computational metaphor. This paper introduces the new concept "bio-fitness" for
evaluating abstraction-strength of a BICM. The concept of bio-fitness would
benefit both computation and biology. An Immune System (IS) is a complex
biological system composed of various biological phenomena. The emerging
field of Artificial Immune System (AIS) exploits the IS mechanisms to abstract
a variety of BICMs. Here we attempt to elaborate the concept of bio-fitness in
the AIS context.

**Keywords:** computation, metaphor, bio-fitness, artificial immune system.

## 1 Introduction

The metaphor abstraction is important stage in bio-inspired computation research
processes. It involves appropriate mapping of mechanisms, structures, and components of counterparts in biology and computation. The immune system is highly complex biological system composed of various explored and unexplored phenomena,
which work in interlinked and cascaded manner to achieve main goal of the system
that are; a) self-non-self discrimination and identifying, and b) combating danger [1],
[2], [3], [4], [5].

In our opinion, a strong bio-inspired computational metaphor (BICM) is one that
works in close concert with its counterparts in biology. It might appear as vital quest
in AIS and bio-inspired computation research that; how strength of BICM could be
defined and evaluated? This paper aims to initiate a scientific discourse in this regard
in the context of AIS research. A biological phenomenon may be used as a basis for
abstracting BICMs with varying strengths. The strength of BICM abstraction may depend upon the abstractor's (person involved in metaphor abstraction) understanding in
biology and computation, and his aesthetic approach towards the field. We propose a
new concept of "bio-fitness" for evaluating abstraction-strength of a BICM, which is
expected to benefit bilaterally for computation and biology.

D.M.A. Hussain et al. (Eds.): IMTIC 2008, CCIS 20, pp. 215–226, 2008.
© Springer-Verlag Berlin Heidelberg 2008

This paper contributes for bio-inspired computation in general and AIS in particular. It gives an overview of Immune Inspired Computational Metaphor (IICM) abstraction in section 2. The proposed criterion of bio-fitness is defined in section 3, and section 4 describes methodology for evaluating bio-fitness. To elaborate the proposed idea, section 5 illustrates bio-fitness of an IICM called DASTON (DAnger Susceptible daTa codON). Section 6 describes some key significances of the proposal for AIS and immunology. Finally, section 7 concludes the paper.

## 2   Immune-Inspired Metaphor Abstraction

The process of metaphor abstraction is central to AIS research. Immune system is highly complex, composed of numerous explored and unexplored biological phenomena, structures, elements. AIS practitioners use their aesthetic approaches to map immune system components and functions to those of computational systems for abstracting computational metaphors. These metaphors result in to novel algorithms, architectures, data-structures, data-types, etc. as components of AIS (see figure 1).

The story of AIS research starts from wet immunology research labs where immunologists perform experiments, in-vitro (in test tubes) and in-vivo (in test organisms), to reveal principles of immune system. Computational researchers can utilize these principles in two ways; a) computational models can be designed to mimic the immunological processes for in-silico (in computers) immunology research or immunoinformatics, and b) novel metaphors can be abstracted and mapped to computational systems called artificial immune systems, see Figures 1 and 2. The AIS metaphor abstraction process may be divided into following three stages.

*Identification of basis mechanisms* – this stage involves the identification of potential mechanisms in immunology and computation that can be used as a basis for the

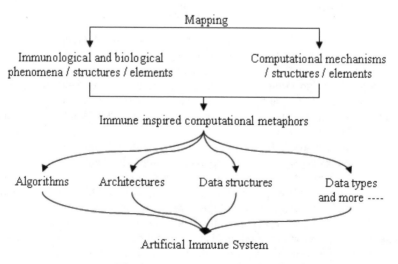

**Fig. 1.** Metaphor abstraction for AIS

metaphor abstraction. An apposite identification would reveal success in subsequent stages. The success in this step requires strong general knowledge of computation, biology, immunology, and existing artificial immune systems.

*Establishing theoretical background* – after identifying some basis mechanisms in computation and immunology, it is important to establish a strong theoretical background for these mechanisms. This may be completed through; a) deep study of mechanisms from literature, b) seeking guidance from experts in immunology and computation, and c) performing some preliminary experimental work.

*Logical mapping* – identification of basis mechanisms and establishment of their strong theoretical background enable one to appropriately map the mechanisms, structures, elements in computation to their counterparts in immunology. This mapping demands aesthetic approach of metaphor designer. The correct mapping must result into strong metaphor. The strength of mapping depends upon the level of completeness of preceding stages.

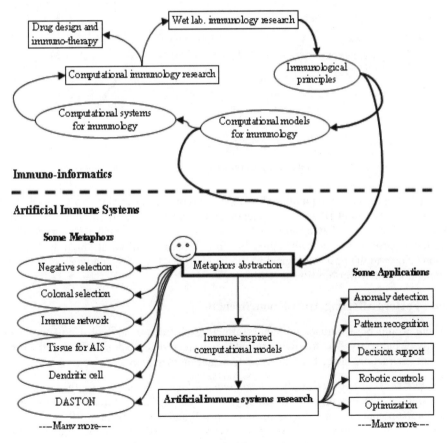

**Fig. 2.** Illustration of AIS research process

# 3  Proposed Criterion for Abstraction-Strength "Bio-fitness"

As described in above section 2, various principles and components of immune system, revealed through in-vivo, in-vitro, and in-silico immunology practices, are exploited to abstract computational metaphors in AIS. It could be an open question for AIS research community that how should "strength of abstraction" for an immune-inspired computational metaphor be defined, and what should be the appropriate method to evaluate this strength. A biological phenomenon (BP) may be used as basis to abstract different computational metaphors with varying abstraction-strengths, low ($CM_L$) to high ($CM_H$), see Figure 3. The abstraction-strength could be high or low (represented by thickness of arrows in figure 3) depending upon various factors (some of the key factors are given in the following subsection 3.1).

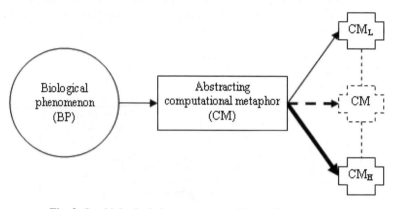

**Fig. 3.** One biological phenomenon resulting various metaphors

In our opinion, a computational metaphor having closer analogy to the basis biological phenomenon (the source phenomenon serving as seed for the abstraction) could be considered having greater "abstraction-strength" or higher "bio-fitness". A computational metaphor with higher bio-fitness ($CM_H$) will sit at "high analogy region" closer to the basis biological phenomenon (BP), Figure 4, and vice-versa for the metaphor with lower bio-fitness ($CM_L$).

## 3.1  Factors Affecting Abstraction Strength

There would be a number of factors affecting abstraction-strength (bio-fitness) directly or indirectly. The following might be considered key factors affecting the process of abstraction, hence bio-fitness of the resulting metaphor;

*Completeness of knowledge about basis immunological phenomena* – there are many unexplored phenomena in immunology; for example; controversial point of views about main goal of immune system, self-non-self and danger-theory, offer many challenges to immunologists [4]. Since, the source of an IICM is some immunological phenomena therefore completeness of the metaphor should depend upon completeness of the respective knowledge in immunology. The AIS practitioners are bound to the knowledge explored by immunologists and immuno- informaticians.

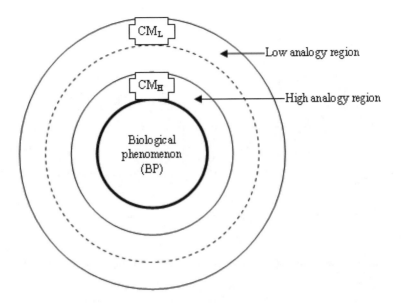

**Fig. 4.** Abstraction strength regions are a biological phenomenon

*Knowledge about computational mechanisms* – for abstracting a metaphor with appropriate strength, AIS practitioner should be well-learned in general computation theory and state of the art computational techniques. A metaphor abstracted with incomplete or improper understanding of computational systems, techniques, and applications might not be able to show good results.

*Aesthetics of the designer* – while mapping knowledge in two distinct fields, personal aesthetics of metaphor designers greatly influence the process of abstraction (the face in Figure 2 represents involvement of human aesthetics). AIS practitioners having same level of knowledge and exploiting same basis mechanisms in immunology and computation might propose distinct metaphors with varying abstraction-strength.

## 4   Proposed Method for Evaluating Bio-fitness

The proposed criterion "bio-fitness" for evaluating abstraction-strength of immune-inspired computational metaphor is supposed to demonstrate the level of analogy between the two fields, immunology and computation. There are a number of interlinked and/or cascaded biological phenomena serving the purpose of immune system [10].

These phenomena can be exploited as basis for abstracting various AIS metaphors. For example, as shown in Figure 5, conceptually immune system may be considered comprising of $n$ explored (regular boundary circles) and unexplored (dashed boundary circles) biological phenomena (BP1, BP2, BP3---- BP$p$, ----, BP$i$, ----, BP$n$). Each, or combinations, of these biological phenomena may serve as basis for abstracting $m$ computational metaphors (CM1, CM2, CM3, -----, CM$p$, -----, CM$j$, -----, CM$m$) for an AIS, where BP$p$ and CM$p$ illustrate any proposed biological phenomenon and

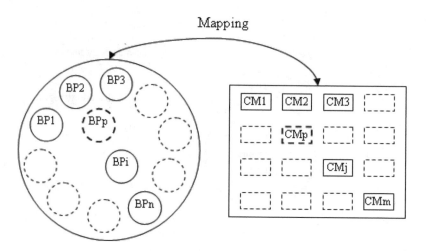

**Fig. 5.** Various biological phenomena and respective computational metaphors

companion computational metaphor respectively. These metaphors would be having varying level of analogies between their counterparts in immunology and computation, i.e. varying bio-fitness levels (as described in section 3).

For evaluating bio-fitness of a computational metaphor CMA inspired from a biological phenomenon BPA, we hypothesize CMA to be compliant with some other biological phenomenon BPB occurring in the vicinity of BPA (see Figure 6). The definition of bio-fitness (abstraction-strength) of a computational metaphor turns out to be

"The compliance of a bio-inspired computational metaphor to biological phenomenon (or phenomena) that has (have) not been directly referred in abstraction process of the metaphor."

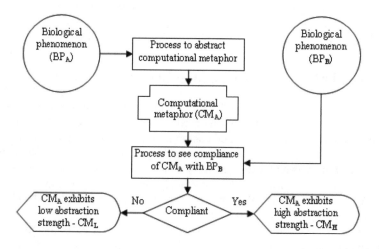

**Fig. 6.** Proposed method for evaluating bio-fitness

The level of bio-fitness of a BICM may be determined through; a) number of compliances with different biological phenomena, and/or b) strength of compliance with a focused biological phenomenon. Measuring bio-fitness level of a BICM is out of the scope of this paper, initially it is important to justify the idea of evaluating bio-fitness, and following sections may help convince the reader in this regard.

## 5   Bio-fitness of DASTON

Hofmyer [6] advises that it is fruitless to propose an idea in isolation or without some practical evidence. The proposal for evaluating bio-fitness of a computational metaphor holds support from our previous work, which proposes DASTON (DAnger Susceptible daTa codON) [7] and shows its conformation to biological phenomenon of polymorphic-susceptibility [8]. Here we briefly overview the process of abstraction and validation of DASTON and method for evaluating its bio-fitness, in attempt to further clarify the idea and justify the worth.

### 5.1   Abstraction and Validation of DASTON

The mapping of a basis immunological phenomenon to computational mechanism is an important step in the process for abstracting immune-inspired computational metaphor. The abstraction of DASTON is not a straightforward mapping of biological and computational counterparts. It involves proposal of a biological phenomenon "danger susceptibility" establishing its base on; danger theory [1-5], infectious disease susceptibility [9-15], and host-pathogen interaction [16-18] (see Figure 7). We aim to present DASTON as a general computational metaphor, therefore attempt mapping Danger Susceptibility to various computational mechanisms, for example; General Processing and Database Query Processing. These mappings lead to following definitions of DASTONs for different cases.

General – The host-data-segments that actively interact with some incident data for producing an intended/expected outcome

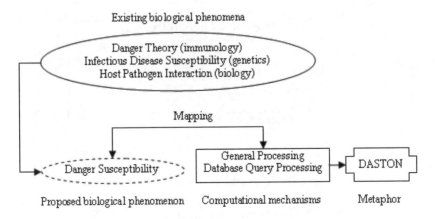

**Fig. 7.** Illustration for DASTON abstraction

Database – The database fields that frequently respond to query-data for returning required information

Intrusion – The host-processes or threads that actively interact with attack scripts to produce malicious activities (respective process logs could also be analyzed to see process behavior)

We have tested the system call benchmark data of the University of New Mexico [19] for the existence of DASTONs. The data is for intrusion detection experiments and contains normal and intrusion-trace system call sequences. Our experiments show that not all but some segments of system call sequences are actively involved to produce intrusive activities. This confirms the existence of DASTONs in system call data and validates our idea (please refer to [7] for further details).

## 5.2  Evaluating Bio-fitness

In order to demonstrate its bio-fitness, the DASTON must comply with some biological phenomenon other than that exploited as basis for its abstraction; it is according to proposed definition of bio-fitness in section 4.

As described in the above section 5.1 and shown in respective Figure 7, abstraction of DASTON is based on a composite biological phenomenon Danger Susceptibility (DS). The biological fact that genetic polymorphism is related to susceptibility [20], [21] has been exploited to prove that DASTON conforms to Polymorphic Susceptibility (PS) [8]; experimental results show that polymorphic-measure of system call DASTONs is higher than other data segments.

The phenomena of Danger Susceptibility and Polymorphic Susceptibility are distinct but occurring in close vicinity in biological system. Hence, compliance of DASTON, abstracted from Danger Susceptibility, to Polymorphic Susceptibility validates its Bio-Fitness, see Figure 8. It also means, as described in section 3, that computational metaphor DASTON exhibits good abstraction strength and closer analogy to the basis biological phenomenon Danger Susceptibility.

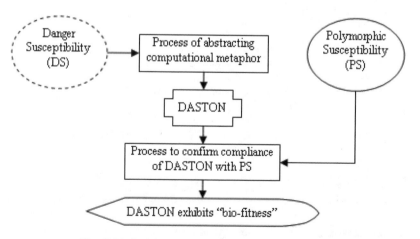

**Fig. 8.** Method for evaluation bio-fitness of DASTON

# 6   Significance of Evaluating Bio-fitness

The BICM may show promising results in computation (exhibit computational-fitness) regardless of their bio-fitness; artificial neural networks, genetic algorithms, and artificial immune systems are among the examples. In the context of bio-fitness these metaphors may be classified as given in Table 1.

**Table 1.** Classification of metaphors

| The bio-inspired computational metaphor exhibiting | |
|---|---|
| Bio-Fitness | Computational-Fitness |
| Yes | Yes |
| Yes | No |
| No | Yes |
| No | No |

Now question raises that if bio-inspired, immune-inspired or hybrid computational metaphor show good performance in computation irrespective of its bio-fitness, then why should we go for evaluating bio-fitness? Certainly, it would be futile to put any scientific effort or proposing a concept without perceiving its benefits for further research and development.

The idea of evaluating bio-fitness is expected to deliver unique benefits to bio-inspired computation in general and to artificial immune system in particular. In addition to computation, the field of immunology might also get some advantages. Following sections describe some key implications of bio-fitness;

## 6.1   Proposing Novel Immune-Inspired Computational Mechanisms

The natural immune system constitutes numerous interlinked and cascaded biological phenomena and structures, which are used as basis for abstracting various immune inspired computational mechanisms (please refer to Figure 5 in section 4). The existing AIS metaphors have exploited only some of the immunological phenomena and components. The components of generalized AIS should be expected to work in concert with their counterparts in immunology. This needs a careful and systematic approach for proposing novel immune-inspired computational metaphors and integrating them appropriately to construct general AIS. For this, the boundary of knowledge should be gradually grown around already tested and validated AIS metaphors.

Since bio-fitness of a bio/immune-inspired computational metaphor tells its close analogy to the basis biological/immunological phenomenon and its compliance to some other neighboring biological phenomena, therefore, evaluation of bio-fitness would effectively help growing boundary of knowledge around the basis biological/immunological phenomenon and propose novel computational metaphors based on vicinal immunological mechanisms. The integration of these metaphors would likely to contribute for developing bio-fit AIS.

## 6.2   Improving Existing Metaphor

The process of research and development always goes for improvements, which is a never-ending effort. The existing metaphors in AIS are based on; existing knowledge in immunology, designers' learning in immunology and computations, and his aesthetic approach (as described in section 3.1). The idea of evaluating bio-fitness of computational metaphors suggests a new improved look for existing AIS metaphors. Bio-Fitness of AIS metaphors might help developing deep context in immunology and computation, so allowing more features for the metaphor. This might inspire AIS practitioners to improve existing metaphors. The improved metaphors might exhibit bio-fitness as well as computational- fitness. It might also help designers to establish reasonable scope of their metaphors and improve process of metaphor abstraction.

## 6.3   Developing Clue for Unexplored Biological Mechanism

The field of immunology is yet to explore numerous phenomena/elements constituting immune system; the mechanisms involved in danger theory perspective could be one of the examples. It would be interesting aspect of evaluating bio-fitness of computational metaphor, with reference to biology in general and immunology in particular, if it helps developing clue for some unexplored biological/immunological phenomena.

The DASTON, computational metaphor, is based on a proposed biological phenomenon danger-susceptibility, which is not apparent in biological or immunological literature but receives strength from various biological and immunological phenomena; danger theory, infectious disease susceptibility, and host-pathogen interaction (see Figure 7, the elliptical shape with doted boundary represents an unexplored biological phenomenon). The bio-fitness exhibited by system call DASTONs (see section) or compliance of DASTON with a well known biological phenomenon, polymorphic susceptibility, gives the clue that basis biological phenomenon, danger-susceptibility, is likely to occur in biological systems. It needs thorough biological testing and validation to appear as general reality.

Similarly, bio-fitness evaluation of other biology/immunology inspired metaphors may provide clue for some other unexplored phenomena. Following would be the steps involved in getting this type of clue.

1  Proposing biological phenomenon, not apparent in literature, based on existing knowledge in the field.
2  Abstracting computational-metaphor from basis biological/immunological phenomenon proposed in step 1.
3  Evaluating bio-fitness of the computational-metaphor abstracted in step 2.
4  If metaphor exhibits bio-fitness then basis biological/immunological mechanism, proposed in step 1, is more likely to occur; else less likely to occur.

Developing these types of indirect clues for unexplored biological/immunological mechanisms through bio-fitness evaluation of computational metaphors should be of valuable assistance for immunologists and immuno-informaticians.

# 7 Conclusions

A bio/immune-inspired computational metaphor (BICM/IICM) is based on biological/immunological phenomena, structures, and elements. The abstraction involves appropriate mapping of the counterparts in computation. The proposed idea of bio-fitness of BICM/IICM is to evaluate strength of abstraction that could also be defined as closeness of analogies in counterparts. An IICM may be called exhibiting bio-fitness if proves compliance with some biological phenomenon in the vicinity but other than that exploited for its abstraction process. The level of bio-fitness may depend upon the number vicinal biological phenomena with which the IICM shows compliance or it could be the strength of compliance to a focused vicinal phenomenon. A BICM/IICM may show its competence for computation (computational-fitness) regardless of bio-fitness, but bio-fitness evaluation is expected to deliver unique bilateral benefits to computation and immunology, for example, exploring novel computational mechanisms, improving existing computational mechanisms, and establishing clue for unexplored biological phenomena. Hence, this paper initiates a distinctive discussion in the fields of bio-inspired computation and Artificial Immune System.

# References

1. Matzinger, P.: The Danger Model: A Renewed Sense of Self. Science 296, 301–305 (2002)
2. Matzinger, P.: The Danger Model In Its Historical Context. Scand. J. Immunol. 54, 4–9 (2001)
3. Gallucci, S., Lolkema, M., Matzinger, P.: Natural Adjuvants: Endogenous Activators of Dendritic Cells. Nature Medicine 5(11), 1249–1255 (1999)
4. Matzinger, P.: The Real Function of The Immune System (last accessed on 06-04-04), http://cmmg.biosci.wayne.edu/asg/polly.html
5. Matzinger, P.: An Innate sense of danger. Seminars in Immunology 10, 399–415 (1998)
6. Hofmeyr, S.A., Forrest, S.: Architecture for an Artificial Immune System. Evolutionary Computation Journal 8(4), 443–473 (2000)
7. Iqbal, A., Maarof, M.A.: Towards Danger Theory based Artificial APC Model: Novel Metaphor for Danger Susceptible Data Codons. In: Nicosia, G., Cutello, V., Bentley, P.J., Timmis, J. (eds.) ICARIS 2004. LNCS, vol. 3239. Springer, Heidelberg (2004)
8. Iqbal, A., Maarof, M.A.: Polymorphism and Danger Susceptibility of System Call DAS-TONs. In: Proceedings of the 4th International Conference on Artificial Immune Systems (ICARIS-2005) (2005)
9. Goldmann, W.: The Significance of Genetic Control in TSEs. MicrobiologyToday 30, 170–171 (2003)
10. Blackwell, J.: Genetics and Genomics in Infectious Disease, CIMR Research Report, (2002) (last accessed on 06-04-04), http://www.cimr.cam.ac.uk/resreports/report2002/pdf/blackwell_low.pdf
11. Coussens, P.M., Tooker, B., Nobis, W., Coussens, M.J.: Genetics and Genomics of Susceptibility to Mycobacterial Infections in Cattle. In: IAAFSC web site 2001 (2001) (publication)
12. Hill, A.V.S.: Genetics and Genomics of Infectious Disease Susceptibility. British Medical Bulletin 55(2), 401–413 (1999)

13. Tavtigian, S.V., et al.: A Candidate Prostate Cancer Susceptibility Gene at Chromosome 17p. Nature Genetics 27(2), 172–180 (2001)
14. Casanova, J.-L.: Mendelian Susceptibility to Mycobacterial Infection in Man. Swiss Med Weekly 131, 445–454 (2001)
15. Denny, P., Hopes, E., Gingles, N., Broman, K.W., McPheat, W., Morten, J., Alexander, J., Andrew, P.W., Brown, S.D.M.: A major Locus Conferring Susceptibility to Infection by Streptococcus Pneumoniae in Mice. Mammalian Genome 14, 448–453 (2003)
16. Weis, J.J.: Host-Pathogen Interactions and the Pathogenesis of Murine Lyme-Disease. Current Opinion in Rheumatology 14, 399–403 (2002)
17. Casadevall, A., Pirofski, L.A.: Host-Pathogen Interactions: The Attributes of Virulence. The Journal of Infectious Diseases 184, 337–344 (2001)
18. Casadevall, A., Pirofski, L.A.: Host-Pathogen Interactions: Basic Concepts of Microbial Commensalism. Colonization Infection and Disease 68(12), 6511–6518 (2000)
19. Intrusion Detection Data Sets (last cited on 01-05-2005), http://www.cs.unm.edu/~immsec/systemcalls.htm
20. Zhou, Y.-S., Wang, F.-S., Liu, M.-X., Jin, L., Hong, W.G.: Relationship between susceptibility of hepatitis B virus and gene polymorphism of tumor necrosis factor-a. World Chin. J. Digestol 13(2), 207–210 (2005)
21. Sabouri, A.H., Saito, M.: Polymorphism in the Interleukin-10 Promoter Affects Both Provirus Load and the Risk of Human T Lymphotropic Virus Type I-Associated Myelopathy/Tropical Spastic Paraparesis. Journal of Infectious Diseases 190, 1279–1285 (2004)

# Analysis of MANET Routing Protocols under TCP Vegas with Mobility Consideration

Razia Nisar Noorani, Asma Ansari, Kamran Khowaja,
Sham-ul-Arfeen Laghari, and Asadullah Shah

Isra University Hyderabad, Pakistan
{razia-nisar, asma_ansari, kamran_khowaja,
shams_laghari}@hotmail.com

**Abstract.** Mobile Ad Hoc Network (MANET) is an autonomous network of mobile nodes in an infrastructure-less environment and has a dynamic topology. It is also called a short-lived network. In order to facilitate communication within the network, a routing protocol is used to discover routes between the nodes. In MANET, temporary link failures and route changes happen frequently. Assuming that all packet losses are due to congestion, Transport Control Protocol (TCP) performs poorly in such an environment. Many TCP variants have been developed for the improved performance of communication in MANET. Simulations were carried out using Network Simulator-2 (NS-2), and the selected MANET Routing protocols. *Ad hoc* On-demand Distance-vector (AODV) and Destination sequenced distance vector (DSDV) were analyzed in accordance with their best performance packet delivery rate, average end-to-end delay, and packet dropping, under TCP Vegas with mobility considerations. The simulation results indicate that AODV has a better throughput performance and low average end-to-end delay compared with DSDV, but AODV suffers from high packet drop.

**Keywords:** MANET, TCP VEGAS, AODV, DSDV, NS-2.

## 1 Introduction

With the ever-increasing demand for connectivity the need for mobile wireless communication is inevitable. The use of portable laptops and handheld devices is increasing rapidly. Most of the portable communication devices have the support of a fixed base station or access points that corresponds to the last-hop-wireless model. This trend can be observed in wide-area wireless cellular systems. However, such a support is not available in settings where access to a wired infrastructure is not possible. Situations like natural disasters, conferences and military settings are noteworthy in this regard. This has led to the development of Mobile Ad-hoc Networks [1].

An Ad hoc network is a dynamically changing networking of mobile devices that communicate without the support of a fixed structure. There is a direct communication among neighboring devices but communication between non-neighboring devices requires a routing algorithm. A lot of work has been done on routing protocols

D.M.A. Hussain et al. (Eds.): IMTIC 2008, CCIS 20, pp. 227–234, 2008.

since they are critical to the functioning of ad-hoc networks [2, 3]. Various routing protocols have been proposed in the literature such as AODV [2] and DSDV [3].

TCP/IP is the standard networking protocol on the internet. It is the most widely used transport protocol for data services like file transfer, e-mail and WWW browser. TCP, primarily designed for wire-line networks, faces performance degradation when applied to the ad hoc scenario. In addition, various routing protocols behave differently over the variants of TCP. It is essential to understand the performance of different MANET routing protocols under TCP variants. In this paper, we have done a performance analysis of two selected MANET Routing Protocols i.e. AODV and DSDV over TCP Vegas with mobility consideration.

The remainder of the paper is organized as follows: section 2 briefly reviews the two distance vector routing protocols, DSDV and AODV; and the working of TCP Vegas; section 3 is used to describe the simulation performance; section 4 covers the discussion of simulation results; while section 5 presents conclusion and future work.

## 2  Related Work

In our study, we have focused our simulations on two MANET routing protocols, namely, AODV and DSDV under TCP Vegas with mobility consideration. This section briefly describes the general working principles behind these two protocols and also the working of TCP Vegas.

### 2.1  DSDV Routing Protocol

In DSDV [1, 4], each node maintains a routing table, which has an entry for each destination in the network. The attributes for each destination are the next hop, metric (hop counts) and a sequence number originated by the destination node. To maintain the consistency of the routing tables, DSDV uses both periodic and triggered routing updates; triggered routing updates are used in addition to the periodic updates in order to propagate the routing information as rapidly as possible when there is any topological change. The update packets include the destinations accessible from each node and the number of hops required to reach each destination along with the sequence number associated with each route.

Upon receiving a route-update packet, each node compares it to the existing information regarding the route. Routes with old sequence numbers are simply discarded. In case of routes with equal sequence numbers, the advertised route replaces the old one if it has a better metric. The metric is then incremented by one hop since incoming packets will require one more hop to reach the destination. A newly recorded route is immediately communicated to its neighbors.

When a link to the next hop is broken, any route through that next hop is immediately assigned infinity metric and assigned an updated sequence number. This is the only case when sequence numbers are not assigned by the destination. When a node receives infinity metric and it has an equal or later sequence number with a finite metric, a route update broadcast is triggered. Therefore, routes with infinity metrics are quickly replaced by real routes propagated from the newly located destination.

One of the major advantages of DSDV is that it provides loop-free routes at all instants. It has a number of drawbacks, however. Optimal values for the parameters, such as maximum settling time, for a particular destination are difficult to determine. This might lead to route fluctuations and spurious advertisements resulting in waste of bandwidth. DSDV also uses both periodic and triggered routing updates, which could cause excessive communication overhead. In addition, in DSDV, a node has to wait until it receives the next route update originated by the destination before it can update its routing table entry for that destination. Furthermore, DSDV does not support multipath routing.

## 2.2  AODV Routing Protocol

The AODV is an on-demand or reactive MANET routing protocol [2, 4, 5]. In AODV, when a source node desires to send a message to some destination node and does not already have a valid route to that destination, it initiates a route discovery process to locate the intended node. It places the destination IP address and last known sequence number for that destination, as well as its own IP address and current sequence number (Broadcast-ID), into a Route Request (RREQ) message. The broadcast-ID and the nodes own IP address, uniquely identifies the RREQ which helps to suppress duplicate RREQ's to flow in the MANET when the same RREQ is received by a mobile node again. After that it broadcasts the route request (RREQ) message to its neighbors, which then forward the request to their neighbors, and so on, until either (a) the destination or (b) an intermediate node with a "fresh enough" route to the destination is found. If neither of these conditions is met, the node rebroadcasts the RREQ.

On the reception of RREQ message, the destination node creates a Route Reply (RREP) message. It places the current sequence number of the destination as well as its distance in hops to the destination, into the RREP, and sends back a unicast message to the source. The node from which it received the RREQ is used as the next hop. When an intermediate node receives the RREP, it creates a forward route entry for the destination node in its route table, and then forwards the RREP to the source node. Once the source node receives the RREP, it can begin using the route to transmit data packets to the destination. If it later receives a RREP with a greater destination sequence number or an equivalent sequence number with smaller hop count, it updates its route table entry and begins using the new route.

In AODV, an active route is defined as a route which has recently been used to transmit data packets. Link breaks in non-active links do not trigger any protocol action. However, when a link break in an active route occurs, a link failure notification is propagated to the node upstream of the break determines whether any of its neighbors use that link to reach the destination. If so, it creates a Route Error (RERR) packet. The RERR packet contains the IP address of each destination that is now unreachable, due to the link break. The RERR also contains the sequence number of each such destination, incremented by one. The node then broadcasts the packet and invalidates those routes in its route table.

There are many advantages of AODV. The number of routing messages in the network is reduced due to its reactive approach that makes it use the bandwidth more efficiently. However, protocol overhead may increase if it is used in highly mobile and heavily loaded networks. Furthermore, due to reactive approach it is more immune to the topological changes witnessed in the MANET environment. As a result, the AODV offers quick adaptation to dynamic link conditions, low CPU processing and memory overhead, low network utilization and determines unicast routes to destinations within the MANET. It also allows mobile nodes to obtain routes quickly for new destinations, and does not require nodes to maintain routes to destinations that are not in active communication. A distinguishing feature of AODV is its use of a destination sequence-number (DSN) that ensures loop freedom. Hence, AODV operates in a loop-free style.

### 2.3  TCP Vegas

TCP-Vegas [6] utilize the congestion avoidance mechanism to avoid packet loss by decreasing its CWND as soon as it detects an incipient congestion. In TCP-Vegas, CWND is determined by difference between expected throughput and actual throughput as follows.

Expected = WindowSize (CWND)/BaseRTT
Actual = WindowSize (CWND)/currentRTT
Diff = (Expected – Actual) x BaseRTT

where BaseRTT is the minimum of all measured RTTs. TCP-Vegas defines two thresholds, namely α and β. If Diff is < α, it considers the absence of congestion and increases its CWND by 1. If Diff is > β, it expects an incipient congestion and decreases its CWND by 1. Otherwise, it keeps the current CWND. For the purpose of retransmitting lost packets, TCP-Vegas maintains fine-grained RTO (Retransmission Time-Out) value for each transmitted packet, which is used to determine the occurrence of a timeout event when a duplicate ACK for a corresponding packet is received. If the timeout event occurs, the TCP sender thinks that the packet is lost and retransmits it without waiting for additional duplicate ACKs.

## 3  Simulation Performance

Performance evaluation was conducted by using NS-2 Simulator [7]. In this paper, "Random waypoint" model was adopted to simulate nodes movement, where the motion is characterized by two factors maximum speed and the pause time. The setup scenario consists of 50 mobile nodes with 10 randomly selected sources for traffic generation and 10 randomly selected destinations for the reception of the generated traffic. Topology is a rectangular area with 1000m x 800m which is standard form for wireless simulations. Data packets can be exchanged between the nodes as they traverse within the hearing range of one another. The traversing speed is set to 20m/s. As nodes move away due to mobility, link breaks and connection deformation occurred. All simulations are run for 300 seconds of simulation time. Simulation parameters are shown in Table 1.

**Table 1.** Simulation Parameters

| VARIABLES | VALUE |
|---|---|
| Simulation time | 300 sec |
| Topology size | 1000 m x 800 m |
| Total nodes | 50 |
| Mobility model | Random Waypoint |
| Traffic type | TCP (Vegas) |
| Packet rate | 4 packets/sec |
| Packet size | 512 bytes |
| Maximum Speed | 20 m/s |
| Number of connections | 10 |
| Pause time | 10, 100,250, 450 sec |
| NS-2 Version | NS-2.28 |

## 3.1 Simulation Results

This section describes the results achieved from the simulations. To analyze the performance of AODV and DSDV under TCP Vegas, 10 mobility scenario files were generated for pause-time 10, 100,250, 450 and 700 for both AODV and DSDV protocols.

There are three performance metrics that are measured in these simulations, namely, packet delivery ratio, average end-to-end delay and Total Packet drop.

### 3.1.1 Packet Delivery Ratio (PDR)

The PDR is defined as the number of received data divided by the number of packets generated. The reason that AODV has the highest PDR is that in AODV, efficacious route cache management is done via a cache entry timeout that ensures that only active routes are maintained in the route cache. This prevents the problem of a stale route entry cache. Also the use of sequence numbers prevents the formation of routing loops; other reason is that AODV is on-demand protocol, searches for route are done only when needed. In critical analysis, Vegas is giving linear delivery rate. The other important point to be noted from (figure 1) is the convergence of AODV protocol. This convergence fact again leads to the betterment of Vegas in different mobile scenarios. In figure-1, Packet delivery ratio starts to fall drastically when mobility increases (pause time 10); that is quite an expected result since DSDV has difficulties to maintain fresh routing information when nodes are moving fast. Practically all discarded packets were dropped because a stale routing table entry forwarded packets over a broken link. DSDV maintains only one route per destination; thus in case of broken link no alternative route can be found. AODV seem to have generally better packet delivery ratio than DSDV that reacts heavily to changes in mobility ratios.

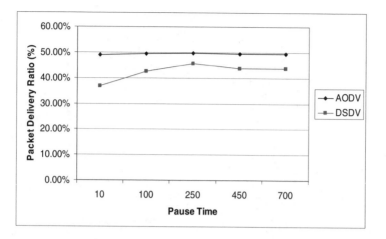

**Fig. 1.** Packet Delivery Ratio

### 3.1.2  Average End-to-End Delay

The End-to-End Delay is defined as the time a data packet takes to travel from source to the destination. Average End-to-End Delay is the delay perceived by all the packets including route acquisition delay. Figure 2 shows, the average End-to end delay for two selected MANET routing protocols namely AODV, and DSDV. It shows variations in the graph, DSDV has highest End-to-End delay (84.53 sec) at pause time 10, because DSDV protocol keeps packets in queues indefinitely until they are delivered to the next hop or the destination node. Therefore, it delivers the older packets rather than the newer ones, and hence there is an increase in average end- to- end delay for DSDV protocol, and having lowest End-to-End delay (25.97 sec) at pause time 250. Whereas if AODV is considered, it gives good performance at pause time 10. and 700, as compared to DSDV. At different pause times AODV performance can be considered on the average as compared to DSDV for average end-to-end delay, the reason for this average delay is because Vegas has better estimation approach and also its capacity to detect any congestion in advance. Vegas attempts to keep the sending rate around a point estimated by the RTT (Retransmission Timeout) samples. The idea is that if increasing the sending rate and the RTT does not increase. However, if the RTT increases as the increase in the sending rate, then more bandwidth can not be achieved, instead just taking up more space in the queues of intermediate than necessary. DSDV has the highest end-to-end delay at pause time 10 due to high mobility and another reason is that DSDV protocol keeps packets in queues indefinitely until they are delivered to the next hop or the destination node. Therefore, it delivers the older packets rather than newer ones, and hence there is an increase in average end- to- end delay for DSDV protocol. As shown in graph (figure-2) at pause times 10 and 450 the DSDV was suffering from the highest end to end Delay as compared with other two protocols (as shown in Figure-2). AODV has somewhat better technique in this regard, since the destination replies only to the first arriving RREQ. This automatically favors the least congested route instead of the shortest route.

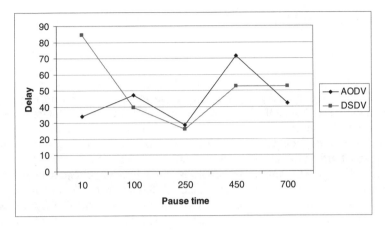

**Fig. 2.** Average end-to-ends Delay

### 3.1.3  Total Packet Dropped

Packet dropping refers to the loss of packet due to no connection availability or nodes are not in the range of each other. TCP Vegas is a TCP congestion control, or network congestion avoidance, algorithm that emphasizes packet delay, rather than packet loss, as a signal to help determine the rate at which to send packets. As shown in Figure-3, AODV has the highest packet drop at all the pause times as compared with DSDV protocol. As shown in figure 3, at pause time 10, AODV has the highest packet drop due to high mobility. Mobility affects the performance of all the proto-cols, link failures can happen very frequently. Link failures trigger new route discov-eries. In AODV, since it is at the most one route per destination in its routing table. The frequency of route discoveries in AODV is directly proportional to the no: of route breaks.

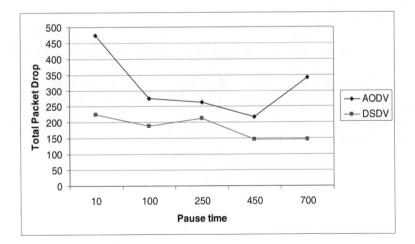

**Fig. 3.** Total Packet Drop

# 4    Conclusion and Future Work

This paper addresses the performance comparison of two table driven protocols namely DSDV and AODV under the TCP Vegas with mobility consideration. From the above performance evaluation of the two protocols, it can be concluded that AODV has a better throughput performance and low average end to end delay as compared to DSDV, but AODV suffers from high packet drop. The future work could be conducted with the analysis of MANET environment under different issues such as node energy consumption, issues of hidden and exposed terminals, and constraints in mobility and traffic criteria. Through the extensive simulations, it is observed that which to select among routing protocols is more important than which to select among variants [6], because the performance difference between variants is not so much outstanding. In future different MANET routing protocols like Optimized link state routing protocol (OLSR), temporarily ordered routing protocol (TORA), Pre-emptive Ad hoc on-demand distance-Vector routing protocol (PAODV) can be added for better analysis evaluation.

# References

1. Royer, E.M., Toh, C.-K.: A Review of Current Routing Protocols for Ad-Hoc Mobile Wireless Networks. IEEE Personal Communications Magazine, 46–55 (1999)
2. Perkins, C.E., Royer, E.M.: Ad-hoc On-Demand Distance Vector Routing. In: Proceedings of the 2nd IEEE Workshop on Mobile Computing Systems and Applications, New Orleans, LA, pp. 90–100 (1999)
3. Perkins, C.E., Bhagwat, P.: Highly Dynamic Destination Sequenced Distance Vector Routing (DSDV) for Mobile Computers. In: Proceedings of the ACM SIGCOMM Conference on Communications Architectures, Protocols and Applications, London, UK, pp. 234–244 (1994)
4. Pandey, A.K.: Study of MANET routing protocols by GloMoSim simulator. International Journal of Network Management, 393–410 (2005)
5. Lu, Y., Wang, W., Zhong, Y., Bhargava, B.: Study of Distance Vector Routing Protocols for Mobile Ad hoc Networks. In: Proceedings of the First IEEE International Conference on Pervasive Computing and Communications, p. 187. IEEE Computer Society, Los Alamitos (2003)
6. Kim, D., Bae, H., Song, J.: Analysis of the Interaction between TCP Variants and Routing Protocols in MANETs. In: Proceedings of the 2005 International Conference on Parallel Processing Workshop (ICPPW 2005) (2005)
7. Fall, K., Varadhan, K.: Ns Notes and Documents, The VINT Project, UC Berkeley, LBL, USC/ISI. And Xerox PARC (2000),
   http://www.isi.edu/nsnam/ns-documentation.html

# CMMI and OPM3: Are They Compatible?

Saman Nazar[1] and Eram Abbasi[2]

[1] Bahria University, Karachi Campus, Karachi, Pakistan
saman.nazar@gmail.com
[2] Bahria University, Karachi Campus, Karachi, Pakistan
eram@bimcs.edu.pk

**Abstract.** In the past, many software companies have reported benefits in productivity and quality by using quality improvement initiatives. CMMI is one such initiative that has gained much success over a long period. OPM3 is a recent addition to the list of maturity models. CMMI is the software process improvement model proposed by the Software Engineering Institute (SEI) and OPM3 is the project management maturity model proposed by the Project Management Institute (PMI). This paper aims to analyze the compatibility of these two models from both theoretical and practical perspectives. As a first step towards the comparative analysis, a detailed one-to-one mapping of all the practices of the two models was carried out. The overlapping and differentiating factors of CMMI and OPM3 were identified. The analysis was undertaken in the hope of finding synergy between maturity models of two different domains proposed by these two world-renowned organizations. In order to validate the results observed, an OPM3 assessment was carried out for a CMMI-appraised software house. The assessment verified the results that were observed during the comparative analysis. The objective of this study was to find out the strengths and weaknesses of each maturity model so that organizations interested in gaining maturity can reap maximum benefit by the application of any one or both of these models. Hence, the objective of the study was successfully achieved.

## 1 Introduction

The past decade or so has seen a tremendous rise in the quest for organizational maturity. Several mature organizations have reported benefits from better performance after reaching a certain maturity level. Credit goes to the family of capability-maturity models developed by the Software Engineering Institute of Carnegie-Mellon University that originally popularized the concept of maturity and process improvement. The project management community was clearly influenced by this concept of achieving maturity and used it to develop its own project management maturity models. The Project Management Institute's Organizational Project Management Maturity Model or simply OPM3 is the most recent addition to this list of maturity models. CMMI (Capability Maturity Model Integration) is a reference model consisting of best practice descriptions for a broad range of engineering activities grouped in a set of process areas. It has two representations: Staged and Continuous. As a descriptive model,

D.M.A. Hussain et al. (Eds.): IMTIC 2008, CCIS 20, pp. 235–242, 2008.
© Springer-Verlag Berlin Heidelberg 2008

CMMI is well suited for appraisal efforts seeking to determine a particular organization's capabilities within the scope of software, systems, integrated product engineering, or acquisition and for guiding the broad direction of process improvement efforts in these areas of expertise [1]. CMMI deals with project management maturity through improvement in 8 process areas. The interactions among the project management process areas are divided in two process area groups:

A. Basic Project Management process areas are:

    a. Project Planning
    b. Project Monitoring and Control
    c. Supplier Agreement Management

B. Advanced Project Management process areas are:

    a. Integrated Project Management for IPPD
    b. Risk Management
    c. Integrated Teaming
    d. Quantitative Project Management
    e. Integrated Supplier Management

OPM3, on the other hand, is industry independent. It seeks to create a framework within which organizations can reexamine their pursuit of strategic objectives via Best Practices in Organizational Project Management. According to OPM3, Projects and Programs are part of a Project Portfolio. It defines organizational project management as the systematic management of projects, programs, and portfolios in alignment with the achievement of organization's strategic goals. It contains a collection of almost 586 generally accepted and proven project management Best Practices (Knowledge), a self-assessment tool to assess organization's maturity against the Best Practices identified (Assessment) and then guidance for possible organizational changes for achieving maturity (Improvement). It complements the widely used Project Management Body of Knowledge (PMBOK) guide [2].

## 2   Model Comparison

Unlike CMMI, OPM3 is in its first public iteration and does not have a market history. Also it is not a final product (e.g., benchmarking, consulting support, and an improved assessment tool are still in infancy). The self-assessment tool is very raw as it has some loopholes which need to be filled. For example: when evaluating practices it either considers that the organization performs the practice or it does not perform the practice. There is no room for partial fulfillment of any practice. Also there are too many Best Practices to choose from. However, on the other hand this is what makes it flexible for use by any organization. It provides a more generic approach towards improving project management leaving it to the organization to decide which Best Practices to focus on first. One major difference between CMMI and OPM3 is that CMMI's process areas are narrowed to system/software engineering perspective while

Table 1. OPM3 and CMMI process categories/ domains

| CMMI process categories | OPM3 process domains |
|---|---|
| Project Management | Project Management |
| Process Management | Program Management |
| Engineering | Portfolio Management |
| Support | |

Note: The information given in the table about CMMI and OPM3 is not comparative.

the Best Practices of OPM3 are more holistic. OPM3 introduces the concept of improving organizational maturity in terms of three domains: Projects, Programs and Portfolios that seem more relevant to an organization. Helping organizations align their projects, programs and portfolios to corporate strategy is the intent of OPM3. CMMI, on the other hand, does not recognize portfolio management. It focuses on project management along with improvement in process management, engineering and support process areas for achievement of an organizational maturity rating.

One major difference in both the models is that in case of CMMI, maturity means that to achieve every next level, different things are measured for compliance. OPM3 treats maturity in a slightly different perspective. Same things are measured every time; it's just the results that improve, indicating increased maturity.

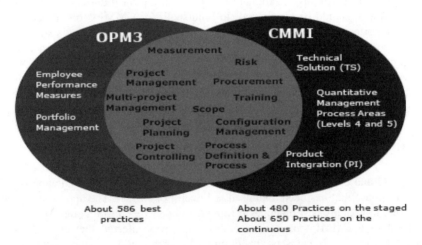

Fig. 1. CMMI and OPM3 overlapping [3]

Figure 1 shows that many practices of CMMI are mapped on to OPM3 even those that are not in its project management process area. This is because the terms of process and project in CMMI and OPM3 are used with slightly different meanings. So even those practices that CMMI has termed to be in process management can be found in OPM3, a project management maturity model.

This brings us to the main difference between CMMI and OPM3 and that is the focus on portfolio management. As its name suggests OPM3 is an "organizational"

**Table 2.** Comparison of terms used by OPM3 and CMMI

| OPM3 terms | CMMI terms |
|---|---|
| Best practices (about 586) | Goals and Practices (about 480 on staged) |
| Key performance indicators | Typical work products |
| Self Assessment Comprehensive Assessment (usually done before improvement) | SCAMPI Appraisal for internationally recognized maturity level ranking (usually after improvement) |
| OPM3 Consultant OPM3 Appraiser | CMMI Instructor SCAMPI Lead Appraiser |

Note: The table gives an overview of the areas where these models map.

project management maturity model so all practices included in this process domain cannot be mapped to any of those in CMMI because CMMI's focus is on achieving success in individual projects through process improvement while OPM3 focuses on achieving organization business/strategic goals through its projects. While CMMI is of the view that consistent success is possible in all projects if the processes have reached a certain maturity level, OPM3 is of the view that projects should be prioritized and aligned with organizational strategy and if a project does not fulfill the success criteria, it should be terminated. In a nutshell, for CMMI "good process" is of crucial importance while for OPM3 "projects supporting organization's strategy" are most important. Hence, OPM3 not only wants good organization-wide processes but also forces organizations to do soul-searching on the kind of projects they take up, always giving due consideration to their ultimate business goals. All practices that are concerned with maintaining cash flows and investments, performing employee performance assessments etc. are out of the scope of CMMI.

# 3   OPM3 Assessment of a Software House in Pakistan

For OPM3 and CMMI comparison, it was deemed ideal to perform an appraisal of a software house that has been appraised at Level 3 of CMMI. The appraisal helped in analyzing how many best practices it followed under the OPM3 model. With this assessment of a CMMI-appraised organization, it was inferred that any CMMI appraised software house can score better on OPM3 continuum if it starts focusing on its Portfolio management because it will be following most of the other practices. The assessment tool provided by PMI was used for this appraisal. There are 151 questions in the tool, which include self-assessment questionnaire. Based on the responses provided by the software house, four types of graphs were developed that represent the responses to these questions.

## 3.1   Results of the Assessment

Figure 2 shows the organization's overall position on a continuum of organizational project management maturity.

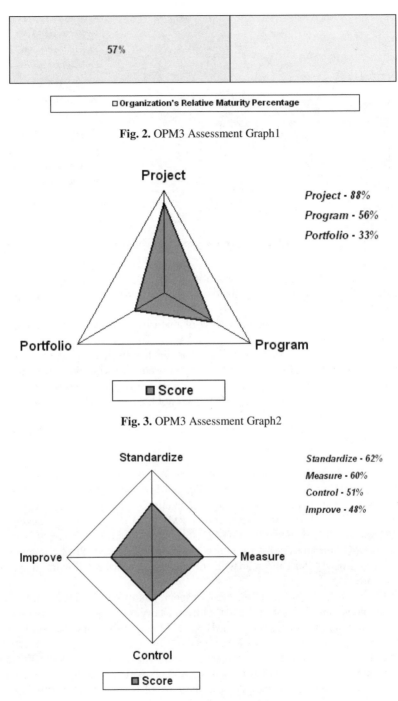

57%

□ Organization's Relative Maturity Percentage

**Fig. 2.** OPM3 Assessment Graph1

Project

Project - *88*%
Program - *56*%
Portfolio - *33*%

Portfolio                    Program

■ Score

**Fig. 3.** OPM3 Assessment Graph2

Standardize

*Standardize - 62%*
*Measure - 60%*
*Control - 51%*
*Improve - 48%*

Improve                    Measure

Control

■ Score

**Fig. 4.** OPM3 Assessment Graph3

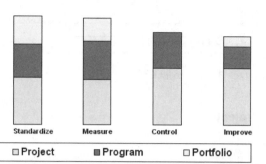

Project Standardize - *83%*
Project Measure - *80%*
Project Control - *100%*
Project Improve - *100%*

Program Standardize - *58%*
Program Measure - *66%*
Program Control - *62%*
Program Improve - *37%*

Portfolio Standardize - *48%*
Portfolio Measure - *40%*
Portfolio Control - *0%*
Portfolio Improve - *18%*

**Fig. 5.** OPM3 Assessment Graph4

Figure 3 shows the organization's maturity in terms of each domain.

Figure 4 demonstrates the organization's maturity in terms of each process improvement stage.

Figure 5 shows the organization's view of maturity in terms of each domain and each stage of improvement.

All these graphs prove that the project domain has been covered really well by CMMI so most of the CMMI practices being followed map with those of OPM3. Program practices too have been followed. The practices that seem to be least paid to heed to are the Portfolio ones since CMMI does not have any such focus.

## 4   Conclusion

The research covered the analysis and comparison of two maturity models namely CMMI and OPM3. CMMI is the maturity model for software process improvement while OPM3 is the maturity model for organizational project management. CMMI does not cover Portfolio management while OPM3 lacks the practices of the technical process areas of CMMI. From the OPM3 assessment of software house it is inferred that any CMMI appraised software house can score better on OPM3 continuum if it starts focusing on its Portfolio management because it will be following most of the other practices.

OPM3, though a useful model in every respect, has its limitations. CMMI has delivered the goods and has stood the test of time as organizations have reported benefits after following the practices of CMMI. OPM3 on the other hand has yet to build a name for it. PMI is keeping in view the needs of the industry and it is expected that OPM3 can very well become the industry standard for achieving organizational project management maturity. If this happens then many software houses are expected to hop on to the bandwagon as happened in the case of CMMI. But all this has yet to materialize. Considering the Pakistani IT industry, this is still a far-fetched idea. No organization took the initiative of CMMI on its own until it was supported by

Pakistan Software Export Board (PSEB). The reason is that the expenditure and the effort involved in going through such huge organizational changing projects requires a strong motivation and incentive, that is so far not available in the case of OPM3. But this does not render the said model as worthless. It is perfectly adaptable and useful to all kinds of projects and organizations including those of the IT industry. All those companies who seek improvement in their project management processes on an organizational level can very well utilize the benefits of this model.

Considering the differences found in the two models, it is recommended that a new model be proposed that includes both the technical features of CMMI and the Portfolio management practices of OPM3. This will help organizations in better aligning their projects to their business strategy while giving due consideration to their technical aspects. This is important because, after all, software companies are business ventures. The trend towards increasing use of IT continues and the challenge remains how to better manage IT projects in order to maximize their economic benefits that is the main business goal. This aspect of meeting business goals is not covered by CMMI. With this aspect included, the new model will be ideal for every organization as it will include both the technical and business knowledge. With this model not only the software processes will improve and the projects will be successfully managed but the organizational project management maturity will also improve.

# References

1. Software Engineering Institute: CMMI staged-version 1.1 (2002),
   http://www.sei.cmu.edu/cmmi/
2. Project Management Institute: Organizational Project Management Maturity Model (OPM3) Knowledge Foundation. Project Management Institute Inc. Newton Square. Pennsylvania 2003. opm3.pmi.org (2003)
3. Philips, B.: CMMI and OPM3 presentation (2005)
4. Bert, D.R., Yael, G.C., Lockett, M., Calderini, S.R., Moura, M., Sloper, A.: The impact of project portfolio management on information technology projects. International Journal of Project Management 23(2005), 524–537 (2005)
5. Bradley, M.: CMMI and OPM3. Montgomery PMI Chapter Meeting, The Cahaba Group (2004)
6. Keuton, T.: CMMI and OPM3 A Powerful Combination for Increasing Organizational Maturity (2005), http://www.projectmagazine.com/v5i4/cmmi.html
7. Cooke-Davies, T.J., Arzymanow, A.: The maturity of project management in different industries: An investigation into variations between project management models. International Journal of Project Management 21, 471–478 (2003)
8. Eman, K.E., Madhavgi, N.H.: Does Organizational Maturity improve quality? IEEE Software, 109–110 (1996)
9. Hillson, D.: Benchmarking organizational project management capability. In: Proceedings of the 32nd Annual Project Management Institute 2001 Seminars and Symposium. Project Management Institute, Nashville (2001)
10. Mendez, Y.B.: Project Management for Software Process Improvement. In: Proceedings of Project Management Institute Global Progress, Prague, Czech Republic, Project Management Institute (2004)

11. Kerzner, H.: Project Management: A Systems Approach to Planning, Scheduling, and Controlling, pp. 313–314. John Wiley and Sons, New York (1998)
12. Software Engineering Institute: CMMI staged-version 1.1, pp. 24 (2002), http://www.sei.cmu.edu/cmmi/
13. Project Management Institute: A Guide to Project Management Body of Knowledge, pp. 220–221. PMI publication Division, Sylvia, NC (1996)
14. Futrell, R.T., Shafer, D.F., Shafer, L.I.: Quality Software Project Management, p. 136. Prentice Hall, Englewood Cliffs (2002)
15. Garcia, S.: How standards enable adoption of project management practices. In: IEEE Software. IEEE Computer Society, Los Alamitos (2005)

# Impact of Cluster Size on Efficient LUT-FPGA Architecture for Best Area and Delay Trade-Off

Khalil Dayo[1], Abdul Qadeer Khan Rajput[2], Bhawani S. Chowdhry[1,3], and Narinder Chowdhry[1]

[1] Department of Electronic & Biomedical Engineering, MUET, Jamshoro, Sindh, Pakistan
[2] Department of Computer Systems & Software Engg. MUET, Jamshoro Sindh Pakistan
[1,3] Postdoctoral Research Fellow, School of Electronics & Computer Science, University of Southampton, UK
krdayo@yahoo.com, bsc_itman@yahoo.com

**Abstract.** The delay of a circuit implemented in a Lookup table (LUT) based Field-Programmable Gate Arrays (FPGAs) is a combination of routing delays, and logic block delays. However most of an FPGA's area is devoted to programmable routing. When these blocks are replaced with logic clusters, the fraction of delay due to the cluster has significant impact on total delay. This paper investigates the impact of logic cluster size when the most favorable LUT size is achieved. As a result, fast and area efficient FPGA architecture can be proposed that can combine the logic blocks into logic clusters. In lookup table FPGA architecture, area and delay are the main factors to be tackled, the best value for each of the parameters depends on complex trade-offs. If an FPGA with smaller LUTs is constructed to minimize the area, the result is poor speed. On the other hand, if an FPGA includes larger LUTs, speed might increase but area is unnecessarily wasted. In this experimental work 20 benchmark circuits were tested to calculate the delay and area metric. Results show increasing logic cluster size has no more effect on delay as well as area, when suitable optimal values of lookup table size (LUT) are established.

**Keywords:** Lookup table, FPGA, CLBs, BLE.

## 1 Introduction to FPGA Architecture

Field-Programmable Gate Arrays (FPGAs) are user programmable digital devices that provide efficient, yet flexible, implementations of digital circuits. An FPGA consists of an array of programmable logic blocks interconnected by programmable routing resources. The flexibility of FPGAs allows them to be used for a variety of digital applications from small finite state machines to large complex systems.

Figure 1 shows a high-level view of an island-style FPGA [1], which represents a popular architecture framework that many commercial FPGAs are based on, and is also a widely accepted architecture model used in the FPGA research community. Logic blocks represented by gray squares consist of circuitry for implementing logic. Logic blocks are also called configurable logic blocks (CLBs). Each logic block is surrounded by routing channels connected through switch blocks and connection

D.M.A. Hussain et al. (Eds.): IMTIC 2008, CCIS 20, pp. 243–252, 2008.

blocks. The wires in the channels are typically segmented and the length of each wire segment can vary. A switch block connects wires in adjacent channels through programmable switches such as pass-transistors or bi-directional buffers. A routing block connects the wire segments around a logic block to its inputs and outputs, also through programmable switches. Notice that the structures of the switch blocks are all identical. The figure illustrates the different switching and connecting situations in the switch blocks (the structures of all the connection blocks are identical as well). In [1] routing architectures are defined by the parameters of channel width (W), switch block flexibility (Fs – the number of wires to which each incoming wire can connect in a switch block), connection block flexibility (Fc – the number of wires in each channel to which a logic block input or output pin can connect), and segmented wire lengths (the number of logic blocks a wire segment spans).

Modern FPGAs provide clusters, embedded IP cores, such as memories, DSP blocks, and processors. Commercial FPGA chips contain a large amount of dedicated interconnects with different fixed lengths. These interconnects are usually point-to-point and uni-directional connections for performance improvement. For example, Xilinx XC4000XL easy path solution, [2] has vertical or horizontal interconnects across 4, 16 or 24 logic blocks. There are dedicated carry chain and register chain interconnects within and between the wire segments around a logic block to its inputs and outputs, also through programmable switches. The figure 1, illustrates the different switching and connecting situations in the switch blocks.

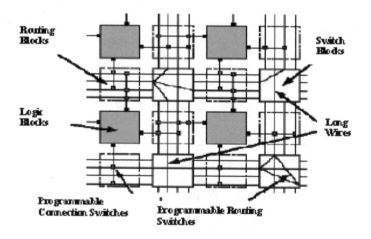

**Fig. 1.** An Island style FPGA Architecture

## 1.1 Basic Logic Element

In general a logic cluster consists of one or more "Basic Logic Elements" (BLEs) connected by fast local interconnect, where the BLE used consists of a 4-LUT and a register. Figure 2 shows a logic cluster consisting of N BLEs and local interconnect. The size of the logic cluster (number of BLEs it contains) used in an FPGA architecture can have a dramatic effect on its area and performance. Previous work [3, 4]

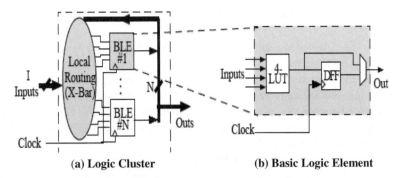

(a) Logic Cluster                    (b) Basic Logic Element

**Fig. 2.** Logic cluster and Basic Logic Element

demonstrated the effect of cluster size on area efficiency. Also, previously it was speculated that as cluster size is increased, circuit speed would improve. While this work reports it as counterintuitive. Figure 2 (a) shows cluster contains (N) basic logic elements fed by (I) cluster inputs. The BLE, illustrated in Figure 2 (b) typically consists of a K-input lookup table and register, which feed a two-input multiplexer that determines whether the registered or unregistered LUT output drives the BLE output.

In this work main focus of the authors is to determine the effect of logic cluster size on circuit speed as well as area, and finding what logic cluster size has the best area and delay trade-off.

## 2  Logic Clustering

Most modern FPGA architectures contain physical hierarchy logic blocks for improved area-efficiency and speed. To implement a design on such architectures, a logic clustering step is typically needed between technology mapping and placement [5]. Logic clustering transforms a netlist of logic cells into a netlist of logic clusters each of which can be implemented using a logic block.

A typical logic block contains N logic cells with I inputs and N outputs. Here, N and I are fixed for the given architecture. There can be other architecture constraints, such as control signals for sequential elements. The logic clustering problem for FPGAs takes as input a mapped netlist and produces a clustered netlist satisfying the cluster parameters. One of the early FPGA clustering algorithms is VPack [3], [6].

The VPack algorithm forms one cluster at a time. At the beginning of each cluster formation, VPack selects as a seed an unclustered logic cell with the most used inputs and places this seed into the cluster. It then calculates the attraction of each unclustered cell to the new cluster. The attraction of a logic cell to a cluster is the number of inputs and outputs that are shared by the cell and the cluster. The cell whose addition doesn't violate cluster constraints and has the largest attraction value will be added to the cluster. The packing process is repeated until we cannot add in new cells to the current cluster. At that time, packing begins with a new cluster. The process terminates when all logic cells have been assigned a cluster. The aim of the VPack algorithm is to minimize the number of clusters needed.

# 3  Trade-Offs in Cluster-Based FPGA

Much of the speed and area-efficiency of an FPGA is determined by the logic block it employs. In a cluster-based FPGA, there are clear trade-offs between cluster size and FPGA speed and area. If a very small logic cluster is used (few BLEs per logic cluster), many logic blocks are required to implement each circuit, and many connections must be routed between the numerous logic blocks. Since routing consumes most of the area and accounts for most of the delay in FPGAs, a small logic block often results in poor area-efficiency and speed due to the excessive routing required connecting all the logic blocks. If, on the other hand, a very large logic block is employed (many BLEs per logic cluster), fewer logic blocks are required to implement each circuit, but the logic block area and delay may become excessive, again resulting in poor area-efficiency and speed. Choosing the best size for an FPGA logic block therefore involves balancing complex trade-offs. Thus it is important to determine the best cluster size for cluster-based architectures. This style of logic block is of interest for several reasons. First, many island style FPGAs, employ cluster-based logic blocks, so research concerning the best size of logic clusters is of clear commercial interest. Also, prior research [3], [7] has shown that increasing the size of logic clusters may be competitive with the speed of FPGA. Consequently it is important to measure the metric of this size. Last and important finding should be determining the impact of clustering when the optimal LUT size is achieved.

# 4  Determination of Best Area-Delay Trade Off

To determine the best area-delay trade-off, authors have arranged a set of 20 benchmark circuits. Table 1 gives the description of these circuits used to measure total delay, intra-cluster delay, inter-cluster delay.

## 4.1  The Delay Metric

The important key metric for FPGAs is their speed measured by the critical path delay. The total critical path delay is defined as the total delay due to the logic cluster combined with the routing delay. Figure 3 shows the geometric average of the total critical path delay across all 20 circuits as a function of the cluster size and LUT size. Observing the Figure, it is clear that increasing cluster size or LUTs, decreases the critical path delay. These decreases are considerable between cluster size = 1 and LUT size = 2 and cluster size = 7 and LUT size = 7. The significant reduction in delay from 35-80 ns to 25-50 ns for different parameters is shown in the figure. There are two trends that explain this behavior. As the LUT and cluster size increases; the average delay decreases.

It is useful to break the total delay into two components: intra-cluster delay (which includes the delay of the muxes and LUTs), and inter-cluster delay. Figure 4 shows the portion of the critical path delay that comes from the intra-cluster delay as a function of K and N.

**Table 1.** Benchmark circuits

| | Circuit Description | | | | |
|---|---|---|---|---|---|
| Circuit | Number of LUTs | Number of Latches | Number of Nets | Number of Inputs | Number of Outputs |
| alu4 | 1522 | 0 | 1536 | 14 | 8 |
| apex2 | 1878 | 0 | 1917 | 39 | 3 |
| apex4 | 1262 | 0 | 1271 | 9 | 19 |
| bigkey | 1707 | 224 | 2194 | 263 | 197 |
| clma | 8381 | 33 | 8797 | 383 | 82 |
| des | 1591 | 0 | 1847 | 256 | 245 |
| diffeq | 1494 | 377 | 1935 | 64 | 39 |
| dsip | 1370 | 224 | 1823 | 229 | 197 |
| elliptic | 3602 | 1122 | 4855 | 131 | 114 |
| ex1010 | 4598 | 0 | 4608 | 10 | 10 |
| ex5p | 1064 | 0 | 1072 | 8 | 63 |
| frisc | 3539 | 886 | 4445 | 20 | 116 |
| misex3 | 1397 | 0 | 1411 | 14 | 14 |
| pdc | 4575 | 0 | 4591 | 16 | 40 |
| s298 | 1930 | 8 | 1942 | 4 | 6 |
| s38417 | 6096 | 1463 | 7588 | 29 | 106 |
| s38584.1 | 6281 | 1260 | 7580 | 39 | 304 |
| seq | 1750 | 0 | 1791 | 41 | 35 |
| spla | 3690 | 0 | 3706 | 16 | 46 |
| tseng | 1046 | 385 | 1483 | 52 | 122 |

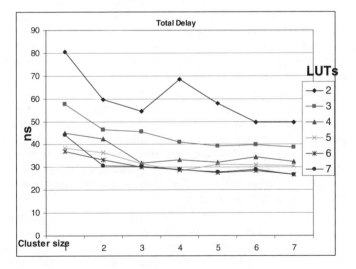

**Fig. 3.** Total Delay in nano-seconds (Cluster Size = 1-7)

It is quite obvious from Figure 4 that the intra-cluster delay decreases as the LUT size increases. This is due to the fact that there is a reduction in the number of BLE levels on the critical path and hence there will be fewer logic levels to implement.

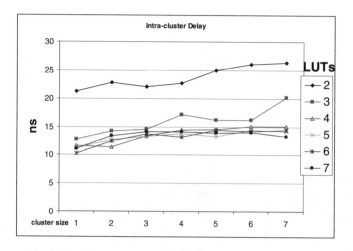

**Fig. 4.** Intra-cluster Delay in nano-seconds (Cluster Size = 1-7)

This will translate into a reduction in intra-cluster delay. Also the second behavior that should be noticed is that the intra-cluster delay increases for any given LUT size as the cluster size is increased, except the cluster size 3 to 7 for LUTs 4-7.

This is because the intra-cluster multiplexers get larger and therefore slower. However, the delay through these multiplexers is still much faster than the inter-cluster delay. As Lookup table (K) increases to 4, there are fewer LUTs on the critical path, and this translates into fewer inter-cluster routing links, thus decreasing the inter-cluster routing delay. Similarly, as N is increased 4 and 5 more connections are captured within a cluster, and again, the inter-cluster routing decreases.

Talking about these trade-offs, it's useful to follow an explicit example. However, it is interesting that increasing cluster size has little impact after a certain point (for N > 3). Figure 6 shows this clearly where for any fixed LUT size, the majority of the improvement in critical path delay occurs as the cluster size is increased from 1 to 3.

Any further increases in cluster size results in a very minimum delay improvement. This behavior suggests that clustering has little effect after a certain point. That concludes, employing larger clusters should always reduce the critical path. Although, the total delay results from figure 4 do not contradict this, what is important here is that how little of an improvement in total delay that is achieved with larger cluster like 6? Further more it is very important to note that cluster size beyond 4 has no improvement in delay for increasing size of LUTs.

Appendix depicts complete graphical view of total delay calculated during the experiments in nano seconds for various sizes of clusters at glance.

The results clearly show the number of levels decreasing with increasing cluster and LUT sizes. But, for any given LUT size it can be seen that most of the reduction in the number of levels occurs as the cluster size is increased from 1 to 3, as the majority of the circuits confirms this behavior shown in Appendix A. Also, recall that the majority of the critical path delay was reduced in the same range. The direct relationship between the number of cluster levels on the critical path and the final total delay is no coincidence. Fewer logic blocks on the critical path leads to improved performance. The main

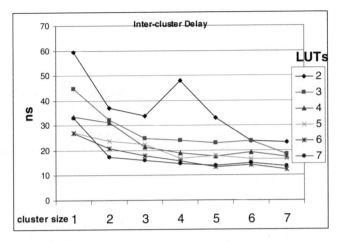

**Fig. 5.** Inter-Cluster delay in nano-seconds (cluster size 1-7)

reason that the total delay did not improve significantly for larger LUTs, as we varied the cluster size from 3 to 7 was that there was no significant reduction in the number of logic block levels.

Without a reduction of the number of inter-cluster levels on the critical path we cannot possibly expect improvements in FPGA performance. Another interesting trend to observe from figure 5 is that increasing the cluster size has less of an effect for architectures composed of larger LUTs.

## 5   Conclusion

Individual as well combined geometric averages of all the tested 20 benchmark circuits show very little effective response when:

➢ 4-5 LUT size is being tested with cluster size 1-7.
➢ Maximum delay reduction is achieved when we test 7-LUT size with 7 cluster size.
➢ However this tendency is quite costly when area factor is being considered, as the best delay optimizing value should be taken under area constrains.
➢ This research concludes that when optimum value of LUT size is achieved, then further increase in cluster size has no much effect on delay.
➢ Any increase in LUT size beyond 4 has little effect on delay, but considerable rise in area.
➢ Clustering technique is useful for smaller LUTs.

**Acknowledgment.** Authors are grateful to Higher Education Commission for funding this project. Special thanks to Nanyang Technological University Singapore, for providing important data and circuits for experiment purpose. Finally authors are grateful to Rastek Technologies and Mehran University for providing excellent environment of research.

# References

1. Brown, S., Francis, R., Rose, J., Vranesic, Z.: Field Programmable Gate Arrays. Kluwer Academic Publishers, Dordrecht (1992)
2. Xilinx Press Release # 0766, Xilinx Easy Path Solutions (2005)
3. Betz, V., Rose, J.: VPR: A New Packing, Placement and Routing Tool for FPGA Research. In: Glesner, M., Luk, W. (eds.) FPL 1997. LNCS, vol. 1304, pp. 213–222. Springer, Heidelberg (1997)
4. Krishnamoorthy, S., Swaminathan, S., Tessier, R.: Technology Mapping Algorithms for Hybrid FPGAs Containing Lookup Tables and PLAs. IEEE Trans. on Computer-Aided Design (2003)
5. Dayo, K.R., Rajput, A.Q.K., Chowdhry, B.S.: Analysis and Comparison of Synthesizing Algorithms for Area and Delay Minimization in FPGA Architecture. Mehran University Research Journal of Engineering & Technology 25(4) (2006)
6. Dayo, K.R., Rajput, A.Q.K., Chowdhry, B.S., Siyal, M.Y.: Investigating the Synthesis Algorithms for Area and Delay Minimization in FPGA Design. IEEEP 46-47, 29–35 (2005)
7. Ohba, N., Takano, K.: An SoC Design Methodology using FPGAs and Embedded Microprocessors. In: Proceedings of the 41st annual Conference on Design Automation, San Diego, CA, USA (2004)

## Appendix: Total Delay in Nano Seconds for Various Cluster Sizes

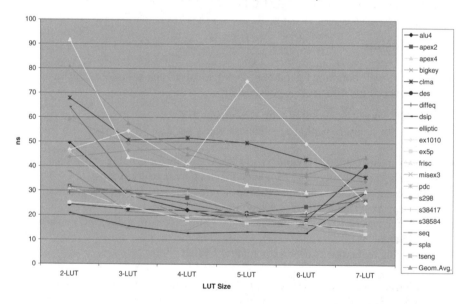

Total Delay in nano-seconds (Cluster Size = 1)

**Total Delay in nano-seconds (Cluster Size = 3)**

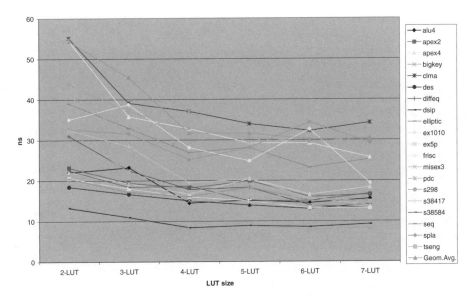

**Total Delay in nano-seconds (Cluster Size = 5)**

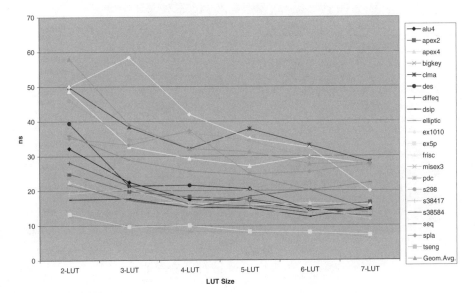

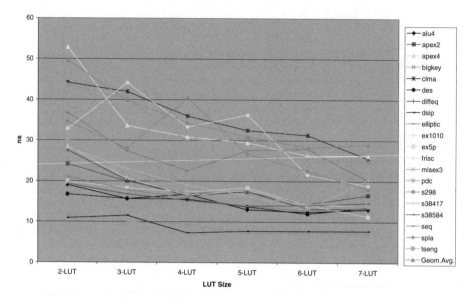

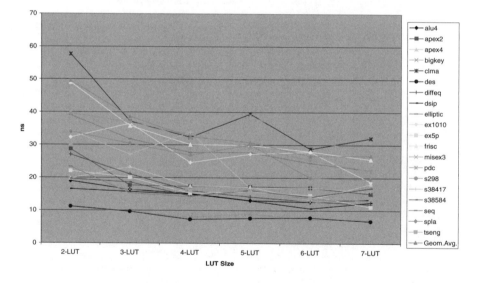

# Effects of Social Status on Foetal Growth Using Obstetric Ultrasonography – A Biometric Approach

Wahab Ansari[1], Narinder P. Chowdhry[2], Kusum Pardeep[3], Aftab A. Memon[4],
Karim Baloch[4], Abdul Qadeer Khan Rajput[4], and Bhawani S. Chowdhry[4,5]

[1] Institute of Information Technology, University of Sindh, Jamshoro, Pakistan
[2] Department of Biomedical Engineering, Mehran University of Engineering & Technology
(MUET), Jamshoro, Pakistan
[3] Department of Radiology, LUMHS, Jamshoro, Pakistan
[4] Institute of Information and Communication Technologies, MUET,
Jamshoro 76062, Sindh-Pakistan
[4,5] Postdoctoral Research Fellow, School of Electronics & Computer Science,
University of Southampton, UK
bsc06v@ecs.soton.ac.uk, chowdhry.np@muet.edu.pk

**Abstract.** The focus of this study is the analysis of biometric data related to the
effects of the foetal growth based on the social status of vegetarian and non-
vegetarian patients referred to clinicians. We have analyzed multiple biometric
parameters such as age, socio-economic status, last menstrual period, and their
associated biometric data. Out of thirty random patients, twelve patients were
selected for investigation. Our results indicate that the growth rate of vegetarians is 56 grams/week, while for non-vegetarians it is 36 grams/week at the
second and third trimester of the middle class patients.

**Keywords:** Biometrics, Medical informatics, Artificial intelligence and Ultra
Sonography.

## 1 Introduction

The bioinformatics is a branch of science, which deals with collecting biological information using computer graphics and modelling for quick visualization and assessment of biological cells for the evolution of growth through a span of time [1]. The
biological images obtained in vitro or vivo are plotted in grey level in between frequency and time. The ultrasound diagnostic is a relatively a new imaging modality,
that of immense value, especially in diagnostic applications in the branches of medicine [2]. It utilizes high frequency sound waves above the range of human hearing, to
the upper limit of above 20 KHz [3]. These images are captured in real time. This
biometric data include Bi Parietal Diameter (B.P.D.) that is proved to be an excellent
parameter for the assessment of gestational age in the second trimester of pregnancy
[4]. The Femur Length (F.L), the linear growth of the femur during pregnancy makes
femur length a useful parameter for assessing gestational age in the second and third
trimester. The other parameters are Abdominal Circumference (A.C) and Foetal
Weight (F.W) during the growth of embryo/foetal in vivo. The Ultra Sonography is

D.M.A. Hussain et al. (Eds.): IMTIC 2008, CCIS 20, pp. 253–260, 2008.
© Springer-Verlag Berlin Heidelberg 2008

painless, portable cheaper and free from ionizing radiations. It is operated dependent [5]. In this context, Artificial Intelligence is being used to gather data through sensors and transducers which converts mechanical energy into electrical energy and then into sound waves [6].

Our objective was to study different patients from the different areas of the region to compare the growth rate of the vegetarian and non-vegetarian patients. This study has been made in the perspective of to identify malnutrition for the patient to attain the American Standards of the foetal weight growth in the vivo.

## 2  Material and Methods

The Ultra Sonography Model-SSD 500 shown in Fig. 1 has grey scale of 64 measures an image of the patient on the basis of sound waves. The interference of sending and reflected waves produces an image of the foetus on the screen under investigation. This is actually a relationship of frequency versus time.

The Tables I and II represents the Biometric Data collected from the patients. This quantitative data is based on the vegetarian and non-vegetarians foetal growth in vivo. The experts of ultrasound mostly collect scan data manually and feed it to the memory for reproducing on the screen for analyzing the image.

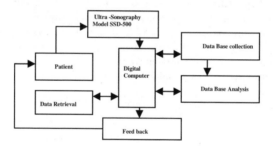

**Fig. 1.** Model presents data retrieval processing and analysis of biometry data

On the other hand other history of the patient is collected by data retrieval system and recorded into the computer. Later on the data is collected and saved in database and through interactive process the output is analyzed for further study of socio-economic status wise growth in the society. In the end for further study for socio-economic feedback is collected for continuous study of biometric data of the foetal growth.

## 3  Findings

The Table I illustrates, the data of twelve random patients out of thirty patients. This is done for the comparison of biometric data on the basis of food habits of the patients. This data is focused on middle class of the patients. The underlying table high lights the biometric data of the various patients who were referred belongs to areas refereed at ultrasound clinic.

**Table 1.** Shows Biometric Data of randomly selected patients

| Patient # | Age | Social Class wise status 1HC 2MC 3LC | Veg./non-Veg Food Habits NV | V | LMP in weeks | BPD (cm) | FL (cm) | AC (cm) | G/A (weeks) | FW (Kg) |
|---|---|---|---|---|---|---|---|---|---|---|
| 1 | 35 | 2 | 0 | 1 | 22 | 4.9 | 3.5 | 15.4 | 20 | - |
| 2 | 25 | 2 | 0 | 1 | 18 | 4.1 | 2.4 | 11.3 | 17-6 | - |
| 3 | 26 | 2 | 0 | 1 | 16 | 4 | 2.4 | 10.4 | 17-5 | 0.194 |
| 4 | 27 | 2 | 0 | 1 | 9 | 4.5 | 3 | 13 | 10 | 0.279 |
| 5 | 35 | 2 | 0 | 1 | 34 | 8.4 | - | 27.5 | 35-4 | 1.9 |
| 6 | 35 | 2 | 0 | 1 | 33 | 8.8 | 6.1 | 27.5 | 34 | 2.1 |
| 7 | 30 | 2 | 1 | 0 | 39 | 9.4 | 7.3 | 32.7 | 39 | 3.1 |
| 8 | 26 | 2 | 1 | 0 | 29 | 7.7 | 5.4 | 22.5 | 30-2 | 1.2 |
| 9 | 28 | 2 | 1 | 0 | 37 | 8.4 | 5 | 22 | 30 | 1 |
| 10 | 27 | 2 | 1 | 0 | 22 | 5.2 | 3.6 | 15.6 | 21-2 | 0.418 |
| 11 | 26 | 2 | 1 | 0 | 34 | 8.8 | 6.9 | 27.6 | 35 | 2.2 |
| 12 | 34 | 2 | 1 | 0 | 34 | 8.1 | 6.6 | 29 | 33 | 2.1 |

## 3.1 Equations

Here are two equations which have been used for the analysis of biometric data for the middle class socio-economic groups.

$$GR_{Avg} = (GR_1 + GR_2) / 2 \qquad (1)$$

where GR stands for Foetal Growth Rate.

$$GR = FW / GA \qquad (2)$$

where, F.W and G.A stands for Foetal Weight and Gestorial Age respectively.

Fig. 2 shows the minimum and maximum age of patients in between 25 to 35 approximately.

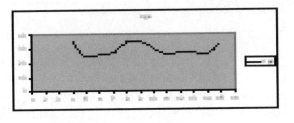

**Fig. 2.** Represents number of patients versus Age

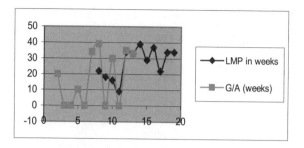

**Fig. 3.** Represents number of patients versus Age

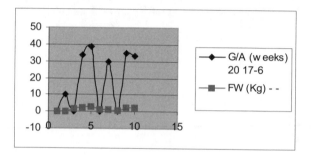

**Fig. 4.** Represents F.W versus G.A

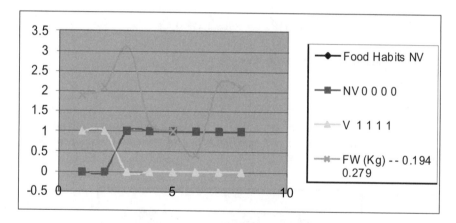

**Fig. 5.** Represents FW versus Food Habits

Fig. 3 shows G.A according to LMP and Ultrasound. The observation shows that there occurs averagely one-week difference. It should be noted that LMP is obtained from the patient, while G.A is measured from the biometric study using Sonography. Generally, the difference belongs to 1 to 3 weeks. Where as, on the basis of the vegetarians and non-vegetarians patients, it comes out to be one-week averagely. Fig. 4 depicts that as G.A increases then the weight increases accordingly.

In Fig. 5, we have analysed the weight of the Foetal Growth with respect to Vegetarian and non-Vegetarian patients. It shows that vegetarian mother grows with some higher weight babies as compared to the non-vegetarian. This is the first time of its kind that biometric study in our country has been undertaken for research.

Fig. 6, shows various Biometric Data collected from the patients under investigations and depicts that younger mother grows healthier Foetal Weight, as compared to old age mother.

Figure [7], [8] illustrates comparative typical Foetal Growth Weight of the (vegetarians) and (non-vegetarians) patients against the International Standard Weights for Foetal in vivo [7].

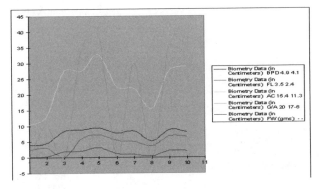

**Fig. 6.** Represents age of patient's vs. different biometric data

**Table 2.** Foetal Weight Compared with American Standards

| Patient No. | Patient Type | | | G/A | Fetal Weight | G/A standard | American Standard | Diff. |
| | Vegetarian | Non-vegetarian | LMP In weeks (Approx) | In weeks U/s (Approx) | (Kg) | (Weeks) | (Mean) (Kg) | Wt (Kg) |
| --- | --- | --- | --- | --- | --- | --- | --- | --- |
| 1 | 1 | 0 | 22 | 20 | 0.305 | 20 | 0.392 | 0.087 |
| 2 | 1 | 0 | 18 | 18 | 0.205 | 18 | 0.259 | 0.054 |
| 3 | 1 | 0 | 16 | 18 | 0.194 | 18 | 0.259 | 0.065 |
| 4 | - | - | - | - | - | - | - | - |
| 5 | 1 | 0 | 34 | 36 | 1.9 | 34 | 2.431 | 0.531 |
| 6 | 1 | 0 | 33 | 34 | 2.1 | 33 | 2.225 | 0.125 |
| 7 | 0 | 1 | 39 | 39 | 3.1 | 39 | 3.552 | 0.452 |
| 8 | 0 | 1 | 29 | 30 | 1.2 | 29 | 1.483 | 0.283 |
| 9 | 0 | 1 | 27 | 28 | 1 | 27 | 1.169 | 0.169 |
| 10 | 0 | 1 | 22 | 21 | 0.418 | 22 | 0.562 | 0.144 |
| 11 | 0 | 1 | 34 | 35 | 2.2 | 34 | 2.431 | 0.231 |
| 12 | 0 | 1 | 34 | 33 | 2.1 | 34 | 2.431 | 0.331 |

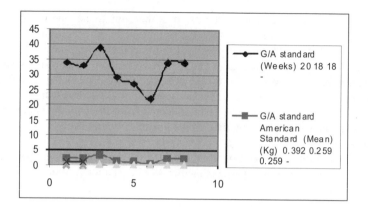

**Fig. 7.** Represents Standard F.W versus Mean across Veg./non-Veg. patients

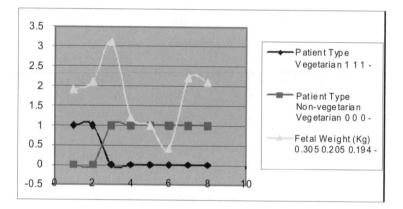

**Fig. 8.** Represents Foetal Weight versus. veg. and non- Veg. patients

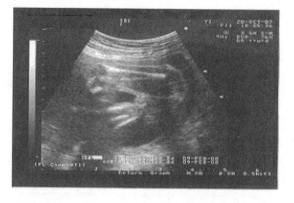

**Fig. 9a.** Graph shows the measurement of femur length

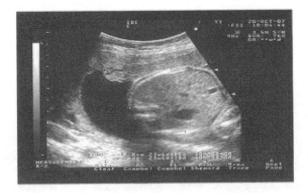

**Fig. 9b.** Graph shows the measurement of abdominal circumference

Our recent study has compared the data of commonly used Foetal parameters such as BPD, AC and FL obtained from socio-economically middle class population. In the third trimester, the growth of the BPD slows down and relies on the other parameters such as Head Circumferences, the Femur Length, the abdominal circumference, and the ratio of head and abdominal circumference (Fig. 9a and 9b). Its values are based on the fact that in the mid trimester is linear and rapid, and the biological variation at each week of gestation is small. A map measure is run round the outer margin of the abdomen, or the sum of 2 measurements taken at right angles to each other on the foetal abdomen is multiplied by 1.57 to obtain the AC.

## 4  Discussion

In the present study, the most of the patients who were referred belong to the middle class towards social status. The study can be further extended to the persons with high Social Economic Group, the private hospitals and to very low social economic group in Rural Health Centres (RHC) and Basic Health Units (BHU). The biometric is not only helps in measuring its results, but its implication towards social setup is great. The sex of the foetus is not under consideration and investigation.

## 5  Conclusion

Our findings indicate that Standard Foetal Weight Growth Rate is substandard as compared to International Standards in our country. The results show that vegetarian's women have more growth rate for their embryo at trimesters as compared to the non-vegetations of the same period for middle socio economic group for closely same G.A and actual age. The Ultra Sonography is considered to be painless and safe for the patients at some stage in collecting the biometric data as compared to other devices. We expect in near future, there will be a very close coordination between various clinical specialities and IT Departments using telemedical links for collection of biometric data from remote areas for the investigations of R and D projects in the field of e-health for the developing countries.

## Acknowledgments

Professor BS Chowdhry acknowledges support of Higher Education Commission, Islamabad for his postdoctoral fellowship at the School of Electronics & Computer Science, University of Southampton, UK.

## References

1. Liebman, M.N.: Bioinformatics. In: Proc. of 23rd Annual EMBS International Conference, Istanbul, Turkey, pp. 4063–4064 (2001)
2. Eyal, S., Ilana, S., Jacques, S.A.: What do Clinical Users Know regarding Safety of Ultrasound during Pregnancy? J. of Ultrasound Med. 26, 319–325 (2007)
3. Eberhard, M.: Ultrasound in Obstetrics and Gynecology, pp. 12–14 (2005)
4. Skovron, M.L., Berkowitz, G.S., Lapinski, R.H., Kim, J.M., Chitkara, U.: Evaluation of Early Third Trimester Ultrasound Screening for Foetal Growth. J. of Ultrasound in Med. 10, 153–159 (1991)
5. Rogers, C.S.: Clinical Sonography, 3rd edn. A Practical Guide. Lippincott Williams & Wilkins (2006)
6. Sonnenwald, D.H.: Applying Artificial Intelligence Techniques to Human Computer Interfaces. IEEE Communications Magazine 26 (1988)
7. Ott, W.J.: The Diagnosis of Altered Foetal Growth: Obstetrics and Gynecology. Clinics of NorthAm. 15(2), 237–263 (1988)

# Bioinformatics Awareness System for Skin Diseases Using Stem Cells Therapy in the Perspective of eHealth

Narinder Chowdhry[1], Wahab Ansari[2], Aftab A. Memon[3], and Fatima Mir[4]

[1] Department of Biomedical Engineering, Mehran University of Engineering & Technology (MUET), Jamshoro, Sindh, Pakistan
[2] Institute of Information Technology, University of Sindh Pakistan
[3] Department of Telecommunication Engineering, MUET, Jamshoro, Pakistan
[4] Research Student MUSTID-MUET, Jamshoro, Pakistan
{chowdhry.np, aftab.memon}@muet.edu.pk, fjancoool@yahoo.com

**Abstract.** In this paper, we have developed a bioinformatics system for skin therapy through stem cell implantation. The purpose is to make use of an eHealth programme. Although some experiments on utilizing universal stem cells have been successfully implemented in the movement of a diseased leg, more experiments will be performed in Pakistan with heart and liver disorders. Our study focuses on the need for physicians or surgeons to induce skin stem cells, and the study of different skin diseases spread through hides and its by-products.

**Keywords:** Bioinformatics, skin therapy, stem cells, biomedical engineering and eHealth.

## 1 Introduction

It has been observed that skin diseases are spreading day by day, particularly in the small cities and generally in big cities of Pakistan. For instance, Hyderabad has a great hide industry, which is popular for its by-products. This has increased economic benefits, but not without the consequences of producing skin diseases. In this municipality, the skin disease is expanding exponentially and the public is unaware of consequences of this threat. However, skin specialists are advising people to take preventive measures in making use of new technologies and services such as skin therapy through stem cells and to establish teleconsultation links via eHealth. As far as skin therapy through stem cells is concerned, the universal stem cells, osteogenic stem cells, mesenchymal stem cells and angiogenic stem cells implantation has been successfully implemented for assisting a patient at Gambat, in the district of Khairpur of Sindh, in increasing the mobility of the diseased lower leg [1]. The patient operated at Gambat had an unhealed fractured leg and had undergone 12 operations in 12 years. His only option was amputation. With this transplant, he is now walking and has 80% regeneration of muscle, bone and cutaneous tissue [2]. This is encouraging that this technique will be beneficial for the treatment of skin diseases. The Government of Pakistan is interested in providing eHealth services to the rural areas people for instant teleconsultation services [3].

D.M.A. Hussain et al. (Eds.): IMTIC 2008, CCIS 20, pp. 261–267, 2008.

Presently, a variety of stem cell applications are under investigation in Pakistan, for example, Thalassaemia, damaged liver recovery, heart and tissue problems. There is an urgent need for skin therapies for growing skin stem cells to remedy spreading skin diseases.

## 1.1 Bioinformatics

Bioinformatics provide information about genomes, which can be exploited with fields of natural sciences including computer science, mathematics, physics and biology [4]. This area of science that collects molecular information at the levels of genes and genome to identify the characteristics of universal stem cells to build and creates 220 basic cells from the human body [5]. This needs scholars in the field of molecular biology such as physicist, computer scientists, engineers, pharmacists and medicine peoples to develop the database for the care of general public use in the perspective of eHealth [6]. Further, it is helpful that DNA sequence data can be collected from internet information at related sites for any tissue or cell required for comparison and analysis of the sample under investigation. This area has specified direction to the network of computer to share genome data for analysis on any particular cell or implantation of results to improve the quality of its use in the unforeseen future. For the analysis and interpretation of genome DNA sequencing databases are given in reference [4], [7] and areas of implementations are clinical disorders, listing disease susceptibility, genetic mutations and polymorphisms. Recently, bioinformatics is being utilized in the field of gene variations and their analysis and prediction of gene and protein structure and function and detection of gene regulation networks and cell modeling [5], [8]. In the future computational biology will be based on clinical level for the treatment of skin care just as repair of hair, kidney and liver disorders.

## 1.2 Origins and Development of Stem Cell

The stem cell is supposed to be transformable to another cell property to acquire the desired cell out of 220 characteristics cell. This can be said as a building block for repairing cells of the body, such as skin, hair, heart, and kidney. According to the categories, stem cells have been classified into three types such as Totipotent, Pluripotent and multipotent stem cells [9]. The origin of the stem cell begun in the mid of the 18th century, when it was known that some cells can produce other cells. Later on, in the beginning of 19th century, it was discovered that some blood cells reproduced other blood cells termed as stem cells. In 1950s in France, radiated people were treated on their bone marrow discovered the immune system in the body. The first identical cell transplant were made in 1960s on identical twins. Next, in 1984 non-identical bone marrow was successfully performed in America. In 1998, James Thompson developed the first embryonic cell and John Gearhart derived germ cells from cells in the fetal gonadal tissue. With reference to [10] the stem cells can be categorized as embryonic and non-embryonic, the first type of stem cells are taken from umbilical cord blood and others are taken from the skin, bone marrow, hair, brain cells and body fat [11]. The work of the famous scientist James Thomason (1998) of the University of Wisconsin at Madison provoked that "The induced cells do the entire things embryonic stem cells do" [10], [12].

The editor is of the view that the treatment of diabetes, heart disease, neurological disorder and Parkinson's and motor neuron disease is hope or hype [12] but he is optimist that research in this area will be fruitful one day. Recent reports discovered [daily dawn] that two groups of scientists have successfully produced stem cell from skin cells, thus it will help foundation cells to make remedy for replacing diseased or damaged tissues or organs. Still it is new but with passage of time it will be a magic bullet of treatment for various deadly and weakening diseases. However, the blank cells of the 220 types of cells can be created if there is no risk of mutation or restriction on ethical grounds in the human body. Further the Japanese scientist claimed that all body cells could be created from stem cells except placenta.

The latest news communicated in daily Jung newspaper dated 8th December, 2007 [13] expresses that stem cell taken from umbilical cord is an ethical problem and the president of America stop funding this research because of opposition from different religious sect. Recently, on the other hand, stem cell taken from skin has less ethical problem and above mention president of America, George has no objection to release funds for stem cells from skins for research. Now stem cell taken from skin cells as devised by Japanese Prof. Shinyayamanak from Kyotd Univ. is encouraging and has know ethical problem as it was take the stem cell from embryo, similarly research performed by American scientists James Thomson, Junyingya has successfully created stem cells having properties of other usefully cells and 14 genes initially and later on basics four genes that can help creating any damaged tissue human body, like heart, kidney, brain, stem cell, these genes has the property when embedded with DNA of skin cell grow and observe the property of the skin where these cells are injected.

### 1.3  Role of Biomedical Engineering

As, the biomedical engineering is playing a vital role in connecting IT-medical science and Engineering together and the immense knowledge in this field have provided a centre of focus at specific field. The bioinformatics is a branch of science, which deals in collecting biological information using computer graphics and modelling for quick visualization and assessment of biological cells, for the evolution of growth through a span of time.

### 1.4  eHealth and Skin Therapy

eHealth is information technology enabled service to the society; the larger network from LAN to WAN has made it possible approach of a patient for a healthcare through computers or mobile devices. The use of eHealth has become an umbrella for bioinformatics, biometry disease diagnosis & prompt health advises through wired and wireless systems. We are expecting for the diagnosis of skin diseases at low level and at high level for genome information collection of every persons keeping record of genetic finger prints at birth and can predict health risk for future use of the concerned person [14]. In this reference a statement of Professor Jean Claude Healy of the world health organization is quoted as that "e-Health is the instrument for productivity gains in the context of existing healthcare systems, but also provides the backbone for the future citizen centered environment". The healthcare is gaining importance because of the fusion of biological sciences, engineering and technologies together for the use of solution of medical problems for general public healthcare.

## 2  Issues and Solutions

### 2.1  Facing Issues and State of Art in Pakistan

The awareness of skin diseases and solution for it through internet is possible by the use of telemedicine and eHealth care. For this purpose clinic in the main city and rural health centres at small cities and villages can play its role through wireless communication systems. The focal point is to make network for biological computational network and keeping databases available for analysis, diagnosis and treatment of patients. As increasing biological and genome information will be widely used in gene therapy for skin diseases at its lowest use level and less harmful for the use of universal stem cells as the case of treatment for mobility of legs for almost incurable diseased leg. This is an issue of the day and we have to take much care for the use of stem cells in the presence of ethical problems. Further issue is that proper measures should be taken for updating the biometric data of the patients for the use of pharmaceutical companies and medical doctors to diagnose and information & communication technologist to design new software to break through for forthcoming problems of the use of genomic data.

### 2.2  Solutions and Planning

Our goal of this research is to review scientific and computational knowledge in the area of bioinformatics, skin therapy through stem cells technology in order to improve the socio-economic effects of the society. On the other hand, this knowledge will be beneficial for patients, physicians or surgeons and pharmaceuticals companies to establish tri wheel connection as a continuous vehicle for the society in the field of e-health. In this context, we have started to develop a prototype based model initially for making interaction between patients, doctors and diagnostic laboratories, Moreover, for patient-physicians consultation is possible through videoconferencing and DSL of 256 Kbps bandwidth, which is now available in big cities through PTCL connections and through remote areas through wireless connectivity.

## 3  Methods

The approach of our study would be creating database of skin diseases, diagnostic system, online consultation for the treatment of skin diseased patients, and readily availability of medicine from pharmaceutical companies. Through teleconsultation, drugs would be provided to the particular patients through pharmaceutical companies on instant service basis. Since, umbilical cord blood stem cell are creating ethical problems so, more safe side is to consider skin stem cell for the treatment of human skin diseases. Our approach is qualitative initially and after, analysis of the quantitative data, the reporting will be made in the future.

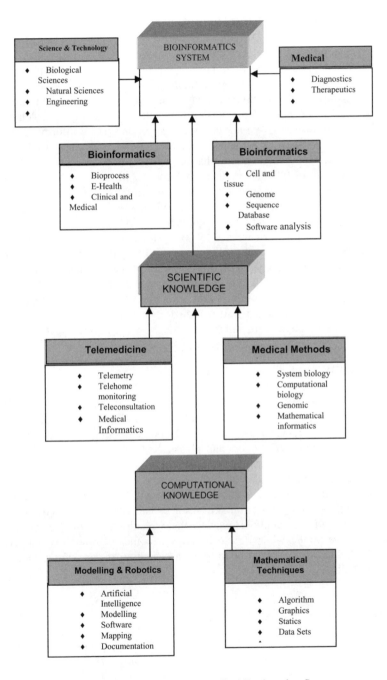

**Fig. 1.** Research Model for Biomedical Engineering System

# 4 Research Model for Bioinformatics System

Fig. 1 presents a proposed model of the bioinformatics research diagram as described below. This research model has three layers; first bioinformatics system's layer consists of four sub links, the science & technology, medical, bioinformatics tools left and bioinformatics tool right. Since we have focused our research of bioinformatics related medical in the field of diagnosis, therapeutics, by using stem cell therapy, especially for skin diseases. We are considering cell and tissue information through genomic data in the form of sequence database for the analysis and awareness of information for patients, doctors, and pharmaceutical companies. On the other hand, the left side diagram of this part by using bioprocess, where we collect eHealth data for clinical and medical use. Secondly, we are expecting scientific knowledge layer, of the system biological knowledge by using computational biology; particularly for the purpose of genome information through mathematical modeling.

On the left side of this part we use telemetry for telemedicine and pharmacy genomic information for pharmaceutical companies to prepare new drugs. On the other hand the algorithms are designed with the assistance of the graphics, static's, artificial intelligence and database for the awareness programs. While, on the other side software could be designed on bioinformatics awareness by using modeling, mapping, genomes and collecting updated information for general use. This model has been designed in the perspective of the reference [15].

# 5 Conclusion

The production of stem cells from skin is encouraging and can safely be used for skin therapy. These emerging technologies including bioinformatics, biomedical engineering, and eHealth in its integrated form, can be a foundation for awareness information for the public, medical practitioners and pharmaceutical companies. In this research paper, a database for genome sequences and skin therapy and stem cells implantation has been discussed. Further, the implementation of teleconsultation has been suggested and the socio-economic impact on society will grow. Thus, more ethnic-free solutions for taking stem cells from skin itself or hair for the growth of desired cell would gradually attract people. However, these resulting curing effects would be beneficial for the public when this technology becomes common. We suggest that patients DNA data banks should be created in big cities of developing countries for ready availability.

# References

1. Rao, I.A.: New Horizons for Stem Cell Science (2007), http://www.pakissan.com/biotech/techarticles/new.horizons.for.stem.cell.science.shtml
2. Nasir, A.: Telemedicine Systems in five more Districts Soon, January 27, 2007, The Daily Dawn Newspaper (2007)
3. Mangi, M.H.: Stem Cell: An Essential Biotechnology Tool. The Daily Dawn News. Science Tech-World (2004)

4. Ardeshir, B.: Clinical Review Science. Medicine and Future. Biomedical J (BMJ) 324, 1018–1022 (2005)
5. Liebman, M.N.: Bioinformatics. In: Proc. of the 23rd Annual IEEE EMBS International Conference, Istanbul, Turkey, pp. 7211–7215 (2000)
6. Chowdhry, B.S., Rajput, A.Q.K., Chowdhry, N.P.: Major Challenges and Opportunities in the Practice of Research and Development in the Area of Bioinformatics. In: 5th Annual International Conference on Role of Medical Universities in Promotion of Research, p. 70. Liaqat University of Medical & Health Sciences (2007)
7. Venter, J.C., Adams, M.D., Myers, E.W., Li, P.W., Mural, R.J., Sutton, G.G.: The Sequence of the Human Genome. Science 291, 304–1351 (2001)
8. Douglas, A.L.: Biological Engineering and Systems Biology new Opportunities for Engineers in the Pharmaceutical Industry. In: Proc. of the 26th Annual International Conference of the IEEE EMBS, San Francisco, USA, vol. 2, p. 5456 (2004)
9. Paulo, A.F., Angus, W.T.: Where are We Going? Stem Cell Technology. Biomedical J (BMJ) 319, 1308 (1999)
10. Nasir, A.: Scientists Transform Human Skin Cells into Stem Cells. November 21, 2007, Daily Dawn Newspaper (2007)
11. http://www.allaboutpopulariissues.org/history-of-stem-cell-research-faq.htm
12. An Editorial. Stem Cell Therapy: Hope or Hype? Biomedical J (BMJ) 319, 1308 (1999)
13. Nayar, Z.: A report: Global Village, 8 December, 2007, The Daily Jang Newspaper, pp. 15 (2007)
14. Richard, J., Richard, K.: Medical and Technological Future Convergence. In: Physicians Computer Connection Symposium, San Diego, California, p. 16 (2006)
15. Helen, K., Adrianne, M.: Recombinant DNA and Biotechnology: A Guide for Teachers, pp. 5–15 (2001)

# How to Build an Open Source Render Farm
# Based on Desktop Grid Computing

Zeeshan Patoli, Michael Gkion, Abdullah Al-Barakati,
Wei Zhang, Paul Newbury, and Martin White

University of Sussex, Department of Informatics
Centre for VLSI and Computer Graphics
Brighton, United Kingdom
{M.Z.Patoli, A.Al-Barakati, M.M.Gkion, W.Zhang, P.Newbury,
M.White}@sussex.ac.uk

**Abstract.** This paper presents an experiment on how to implement a Grid-based
High Performance Computing solution using existing resources typically available in a teaching or research laboratory. A cost-effective solution is proposed
based on open source software components, and, where appropriate, our own
software solutions, for large scientific applications in the public sector such as
universities and research institutes. In such institutions, classical solutions for
HPC are often not affordable, yet they usually have at their disposal a large
number of machines that can be utilised. The Department of Informatics at University of Sussex, for example, has just installed 150 new Core2 Duo machines
across 3 laboratories. By scaling this number up across the whole University, it
can result a large potential computing resource for utilization. Typical processor
usage rates are often somewhere between 10% and 20% (i.e. user-generated
processes) for most machines. This paper proposes a solution that exploits the
remaining 80% to 90% processor power through consumption of available
computer idle time without disturbing current users. To achieve this goal, the
open source Condor High Throughput Computing software was selected and
implemented as a desktop Grid computing solution. This paper presents our experiences in finding a solution so that other institutions can develop similar
Grid solutions for their own large scientific experiments, taking advantage of
their existing resources. The implementation of our solution is analyzed in the
context of building a render farm.

**Keywords:** CPU scavenging, Desktop Grid computing, Render farm, Open
Source, Blender Render Farm.

## 1 Introduction

Many research institutions such as universities, and colleges, etc. have, or would like
to implement, projects (research or students projects) that require high performance
computing (HPC) environments. However, due to limited funding opportunities, it is
not efficient to afford such an investment in classical HPC solutions such as those
offered by IBM, Sun, Intel, ClusterVision and others. On the other hand, most institutions probably have many hundreds of Pentium IV, Dual core, or Core2 Duo

D.M.A. Hussain et al. (Eds.): IMTIC 2008, CCIS 20, pp. 268–278, 2008.

machines installed on their campuses that are used for limited computing operations, i.e., programming small applications in teaching laboratories, working on office automation types of jobs, e.g. word processing an essay. For the analysis of these resources' usage against their cost, it is important to take under consideration that the computing resource is more expensive than the actual computing utilization by a considerable degree. Therefore, it would be useful to get better value for money by utilizing these machines more effectively.

There are several ways to achieve this in a university environment. First of all, it is important to make sure that the necessary machines are available 24/7 to students. However, this does not guarantee 100% utilization and involves other costs such as environment security needed to ensure open access to laboratories 24/7; this is one end of the utilization scale. Second, different ways of consuming unused process power during machine idle times can be applied. There are also two issues to consider here; a) the cost of developing an infrastructure (e.g. networking, software licensing, technical expertise, etc.) to run a scheme to recover such machine idle time, b) the choice, design and implementation of a solution – in short there is a need to implement a solution, and then evaluate the results.

In this paper (at this stage), the actual implementation issue is more crucial because we will run the solution within the confines of the research laboratory utilizing some 40+ existing machine (private network), and thus should have no impact on external resources. All researchers will manage and run the desktop Grid computing solution for their own research use. In order to have an implementation focus, and because the research group is concerned with research utilizing computer graphics, it was to decided to use as an example HPC application the need for a render farm to process large computer graphic animations.

Building large-scale animations needs three main components:

1. 3D authoring package such as 3ds max, Maya or Blender to build the computer animation
2. cluster of computers on which to distribute the animation rendering jobs.
3. job submission application to distribute and recover the rendered results.

Based on open source software as much as possible (with any extra software designed by our self being made available as open source), Blender [3] was selected as our 3D authoring package and the Condor High Throughput Computing solution [8] for implementing the animation rendering computing solution. Finally, the effort of designing a web service based solution for linking Blender to Condor is in process.

Because we intend to utilize a number of 40+ machines (20 old machines organized as a cluster, 8 newer machines organized as a cluster and the rest being machines used for other individual research purposes) the solution was applied as a desktop Grid computing environment that employs CPU cycle scavenging. This is a process that allows developing an HPC environment by installing a small piece of software on each machine, which should always be running, and monitoring for the CPU to become idle. When it is idle, that computer's power can be used for many other applications, and when the user starts to work on their computer, the software immediately suspend the Grid application to allow the computer users foreground application to retake control of the computers resources [1].

In addition, the application software needs to be deployed on each executor machine in the desktop Grid computing environment. During the specific experiment, a complex 3D animation of a Roman Villa in Blender was developed and rendered on the Condor based Grid.

This paper is organized as follows: section 2 describes some background to this experiment, section 3 describes how to set up the desktop Grid computing environment, section 4 discusses briefly the tools used to create the animation file, section 5 discusses how to make the job submission for the animation to the desktop Grid computing environment, and section 6 discusses the results followed by some conclusions in section 7.

## 2  Background

### 2.1  Related Work

There have been several research projects related to CPU scavenging, such as SETI@Home which is well known as an experiment based on volunteer computing, developed at the University of Berkley for the analysis of radio telescope data using idle machine power from volunteer users on the Internet. This works by allowing the user to install the BONIC [4] software on their computer so that when their machine is idle it can be used as a part of distributed computing environment [2]. Other researchers have completed similar type of work using Sandbox technology where they could create a virtual machine with Linux OS on physical machine with Windows OS and install a piece of software like a screen saver to harvest the idle CPU cycles [1]. This experiment uses a similar approach for absorbing idle time. However, it is based on Condor and other freeware and open source tools for developing a local desktop Grid computing environment utilizing the institution's existing resources for the institutions benefit. It is worth noting that there are many commercial and open source solutions for implementing a desktop Grid solution, and one has to make a choice based on the advantages and disadvantages of each solution with respect to their Grid application. For example BONIC could be selected for building a render farm, but there is so far little use of BONIC in this application area. On the other hand, there are commercial versions of render farms that utilize Condor. Given that commercially Condor is used in this field, it seems reasonable to see if we can build an open source version.

The basic difference between the specific desktop Grid computing environment and solutions such as SETI@Home, which are based on volunteer computing, is that the team decided to build a solution for their institutions own benefit, focused on running their own large scale applications, on computing resources managed similarly by the team thus trusted, i.e. not needing any redundancy, and this solution does not need any foreground software (e.g. a special screen saver) to execute the application, and the installation of the Grid solution can eventually be automated [4]. Other research work being done that is similar to this approach is that done by Monash University where they have implemented the Monash Sponge facility. Here, they use the Condor like us for harvesting the idle CPU cycles to execute C, C++, FORTRAN 77 and Globus jobs [5].

## 2.2  Alternative Rendering Solutions

There may be many solutions for any rendering animation problem but the adoption of the right solution depends both on requirements and budget. If it can be affordable, existing online commercial render farms can be used like: Rendercore [7], ResPower super / farm rendering [6], and many others, who offer their services for hundreds or thousands of dollars. If, of course, the results of the selected rendered animation are not satisfactory, it is necessary to re-submit them, simply to pay them again. The user usually submits the files, and subsequent results can be downloaded within the next few hours—cost being related to rendering time in some way. Another solution, rather than using a commercial render farm for multiple animations, it could be achieved by buying a latest Blade system from Intel, AMD, IBM or BOXX Technology and installing a rendering solution on it. This solution is cost effective and requires many thousands of dollars. Other solutions include setting up a 3ds max Backburner based render farm on existing machines [15]. However, we have tried the Backburner render solution in a teaching lab and it seemed that as soon as a job is submitted the server.exe software that was installed as a service became a foreground service when it was executed, and did not seem to scavenge idle cycles—it is worth noting that BOXX Technologies offer a dedicated render farm Blade based solution with 3ds max's Backburner.

## 2.3  Why Use Condor?

As mentioned above, there are many solutions for implementing a desktop Grid computing environment. However, Condor Render was preferred for the Grid based render farm for several reasons: first, it consists of a cost effective solution with zero software licensing cost (which eliminated all commercial options, obviously); second, due to the fact that most of the public sector institutions like universities use Windows operating systems on most machines (but a few labs use Linux), it was important to develop a Grid solution that operated on Windows, but could also include Linux if necessary; third, many existing 3D authoring tools like: 3ds max and Maya also work on Windows only, and others like the open source Blender work on both Windows and Linux.; and finally, as mentioned above, some early indication that the Grid software chosen was adaptable to the render farm requirements was useful. This last requirement was actually the most important factor because as there is a commercial version of a render farm that uses Condor called GDI|Explorer and a freeware version called GDI|Queue (albeit using 3ds max 6 – which is well out of date).This was the main reason to conclude to the solution of Condor for this experiment.

   Taking into consideration that Condor operates on many platforms including Windows and Linux, team's department uses 3ds max and Maya in most cases (although for a completely open source solution Blender was preferred for this experiment), and there were sufficient freeware (e.g. GDI|Queue, Blender_Render) and open source (e.g. Condor Render for Maya) utilities to link the two (3D authoring tool and compute clusters) together in order to produce a first prototype solution. This initial prototype solution was tested and the results are discussed below. The next step is to build the functionality given by these separate utilities (e.g. GDI|Queue, Blender_Render, etc.) into a single web services based application to link Blender (or 3ds max or

Maya) with the Condor render farm. It is also worth noting that a service orientation approach was adapted for the design of this desktop Grid computing environment, and that the latest version of Condor uses web services.

## 3  Setting Up the Desktop Grid Computing Environment

In this desktop Grid computing environment for initial testing purposes, it was used a limited number of dedicated machines configured with Condor and Blender applications running on them. For the future we plan to add 3ds max and Maya as 3D authoring packages to the Condor render farm, and incrementally add extra machine to the compute cluster.

### 3.1  Hardware Specification

For this initial Grid setup, six old machines discarded by the department of Informatics teaching labs were used; four are Pentium IV Duo Core with the clock speed of 2.80 GHz and two are Pentium IV single core machines with clock speed of 2.80 GHz, all with only 512MB RAM to act as compute or executer machines. In addition, an old Xeon server 2.40 GHz machine, which was already available in lab with 1GB of RAM, was used. These were connected with an old 32 port router discarded from the same teaching lab and some of the UTP cables. The NIC cards in all systems were 100Mbps so they operated on 100Mbps network. All equipments used in this experiment were no longer used by department, thus there effective cost to the research experiment was zero.

**Fig. 1.** Image showing the 7 machines used for the experimental desktop Grid

### 3.2  Desktop Grid Computing Environment Render Farm Setup

The setup of the desktop Grid render farm solution required three main components: a **Central Manager**, a job **Submitter** and multiple **Executers** or servers. To test the Condor Grid middleware it was possible to install all components on a single machine. But for an actual render farm solution, more than one executer could be used.

For the first test, six old machines were used as described above for the **Executers**. The functionality of each component can be described as:

1. The **Central Manager** of the desktop Grid is a machine that holds the scheduler for the Grid that will distribute the render jobs to different executers (sometimes called servers). To operate Condor, the user needs to install the Negotiator, Collector and Master components on the machine acting as the central manager.
2. The job **Submitter** is the machine that can submit the render jobs to desktop Grid. To operate Condor correctly a job submitter machine must have the Scheduler and Master components installed.
3. The **Executer** machines each act as a server that gets the job from the Central Manager and executes them. These machines must have Start and Master components to provide the services. These machines must also have the application component installed, e.g. the Blender rendering component.

In the actual setup, a dual processor server was used to act as both the Central Manager and job Submitter, and the other six machines for Executers. Adding more Executers could be easily done by adding more machines to the desktop Grid. For example, there were 26 more ports left on the network switch.

In this experiment, it is worth considering issues such as space resources to house lots of old machines, or power consumption, or heat generated by lots of machines, etc. This is part of running the infrastructure issue. However, we happen to have spare rooms, which could house many old machines which are fully networked in the research lab space. Using old machines (that the department discarded after 4 years for whatever reason) for building a render farm in this way was a cost effective solution because while the department may not have the resources to continue maintenance on them, research students may be able to do this more effectively because they do not have any quality of service issues to worry about—if a machine in the desktop Grid dies, replacement is not crucial and its repair can often be effected by cannibalizing other discarded machines. Also, when a machine finally and completely dies, it can usually be replaced with another one from the departments 'discarded pool' of machines when they renew teaching lab machines on a cyclical basis.

This approach has 'green computing benefits' because your institution can get more use out of its older machines by passing them on from teaching labs internally to the institute rather than passing them on to other outside organizations, which is common practice. Machines greater than 4 years old generally have limited value, but the disposal costs would have to be borne by the giving or receiving organization, and nowadays this in not inconsequential due to environmental restrictions placed on disposal of electrical equipment. Institutions', when they pass machines on, will thus incur costs in this process, so it may be better to literally 'run them into the ground' thus getting maximum use of the processing power before incurring unavoidable disposable costs.

### 3.3  The Condor Software Installation

The installation of condor on the Windows operating system is not as difficult as other Grid tools, which is another reason for choosing Condor. However, we also tried out Globus (Linux based) and the 3ds max Backburner solution mentioned above. Condor can be downloaded free of cost with its instruction manual, etc. from University of Wisconsin website for different operating systems [8]. The condor has nice user-friendly installation wizard which takes no more than five minutes to install on single machine and then some changes must be made in the Condor_config file for proper configuration. It is important that the following changes are applied in the Condor_config file after installing it on Windows.

1. Select TESTINGMODE instead of UWCS mode set in default.
2. Collector name should be the name of central manager.
3. Hostallow_Read should be set to all names of machines which can view the status of computers' pool.
4. Hostallow_Write should be set to all names of machines which can join the pool and can submit the job to it.
5. Comment out the fields NEGOTIATOR_PRE_JOB_RANK. NEGOTIATOR_ POST_JOB_RANK, and MaxJobRetirementTime.

### 3.4  Software Requirements for Rendering

For a complete open source solution, each machine acting as an Executer must have Blender installed on it. However, for rendering 3ds max files, it would be possible to install 3ds max Backburner, etc., on each Executer.

## 4  3D Animation

For our solution we used 3ds Max, Maya, and Blender for developing complex 3D models and animations. For example, Figure 2 illustrates the use of Blender to create initial animation file for the Roman villa. During the initial setup of the render farm, it

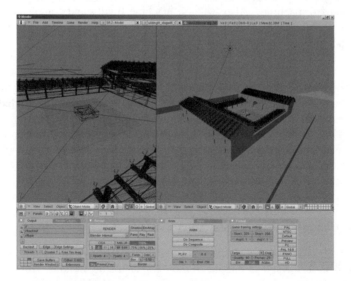

**Fig. 2.** The Roman villa using Blender 3D application

was tested a complex 3D animation of a Roman British building excavated at Fishbourne Roman Palace that was developed in 3ds max and exported to Blender via COLLADA [16] for further processing as a Blender animation file. This small test animation has around 400 frames that took 7 hours on a Pentium IV machine to render. No advanced lighting effects and illumination were included at this stage. On this experimental render farm described above it took 55 minutes to render the animation.

# 5 Job Submission

Building a desktop Grid computing environment for implementing a render farm application was previously a difficult task, but rapid development in open source Grid software is making this task easier nowadays. However, a Grid based render farm application still requires a job submission tool along with the actual animation files. No integrated solution was found for this task, instead relying on several utilities – future work will be to build an integrated job submission application based on web services. Below we describe how to run the job submission and other possible ways to do the same.

## 5.1  Writing a Submission File

A Grid job can be submitted along with a job submission file which defines the type of submission and environment variables, input and output parameters, and other arguments; an example is shown below. This file can be written manually by the user (assuming they know what they are doing).

This file contains around 10 parts, see bellow, which was really hard to write manually, especially when dealing with thousands of frames to distribute them on a large number of Executers. An alternative way of manually writing these files is to use a collection freeware tools to achieve this like Blender_render [9], Condor_render [10], and GDI I Queue [11], that allow users to write Grid job submission scripts and automatic submission for 3D authoring packages tools like: Maya, 3ds max and Blender. Again, open source software for this experiment was used, i.e. Blender_render, to generate the submission script initially. However as already mentioned, we plan to write our own solution for generating this script in order to achieve additional functionality that Blender_render does not include.

Example of part 1 of sub file for rendering roman villa animation on Condor Grid.

```
# # Part 1
#
universe = vanilla
# update the path as required -
environment = path=C:\Program Files\Blender
Foundation\Blender;c:\windows;c:\windows\system32
should_transfer_files = Yes
transfer_files = always
when_to_transfer_output = on_exit
transfer_executable = False
transfer_input_files = villa_stage5.blend
```

```
Executable = "C:\Program Files\Blender
Foundation\Blender\blender.exe"
Log = log.txt
Output = out.txt
Error = err.txt
Arguments = -b villa_stage5.blend -s 1 -e 40 -a
Queue()
```

## 5.2  Submitting the Job

The job is submitted to the Condor Grid using either the Condor_submit command, a Command Line Interface (CLI) command, or using some other web portals or desktop interfaces. In this case, CLI was used to submit the .sub file with animation files to the Condor Grid, i.e. Condor_submit rom_villa.sub.

## 5.3  Rendering

The script generated by Blender_render is then submited to Condor using Condor_submit command from the Submitter machine. The Central Manager then distributes the animation frames to all executers according to .sub file originally submitted. All the Executers start their execution either immediately if the machine is dedicated, or they start after the machine becomes idle if it is non-dedicated. For this experiment, all dedicated machines have been used, but some non-dedicated machines have been also tested. After the rendering, the result is stored on the same machine where they are originally executed, thus an administrator has to get the results manually across the network file sharing system—this is of course one of the functionalities that needs to be automated to return the results back to the Submitter and the Central Manager.

# 6  Results

The animation was submitted to a single computer first and then to the experimental desktop Grid computing solution by chopping it into the 10 equal pieces, i.e. 40 frames per processor core, as there were ten cores from six executers in our Grid. When the same animation was executed on one of the executer machines with only a single core it rendered the animation in around 7 hours, but the same animation was rendered on the complete Condor desktop Grid within 55 minutes, this saved around 6 hours and 5 mins for a small animation, proving the render farm to be 88% faster than a normal machine with one core.

The important thing to note is that two single core machines completed their 40 frames in 40 minutes, though the dual core machines took around 55 minutes to render 80 frames each; of course other factors such as double memory for single cores and some other resources inside the CPUs will have an effect. Better performance of the Grid based render farm depends on better distribution of jobs on each machine; if we submit the same animation by chopping it into 20 frames then we may get much better performance. But if we divide the job in more small pieces e.g. 1 or 2 frames per job, then we may have to consider the network delay issues as the frame execution time is less than 1 minute only like in this case with not so heavy illuminations and other light effects, hence its transfer time will increase its throughput time. But in case, we have an animation where each frame has high volume of illumination or light effects then rendering time for each frame could be in hours, in this case it will better to distribute the jobs in as small pieces as possible, e.g. 1 to 5 frames per job.

## 7  Conclusion and Future Work

We have proved that it is possible to build a desktop Grid computer environment for almost zero cost by utilizing the departments discarded machines. On this desktop Grid computing environment we have implemented a first prototype version of an open source render farm that can easily be expanded with the addition of extra exe-cuter machines as they become available from the departments discarded pool of machines. This is a very effective way of getting extra value for money out of older machines and is a green computing way to use older machines rather than immediate disposal (although eventual disposal cannot be put off forever). We have further proven that it is possible to build an open source render farm with little difficulty. We also have further work to do, namely the implementation of a more integrated open source job submitter application based on web services that determines the machine resources and availability at run time. The next version of our desktop Grid comput-ing environment will deploy the latest Condor software that is based on web services – thus implementing service-orientation. Use of other 3D authoring tools like 3ds max and Maya is also planned. Further down the line, we will be looking at other large scale computing applications that may benefit from the use of a desktop Grid comput-ing environment. Another automatic functionality for the 'job submitter' application would be to determine the machines resources for better load balance, between machines and network bandwidth.

## References

1. Andersen, R., Vinter, B.: Harvesting Idle windows CPU cycles for Grid Computing. In: Proceedings of The International Conference On Grid Computing & Applications (GCA 2006) (2006)
2. Anderson, D.P., Cobb, J., Korpela, E., Lebofsky, M., Werthimer, D.: SETI@home An Experiment in Public-Resource Computing, Communications of the ACM, Vol. 45(11) (2002), http://www.seas.gwu.edu/~jstanton/courses/cs235/papers/p56-anderson.pdf
3. Blender free open source 3D content creation suite, http://www.blender.org/
4. Berkeley Open Infrastructure for Network Computing, http://boinc.berkeley.edu/
5. Jenkins, G., Goscinski, W.: User Guide for the Monash SPONGE Facility, http://www.monash.edu.au/eresearch/activities/sponge-userguide.pdf
6. ResPower Super/ Farm Unlimited Rendering, https://www.respower.com/
7. Remote Rendering Services – RenderCore, http://www.rendercore.com/ rendercoreweb/priceInfoView.do
8. The condor project, University of Wisconsin, http://parrot.cs.wisc.edu/
9. Condor User forum, https://lists.cs.wisc.edu/archive/condor-users/2006-July/msg00173.shtml

10. GDI | Queue, Distributed Rendering Made easy,
    http://www.cirquedigital.com/products/gdiqueue/index.html
11. Rendering Maya project with condor,
    http://www.cs.wisc.edu/condor/tools/maya/
12. Delaitre, T.: Technical Report Experiences with submitting Condor jobs to the Grid
    (2004), http://www.cpc.wmin.ac.uk/papers/reports/condor.pdf
13. Thain, D., Tannenbaum, T., Livny, M.: Condor and the Grid, http://media.
    wiley.com/product_data/excerpt/90/04708531/0470853190.pdf
14. condor version 6.8.7 Manual, Condor Team, University of Wisconsin–Madison, http://
    www.cs.wisc.edu/condor/manual/v6.8.7/condor-V6_8_7-Manual.pdf
15. How to Build A Renderfarm (without pulling your hair out),
    http://www.boxxtech.com/downloads/Community/Rendering.pdf
16. COLLADA.org community,
    http://www.collada.org/mediawiki/index.php/Main_Page

# Implementation Challenges for Nastaliq Character Recognition

Sohail A. Sattar[1], Shamsul Haque[1], Mahmood K. Pathan[1], and Quintin Gee[2]

[1] Computer Science & Information Technology
NED University of Engineering & Technology, Karachi, Pakistan
{sattar, pvc, deansth}@neduet.edu.pk
[2] School of Electronics & Computer Science
University of Southampton, Southampton, UK
qg2@ecs.soton.ac.uk

**Abstract.** Character recognition in cursive scripts or handwritten Latin script has attracted researchers' attention recently and some research has been done in this area. Optical character recognition is the translation of optically-scanned bitmaps of printed or written text into digitally editable data files. OCRs developed for many world languages are already in use but none exists for Urdu Nastaliq – a calligraphic adaptation of the Arabic script, just as Jawi is for Malay. Urdu Nastaliq has 39 characters against Arabic 28. Each character then has 2-4 different shapes according to its position in the word: initial, medial, final and isolated. In Nastaliq, inter-word and intra-word overlapping makes optical recognition more complex. Character recognition of the Latin script is relatively easier. This paper reports research on Urdu Nastaliq OCR, discusses challenges and suggest a new solution for its implementation.

**Keywords:** Nastaliq, Script, Cursiveness, Font, Ligature.

## 1 Introduction

A single script with its basic character shapes is adapted for writing in multiple languages e.g. Roman script for English, German and French while Arabic for Persian, Urdu, Malay, Uygar, Kurdish, Sindhi, etc. Nastaliq is a beautiful calligraphic and most widely used style of writing Urdu using an adapted Arabic script that has 39 characters as against 28 in Arabic. In Urdu, many character shapes have multiple instances. The shapes are context-sensitive too – character shapes changing with changes in the preceding character or the succeeding one. At times even the $3^{rd}$, $4^{th}$ or $5^{th}$ character may cause a similar change as depicted in an n-gram model in a Markov chain.

Research in Urdu text recognition is almost non-existent; however, considerable research has been done on Arabic text recognition that uses the Naskh style of writing. The Arabic script is considered to be a difficult one with a much richer character set than Latin; the form of the letter is a function of its position in the word: initial, medial, final or isolated. It changes its shape depending upon its position, and each shape has multiple instances, words written from right to left (RTL) [1]. Arabic characters have

D.M.A. Hussain et al. (Eds.): IMTIC 2008, CCIS 20, pp. 279–285, 2008.
© Springer-Verlag Berlin Heidelberg 2008

features that make direct application of algorithms for character classification in other languages difficult to achieve, as the structure of Arabic is very different [2].

The Latin, Chinese and Japanese scripts have received ample research and work has been done on the optical recognition of these scripts. Compared to this only few papers have specifically addressed the recognition of Arabic text and languages using the Arabic script like Urdu, Farsi and Malay, for various reasons. One of them is the complexity of the Arabic script itself while a lack of interest in this regard accounts for another [3]. Khorsheed *et al.* [4] presented an approach in which the system recognizes an Arabic word as a single unit using a Hidden Markov Model. The system depends greatly on a predefined lexicon, which acts as a look-up dictionary. All the segments in a word are extracted from its skeleton, and each of the segments is transformed into a feature vector. Then each of the feature vectors is mapped to the closest symbol in the codebook. The resulting sequence of observations is presented to a Hidden Markov Model for recognition [4]. Malik *et al.* [6] proposed a system that takes online input from the user writing the Urdu character with the help of stylus pen or mouse, and converts handwriting information into Urdu text. The process of online Handwritten Text Recognition is divided into six phases, each of which is implemented using a different technique depending upon the speed of the writer and the level of accuracy.

Fanton [5] has discussed the features that Arabic writing has, and identified the fact that these features impose computational overload for any Arabicized software. He also noted that the way in which Arabic is printed imitates handwriting. He pointed out that Finite State Automata give an efficient solution for the translated problems, which can be formalized as regular languages.

Chen *et al.* [7] addressed the problem of automatic recognition of an image pattern without any consideration of its size, position and orientation. In this regard, the extracted image features are made to have properties that are invariant with image transformation including scale, translation and rotation. They approximated the transformation by affine transformation to preserve collinearity and ratios of distances.

# 2   Urdu Nastaliq Character Set

Urdu uses an extended Arabic adapted script; it has 39 characters as against Arabic 28. Each character then has 2-4 different shapes depending upon its position in the word; initial, medial or final. When a character shape is written alone, it is called an *isolated* character shape. Each of these initial, medial and final character shapes can have multiple instances, the character shape changing depending upon the preceding or the succeeding character. This characteristic of having multiple instances of these character shapes is called *context sensitivity*. A complete language script comprises an alphabet and style of writing. Urdu uses an extended Arabic script for writing. It has two main styles, Naskh and Nastaliq. Nastaliq is a calligraphic, beautiful and more aesthetic style and is widely used for writing Urdu.

## 2.1  Nastaliq Script Characteristics

Nastaliq script has the following characteristics:

- Text is written from right to left
- Numbers are written from left to right

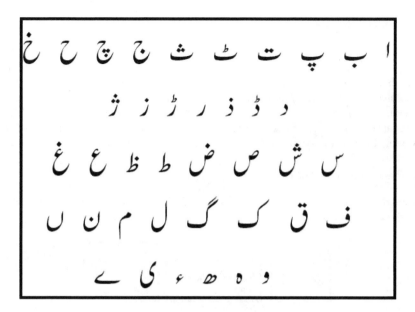

**Fig. 1.** Urdu Alphabet

**Fig. 2.** Urdu Numerals

- Urdu Nastaliq script is inherently cursive in nature
- A ligature is formed by joining two or more characters cursively in a free flow form
- A ligature is not necessarily a complete word, rather in most of the cases a part of a word, sometimes referred to as a sub-word
- A word in Nastaliq is composed of ligatures and isolated characters
- Word forming in Nastaliq is context sensitive i.e. characters in a ligature change shape depending upon their position and preceding or succeeding characters.

## 3   Problems Pertaining to Nastaliq Script

Here we describe the complexities of Nastaliq script with particular reference to optical recognition of printed text.

### 3.1  Cursiveness

Nastaliq, with its inherent cursive nature, makes a complex script. A single word in the script can comprise several ligatures formed in turn by combining several characters cursively joined together, along with isolated characters.

## 3.2  Context Sensitivity

Ligatures in Nastaliq are unique combinations or units of characters that change their shape according to their position within the unit. An initial "BAY",for example, which is the second character in the alphabet, is quite different from medial, final or an isolated one. Added to this is the dependence of each character on the preceding or succeeding characters it joins with. A character might take as many as 20 different shapes according to the character it is joining with. Sometimes even the $3^{rd}$, $4^{th}$ or $5^{th}$ preceding or succeeding character may initiate a change in shapes.

## 3.3  Position/Number of Dots

Several Urdu characters (17 out of 39) are differentiated by the presence of dots placed over, below or within them.

**Fig. 3.** (a, b, c) Dots

Three situations of ambiguity arise because of this. In the first instance, one character may have a dot while the other does not Fig 3(a). In the second case, two similar characters have different numbers of dots to distinguish their different sounds Fig 3(b). Lastly, two characters may be different only because of the difference in the position of dots Fig 3(c).

## 3.4  Kerning

For better visual appeal, the space between pairs of characters is usually reduced or kerned. If we attempt to reduce the natural space between two characters, it causes a slight overlapping of the characters making them less identifiable and more difficult to differentiate. When scanned, the lack of white pixels between the two characters makes them read as a single continuous character shape.

Although the problem appears infrequently in Latin script, e.g. *fi*, it is a common one in Nastaliq, because kerning implies further cursiveness, overlapping characters in numerous possible ways.

## 4  Nastaliq Character Recognition

Character recognition of Arabic Naskh script is relatively more difficult than Latin script. However, character recognition of Nastaliq script is even more complicated due to the inherent characteristics of this writing style.

In Latin script character recognition there are three levels of segmentation, namely: lines, words and characters. First, the text region is segmented into lines of text, then

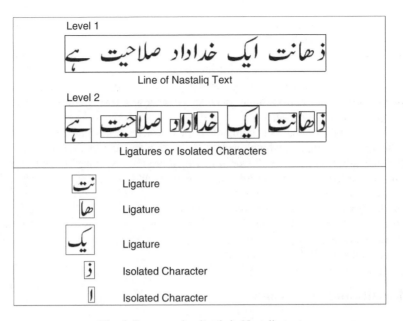

**Fig. 4.** Segmentation levels in Nastaliq text

each of the lines of text is divided into words, and then each word into its constituent characters.

Character recognition in Nastaliq script has only two levels of segmentation; first the text is segmented into lines, then the lines into ligatures and isolated characters.

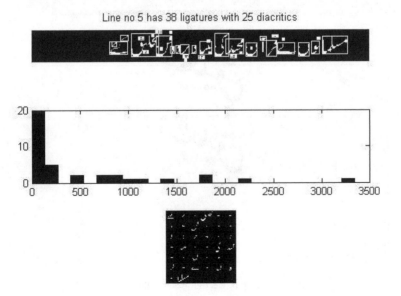

**Fig. 5.** Ligatures and isolated characters identified

Further segmentation of ligatures into character shapes has more challenges and is an open field for research.

Our proposed system for Nastaliq character recognition does not require segmentation of a ligature into constituent character shapes, but rather requires only two levels of segmentation. It then uses cross-correlation for recognition of characters in the ligatures line-by-line, writing their character codes into a text file in sequence, as the character is found in a ligature.

As the recognition process is completed, the character codes in the text file are given to the rendering engine, which displays the recognized text in a text region.

The segmentation phase is shown in Fig 5, where the text image is first segmented into lines, and then each of the lines into ligatures and primitives. Here we call isolated characters and diacritics *primitives*.

The horizontal axis measures the number of black pixels, while the vertical axis unit is number of elements, either ligatures or diacritics. Fig 5 also shows that the text line under consideration is 5th in the text image and is further segmented into a total of 63 elements having 38 ligatures and 25 diacritics.

## 5   Results and Discussion

We used Matlab for rapid prototyping and experimentation; however the Urdu Nastaliq character recognition system was developed in Microsoft VC++ 6.0.

We performed experiments on a small subset of Urdu Nastaliq words keeping the same font size, and the results are very encouraging.

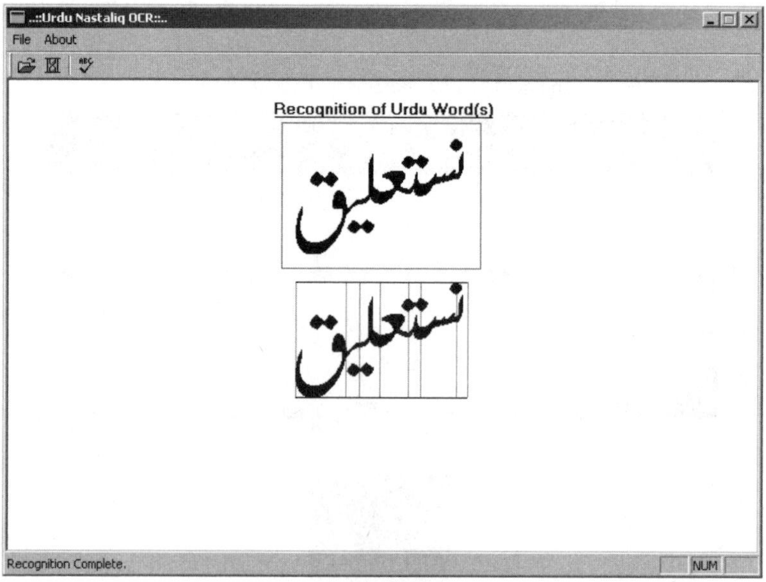

**Fig. 6.** Recognition result

Our Nastaliq character recognition system requires a character-based True Type Nastaliq font, and an image of Nastaliq printed text written with the character based True Type Nastaliq font.

After the segmentation is completed, the isolated character shapes and the ligatures have been identified. In the recognition phase, the True Type Nastaliq font file is loaded into the main memory, and each of the character shapes in the font file is matched with the shapes identified in the text image using cross-correlation for recognition of character shapes, line-by-line, writing their character codes into a text file in sequence as the character is found.

# 6  Conclusion and Future Work

In this paper, we discussed the complexities of the Urdu Nastaliq style of writing, which uses an adapted Arabic script and poses more challenges for the implementation of character recognition than the Arabic script which uses Naskh style of writing. We also discussed our new Urdu Nastaliq character recognition system with its results on a small subset of Urdu Nastaliq words.

For the future, we plan to extend our system to include a greater set of Urdu Nastaliq words.

# References

1. Hadjar, K., Ingold, R.: Arabic News paper segmentation. In: Seventh International Conference on Document Analysis and Recognition (ICDAR 2003), Edinburgh, Scotland, vol. 2, pp. 895–899. IEEE Computer Society, Los Alamitos (2003)
2. Altuwaijri, M.M., Bayoumi, M.A.: Arabic Text Recognition Using Neural Networks. In: International Symposium on Circuits and Systems (ISCAS 1994), vol. 6, pp. 415–418. IEEE, London (1994)
3. Syed, M.S., Nawaz, N., Al-Khuraidly, A.: Offline Arabic Text Recognition System. In: International Conference on Geometric Modeling and Graphics (GMAG 2003), pp. 30–35. IEEE, London (2003)
4. Khorsheed, M.S., Clocksin, W.F.: Structural Features of Cursive Arabic Script. In: 10th British Machine Vision Conference, Nottingham, UK, pp. 422–431 (1999)
5. Fanton, M.: Finite State Automata and Arabic Writing. In: Workshop on Computational Approaches to Semitic Languages (COLING-ACL 1998), University of Montreal, Canada (1998)
6. Malik, S., Khan, S.A.: Urdu Online Handwriting Recognition. In: IEEE International Conference on Emerging Technologies, Islamabad, Pakistan, pp. 27–31 (2005)
7. Chen, Q., Petriu, E., Yang, X.: A Comparative study of Fourier Descriptors and Hu's Seven Moment Invariants for Image Recognition. In: IEEE Canadian Conference on Electrical and Computer Engineering (CCECE-CCGFEI), Niagara Falls, Canada, pp. 0103-0106 (2004)

# Enhancing Business Decisions with Neurofuzzy Technology

Eram Abbasi[1] and Kashif Abbasi[2]

[1] Bahria University, Karachi Campus, Karachi, Pakistan
eram@bimcs.edu.pk
[2] GSESIT, Hamdard University, Karachi, Pakistan
k.abbasi@bigfoot.com

**Abstract.** In the modern era of globalization and dynamism, organizations have to make intelligent but complex decisions. Computer applications are being used to aid in this complex decision-making process, but there is one big limitation to it. Computers lack fair judgment and expertise that a human expert possesses. In order to overcome this problem, soft computing techniques can be used. This paper discusses potential uses of a technique known as 'Neurofuzzy Technology' in improving decision-making. To establish a basis for conceptual understanding, the paper discusses Fuzzy Logic, Neural Networks and Neurofuzzy technologies in general, and then proposes how Neurofuzzy Technology can be used for decision-making. With these theoretical bases, a model is developed to demonstrate the usefulness of Neurofuzzy Technology in business decisions. This model is developed from a sample data file using Fuzzy Technology. The data is later used to train the system by using Neural Net technology. The trained data is evaluated to check the performance of the system in terms of accuracy of results obtained.

**Keywords:** Neural Networks, Fuzzy Logic, Neurofuzzy Technology.

## 1 Introduction

Computers are great when it comes to the manipulation of data and performing mathematical operations, but when it comes to fair judgment and expertise, human experts remain unchallenged. In order to improve the man-machine co-ordination process and to bridge the gap between human reasoning and computer operations, soft computing technologies can be used. Soft computing is a combination of various computing technologies such as Neural Networks, Fuzzy Logic and Genetic Algorithms. These technologies can further be integrated as hybrid systems, to enhance their performance.

## 2 Fuzzy Logic

Fuzzy logic is a remarkable technology with a wide variety of applications in industrial control and information processing. Fuzzy systems deal with how human beings

D.M.A. Hussain et al. (Eds.): IMTIC 2008, CCIS 20, pp. 286–294, 2008.

handle imprecise and uncertain information. The technology was introduced by Zadeh [1] as a means to model the uncertainty of natural language.

Fuzzy logic is a superset of conventional (Boolean) logic [11] that has been extended to handle the concept of partial truth. Truth values between "completely true" and "completely false". The transition of truth values from "completely true" to "completely false" can easily be exhibited through fuzzy sets as compared to crisp sets. For example; a set 'A' of old ages

$$\{ A \mid A >= 35 \} \tag{1}$$

Equation (1) states that ages greater than or equal to 35 are regarded as 'old age' and are members of set 'A'. All other ages below 35 are not considered as 'old age' and are non-members of 'A'. In this case, a person aged 34.8 is not considered as a member of set 'A', but in actual world a person of age 34.8 can also be considered already old. Such imprecision and uncertainty of terms as well as flexibility in truth values can be handled in fuzzy sets but not in crisp sets. In figure 1 [10] the crisp set 'A' of 'old ages', shows that ages 35 years or below are not members of the set 'A'.

Figure 2 [10] shows a fuzzy set 'A' of 'old ages', where ages below 35 years are still members of the fuzzy set 'A' but their degrees of membership are lower. Generally, membership degrees in the fuzzy set of old ages will decrease as ages decrease.

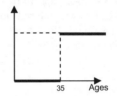

**Fig. 1.** Crisp set 'A' showing ages below 35 years are not members of the set

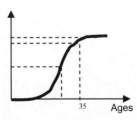

**Fig. 2.** Fuzzy Set 'A' showing ages below 35 years are members of the set

Just as numerical variables take numerical values, similarly in fuzzy logic, linguistic variables take on linguistic values. Linguistic values are words (linguistic terms) with associated degrees of membership in the set. Linguistic variables take on values defined in its set of linguistic terms. Linguistic terms are subjective categories for the linguistic variable.

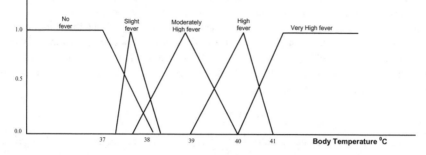

**Fig. 3.** Sample Linguistic Variables, Linguistic Terms, Fuzzy Sets, Membership Functions

Each linguistic term is associated with a fuzzy set [10], each of which has a defined membership function (MF). Figure 3 shows a fuzzy set with linguistic terms and membership functions. Membership functions are subjective measures for linguistic terms and are not probability functions.

## 2.1   Fuzzy System Model

Figure 4 shows a general fuzzy system model. In this model, once the linguistic variables are identified, their crisp values are entered into the system. The fuzzification module then processes the input values submitted to the fuzzy expert system where, crisp inputs are fuzzified into linguistic values to be associated to the input linguistic variables [11].

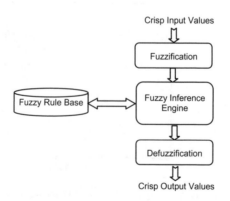

**Fig. 4.** Fuzzy System Model

The system uses a rule base containing fuzzy IF-THEN rules to derive linguistic values for the intermediate and output linguistic variables. Once the output linguistic values are available, the defuzzifier produces the final crisp values from the output linguistic values.

## 3   Artificial Neural Networks

Artificial Neural Networks (ANN), or simply Neural Networks, can be defined as large sets of interconnected units that execute in parallel to perform a common global task. These units usually undergo a learning process, which automatically updates network parameters in response to a possibly evolving input environment. ANNs are developed on the concept of Biological Neural Networks (BNN) and work on the principles of BNN.

### 3.1   The Biological Neural Network

The basic unit of a biological neural network is called a biological neuron, which is shown in figure 5. The biological neurons [9] have a cell body called the soma, and root-like structures called dendrites through which input signals are received. These dendrites connect to other neurons through a synapse. Signals collected through the numerous dendrites are accumulated in the soma. If the accumulated signal exceeds the neuron's threshold, then the cell is activated. This results in an electrical spike that is sent down the output channel called the axon. Otherwise, the cell remains inactive.

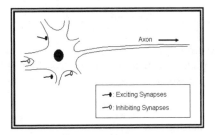

**Fig. 5.** A Typical Biological Neuron

### 3.2   Modelling the Biological System

The artificial neural networks are developed on the concept of the biological neuron but the structure is still far more different. Some of the major differences in the biological system and the artificial neural networks are:

- The artificial neural networks have a very small fraction of the number of neurons as compared to the number of neurons in the brain.
- Similarly, the interconnectivity of artificial systems is far more less than the massive interconnections in the brain.
- Neurobiological evidence indicate that a given neuron is either inhibitory to all neurons to which it is connected or excitatory to all. Whereas, in many artificial neural network models, a mixture of inhibitory and excitatory connections emanating from a single node is allowed.
- There are several types of neurons in the human nervous system, while most artificial neural network models use a single type of node.

## 3.3  Features of Artificial Neural Networks (ANN)

There exists various type of ANN based on their architecture, learning techniques used, and the tasks that they perform. Yet, some basic characteristics that can be commonly considered are:

- The basic component of the ANN is a single unit which is referred to as a 'node' or 'neuron'. Mostly these nodes not only look alike but also behave in a similar fashion.
- All neural networks are composed of nodes that are interconnected with each other. Anything happening to one node does not affect the action of another node, until they are directly connected to each other. Any given node's level of activation and its output depend on its current state and the outputs of the other nodes to which it is directly connected.
- In almost all ANN, the nodes mostly execute in parallel. To a certain degree in some ANN sequential execution is also observed.
- The learning process in ANN takes place in several small stages, until the final result is obtained.

## 3.4  Learning in ANN

There are two major categories of learning modes observed in ANN:

### 3.4.1  Supervised Learning
In this method user-supplied information is provided along with the training pattern. This guides the NN in adjusting its parameters. This method is also referred to as 'learn-by-example' methods.

### 3.4.2  Unsupervised Learning
In this method, user-supplied information is not initially supplied with the training pattern. Rather this information is later on used to interpret the **results.**

# 4  Neurofuzzy Technology

In order to enhance Fuzzy Logic systems [5] with learning capabilities, Neural Net technologies can be integrated with fuzzy systems. This new technology is termed as "Neurofuzzy Technology"

Fuzzy Technology provides a simple way to draw definite conclusions from vague, ambiguous or imprecise information. It helps in decision making with its ability to work from approximate data and find precise solutions. On the other hand, Neural Network (NN) is a technology that learns from the underlying relationship of data. Neural Network posses self-learning and self-tuning capability hence, can be applied to model various systems. As NN are unable to handle linguistic information, therefore, integrating Neural Nets with Fuzzy Logics give a powerful solution to decision making.

In a NeuroFuzzy system, an expert's knowledge is converted to symbolic form [6] and then used to initialize the neural network. Knowledge is then obtained from the output of the neural network. This output is obtained as fuzzy logic representation as an output of the neurofuzzy system.

# 5  Modeling Customer Selection System for Target Market

In order to ascertain the usefulness of Neurofuzzy Technology, a small model application has been prepared. This application is about "Customer Selection for Target Marketing in a business environment".

Selecting a customer to target for direct marketing of goods/services is a task of careful analysis. In a business environment, in order to make direct marketing activities more efficient, it is necessary to have a clear definition of the target customers. One possibility to make sure that all the target people are selected is to make the selection criterion more liberal. However, the disadvantage of this solution lies in the fact that there will be too many people selected who are not supposed to be in the original target group. In other words, if the conditions are too liberal, the selected group will be much bigger than desired, leading to waste of time and other resources. Similarly, if the criterion is too strict, then there are chances of loosing some target people. Thus to handle this problem, this case study is being carried out using Neurofuzzy Technology. This model application, integrates neural network technologies to train fuzzy logic systems. This model has been greatly simplified just to serve the purpose of showing the integration between Fuzzy Technology and Neural Networks

The model application is developed by carrying out the following steps:

- Obtaining training data
- Creating an "empty" fuzzy logic system
- Entering all existing information in the solution
- Configuring the NeuroFuzzy Module
- Training the sample data
- Evaluating the systems performance

## 5.1  System Description

This model has been developed for selecting customers for a certain product or service. Here, the linguistic variables and the rules of the fuzzy logic system describe the target group for the product. Then the fuzzy logic system assesses to what degree a customer satisfies the criteria of the target group.

## 5.2  System Structure

The system structure identifies the fuzzy logic inference flow from the input variables to the output variables. The fuzzification in the input interfaces translates analogue inputs into fuzzy values. The fuzzy inference takes place in rule blocks that contain the linguistic control rules. The outputs of these rule blocks are linguistic variables.

## 5.3  Variables

Linguistic variables are used to translate real values into linguistic values. Linguistic variables are defined for all input, output and intermediate variables.

Figure 6 shows the complete model with input and output variables. In this model, 6 input variables and 3 output variables have been used. The 6 input linguistic variables used in this system are:

- Age
- Number of children
- Demographics
- Annual income spending
- Marital status
- Savings

The 3 output linguistic variables used in this system are:

- Financial status
- Personal information
- Selected customer

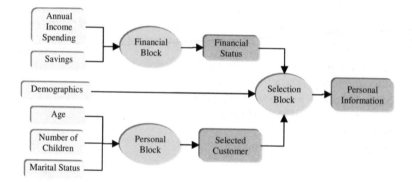

**Fig. 6.** The model showing variables and rule block

## 5.4 Rule Blocks

The rule blocks contain the control strategy of the fuzzy logic system. Each rule block confines all rules for the same context. In this model, 3 rule blocks have been designed containing total 42 rules. The 3 rule blocks are:

1. *Financial Block*: with 2 inputs and 1 output
2. *Personal Block*: with 3 inputs and 1 output
3. *Selection Block*: with 3 inputs and 1 output

## 5.5 Neurofuzzy Design

The application is developed from a sample data file using fuzzy technology. After creating the fuzzy system from a sample data file, the generated system is then trained with the neural networks.

### 5.6  Neurofuzzy Training of Fuzzy Systems

For the Customer Selection example, all rules and the output variable have been selected for training.

### 5.7  Evaluation

After completion of the actual training, the resulting fuzzy logic system was tested various scenarios based on customer information. The result showed that 80% of the customers selected by the system were same as the customers selected by a human being. The result did not show 100% improvement but the trained data showed significant improvement in the system. As the results were satisfactory, therefore, further training was stopped. Though the data set used for training the system was small and training was carried out for a limited time, still significant results were obtained. In actual world if more training is provided then it is expected that the system would be able to show better results. Hence, it is concluded that applying neurofuzzy technology in decision-making process can reap into better decision in lesser time.

## 6  Conclusions and Recommendations

The aim of this study was to find out how Neural Networks and Fuzzy Technology work and how these technologies can be integrated to develop Intelligent Business Systems that can help in intelligent decision-making.

Neural networks are important for their ability for adaptation and learning, fuzzy logic for its exploitation of partial truth and imprecision. As both these technologies offer their own benefits, it has been emphasized to develop a partnership among both the methodologies and make them work together.

The model discussed in the paper, related to "Customer Selection for Target Marketing" reveals that integrating Fuzzy and Neural Technologies results in better management and analysis of data, which in turn result in better decision making. Training fuzzy rule sets with neural technology resulted in far more efficient results. Neurofuzzy Technology facilitated to manage imprecise, partial, vague or imperfect information. Fuzzy logic enables the use of experience and experimental results to deliver solutions that are more efficient. It does not replace or compete with conventional control techniques. The 'Customer Selection for Direct Marketing' example demonstrates that fuzzy logic delivers a transparent, simple solution for a problem that is difficult to solve using conventional techniques. Thus, it has the ability to mimic human decision-making process.

Non-availability of proper software to develop Neurofuzzy application in this region was a major problem that was faced in developing the system. Easy availability of proper software and sufficient time are the major requirements for learning this technology and develop highly professional systems.

Studying various applications of Neurofuzzy Technology reveals that the technique is very successfully applied in various fields. However, numerous examples of systems using Neurofuzzy Technology are recorded, yet creators of these systems have indicated that that still a great deal of research needs to be done in this area. More advanced Neurofuzzy techniques are yet to be used in creating a truly intelligent system.

Communication and managing intelligent business systems is one of the key areas of research of the current age. Most research work being done these days is either directly addressing communication technologies or are one way or other related to this area. The Internet in recent years has developed from simple text pages to high level of interactivity that exists now through forms, JavaScript and Java applets. Intelligent Automated negotiation systems using soft computing can very well become the next step on the road. Since the volume of information available on the Internet is so vast it is becoming increasingly difficult to find what is being searched for. Soft Negotiation techniques are in their infancy, but this is an area for intense development. Similarly, soft computing can be utilized in other business areas, such as Customer Relationship Management, in Medicine for diagnosis and prescription purposes, Data Mining Technologies.

# References

1. Zadeh, L.A.: Fuzzy sets. Information and Control 8(3), 338–353 (1965)
2. Azvine, B., Azarmi, N., Tsui, K.C.: Soft computing - a tool for building intelligent systems. BT Technology Journal 14(4), 37–45 (1996)
3. von Altrock, C.: Fuzzy Logic and Neurofuzzy Applications in Business and Finance. Prentice Hall, NJ (1996)
4. von Altrock, C.: Applying fuzzy logic to business and finance, optimus, 2/2002, pp. 38–39 (2002)
5. von Altrock, C.: Practical Fuzzy-Logic Design - The Fuzzy-Logic Advantage. The Computer Applications Journal Circuit Cellar INK 75 (1996)
6. Carpenter, G.A., Grossberg, S.: Fuzzy ARTMAP: A Synthesis of Neural Networks and Fuzzy Logic for Supervised Categorization and Nonstationary Prediction. In: Fuzzy Sets, Neural Networks and Soft Computing, Van Nostrand Reinhold (1994)
7. Juang, C.-F., Lin, C.T.: An online self-constructing neural fuzzy inference network and its applications. IEEE Transactions on Fuzzy Systems 6(1), 12–32 (1998)
8. Zlotkin, G., Rosenschein, J.S.: Mechanism design for automated negotiation, and its application to task-orientated domains. Artificial Intelligence, 86(2), 195–244 (1996)
9. FuzzyTECH User's Manual and NeuroFuzzy Module 5.0, INFORM GmbH/Inform Software Corp (1997)
10. Jang, J.-S.R., Sun, C.-T., Mizutani, E.: Neuro-Fuzzy and Soft Computing. Prentice Hall, Englewood Cliffs (1997)
11. Zadeh, L.A.: The Roles of Fuzzy Logic and soft computing in the conception, design and deployment of intelligent systems. In: Nwana, H.S., Azarmi, N. (eds.) Software agents and soft computing: concepts and applications, pp. 183–190. Springer, Heidelberg (1997)

# A System Design for a Telemedicine Health Care System

Asadullah Shaikh[1], Muhammad Misbahuddin[1], and Muniba Shoukat Memon[2]

[1] Department of Software Engineering and Management,
I.T University of Goteborg, Sweden
shaikh@ituniv.se, misbahud@ituniv.se
[2] Department of Computer System and Software Engineering,
Mehran University of Engineering and Technology
Jamshoro, Pakistan
hina04_muet@hotmail.com

**Abstract.** Telemedicine systems have been steadily increasing in number and scale over the past decades. There is an intensification of the need for telemedicine in the era of national health care systems. The increasing size of distributed telemedicine systems creates a problem of data integration, vendor lock-in and interoperability. This paper discusses a telemedicine system architecture, which is being built as a Service Oriented Architecture (SOA), because we believe that by the adoption of SOA, several problems in telemedicine systems can be resolved. Currently, hospitals become limited to a single vendor because of the introduction of new proprietary technology. We present an overview of such an architecture, which draws the attention of readers towards the solution of users' problems. In general, our proposed solution emphasizes a web services solution.

**Keywords:** Tele-Wound, Telemedicine.

## 1 Introduction

Telemedicine is a part of telehealth, which is based on the technologies by using telecommunication for the interaction between health professionals and patients in order to execute medical actions at distance. Telemedicine is a very vast field in today's world, which is widely used to reshape the systems in the health care. Telemedicine is helping different healthcare system to solve the problems in several ways.

Telemedicine applications need to run in heterogeneous computing environments. Data integration, vendor locking and interoperability are of major concern in telemedicine applications [1]. In the current situation, most of the telemedicine systems are designed by the consideration of today's needs of companies. Due to this, lot of problems is occurring in the area of telemedicine.

## 2 Background and Motivation

Most of the telemedicine systems vary from each other. Also the infrastructure of current telemedicine system is quite different then previous ones. It means that the

D.M.A. Hussain et al. (Eds.): IMTIC 2008, CCIS 20, pp. 295–305, 2008.
© Springer-Verlag Berlin Heidelberg 2008

client should be limited to the same vendor in order to maintain the integration and interoperability between the old and new telemedicine system. This includes all these things including those vendors who cannot adopt the new technology, which requires a heavier investment for the integration to previous systems. Over the past decades, the cost of integration of the telemedicine system has raised many folds. Due to that the implementation became more difficult. Here in this regard many issues are raised but still there is a matter of vendor locking and interoperability, which stands on top in few industries. Even though there is always a query which is being raised by most of the industries while working on each telemedicine system components that could our system be interoperable in future [2]?

Now the question is this how the user can be sure that his telemedicine system will not create the problem of interoperability along with vendor locking and data integration in future. What will happen when he wants to integrate his one telemedicine component into other? With the consideration of all aspects, we are going to give an example design of a telemedicine system, which suggests a few solutions to those people who are already working on this area or indented to work in future.

The main theme of this paper is to show exemplary system architecture of telemedicine system along with the implementation. The lack of previous published system architectures in this subject further motivates the study as well as the importance of our design choice in telemedicine systems. The results have the possibility to affect the ongoing discussions about the research field of telemedicine system by presenting a system with a design description of vendor locking, data integration and interoperability.

Another motivation for this type of study is to raise the interest of all those persons, who are afraid of not having the vendor support in the area of telemedicine. Telemedicine systems are most discussed and debated topic in today's society.

## 3 Implemented Proposed Architecture

In the telemedicine world, many telemedicine applications exist in the environment and of similar architecture have been proposed for ambulatory of patient monitoring but our Tele-Wound$^{TM}$ application is a sort of different application influenced by Services oriented Architecture. The targeted audience of this application will be leg ulcer patients and the main purpose of this application is to serve all those patients who cannot move freely. Their pictures of infected areas will be fetched through camera (cell phone camera or digital camera) by the nurse from their location, after fetching the picture he/she will send those pictures into our application server along with little description of patient.

This Web based architecture of Tele-Wound$^{TM}$ is utilizing telecommunication (GSM/3G) and internet technology and their major components/modules are more dependent on those technologies. This architecture is also confirming and assuring the availability of medical data from the other resources. Special lists in the doctor's end would be needed with internet access and a browser so that they can help out patients from any place or city or even country for serving the patients/nations.

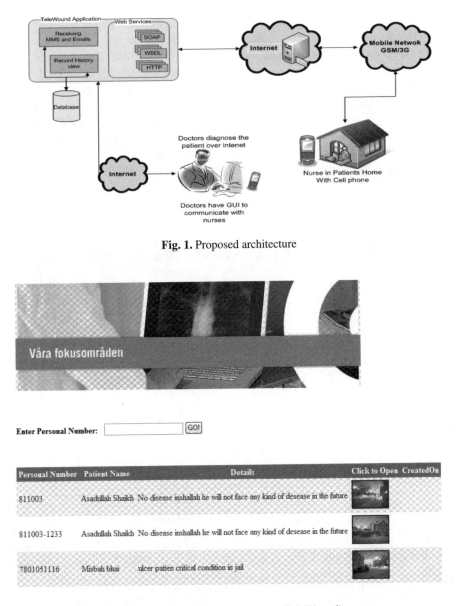

**Fig. 1.** Proposed architecture

**Fig. 2.** Screen shot of application (TeleWound)

The above context diagram (Figure 1) presents the sending and receiving MMS and E-mails from the nurse or patient's end. The MMS messages will consist of patient's picture fetched by the nurse(s), which will be stored on database at the clinic/hospital, where the patient's responsible doctor can work on the patient's records and make prescription on the basis of received data. The application server may cater to many patients concurrently but it depends on the availability of the doctors/specialist/consultants having different areas of expertise. The specialist may use any computer with Internet access to offer his or her expertise. Standard browser interface is required to connect the application server for accessing the patient's data and giving feedback to them through text, voice or video chat or SMS messages. In this way the patient can save an unnecessary trip to the doctor or one can ask the patient to come for a visit if the ulcer/disease is worsening. Figure 2 is showing the screenshot of the main interface of our implemented Tele-Wound$^{TM}$ architecture' which comprises of three main columns named as "Personal Number", "Patient's name" and "Details of disease". Each field has CRUD service and Figure 3 follows the same sequence while indicating the start machine wizard in order to get the data from the server.

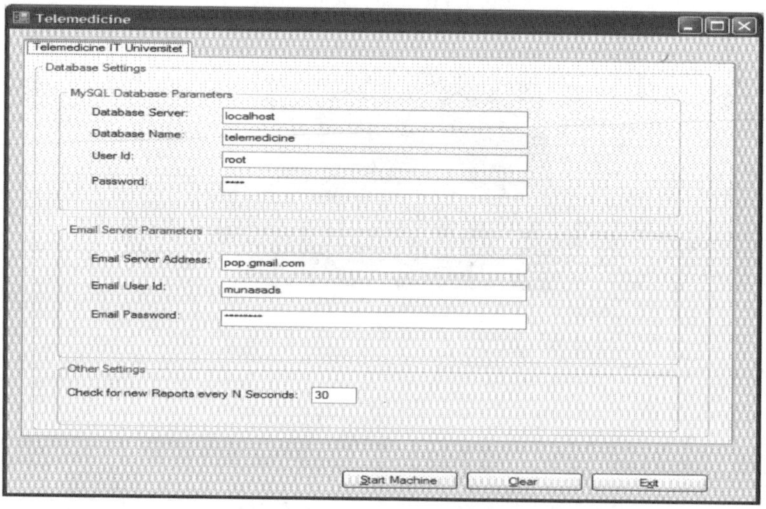

**Fig. 3.** Screenshot of getting data

## 3.1 Design Description

The designing of architecture of Tele-Wound$^{TM}$ application will be SOA based architecture. Figure 4 shows the system components of Tele-Wound$^{TM}$. Doctor's component represents the doctor's activities in the application. This component also contacts the external email server for receiving emails. The method/service CRUD_GetDataset() request the email from the email pop3 server and give it back to doctors component where the doctor's component receives all emails/MMS and separates patient's images and text descriptions from each other in order to save them into database. This component also requires patients records so the doctors are able to

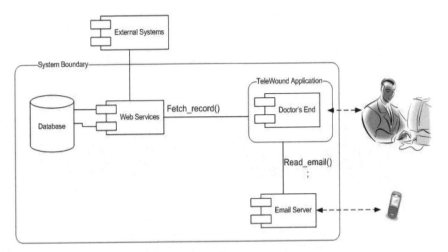

**Fig. 4.** Component Diagram of Tele-Wound $^{TM}$ application

treat their patients after viewing and analyzing their previous history records, CRUD_GetDataResponse() will fetch the records from database to the doctors components by using web services.

Below are the CRUD services used in Tele-Wound$^{TM}$ along with the CRUD type and description.

### 3.2 Implementation of Web Services

The application will communicate with web services. Web services technologies can be used to create a contract between different software systems [3]. We have planned to use different web services. The main advantage of this approach is that it allows organizations to cost-effectively utilize reusable component without adopting expensive technique and technologies.

Here we used different web services in Tele-Wound TM application, these web services are allowing to communicate and interchange the data between similar or dissimilar telemedicine systems. Simple Object Access Protocol (SOAP) has been used as an XML based message-binding protocol, the primary protocols of SOAP are HTTP and HTTPS, and POP3 is used for communicating with email server. Here SOAP defines how messages can be well ordered and processed by system that facilitates cross platform independency in different programming language or platform, and thus interoperability between applications written in different programming languages and running on different operating systems can be achieved.

Figure 5 shows that how WSDL interchange different platforms, share the web services contract and communicate using SOAP over HTTP. This will help to resolve the problem of data interchange from one platform to another. To make it possible, both ends need to understand XML to interact data within different platform. By

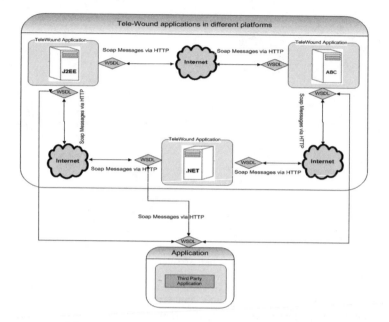

**Fig. 5.** *Tele-*Wound applications in different platforms

considering the figure 5, we can resolve the interoperability problem in cross platforms within the similar telemedicine systems and if two or more different telemedicine applications from different vendors use these services then the application will be able to communicate and interchange their data with each other. By this process the user will not be dependent on a single vendor because, if the technologies are changed then user will be able to easily change his telemedicine application according to new requirements.

### 3.3   Interoperability through Web Services

Interoperability, that is ultimately required now a days can be achieved in our system by adopting the services oriented architecture. Interoperability is one of the important achievements that are gained from implementing those web services according to the services oriented interoperability. It is difficult to integrate two or more dissimilar system, but a web services layer between application and client can significantly remove these difficulties. Figure 5 shows two dissimilar applications communicate through XML based SOAP message over HTTP and they communicate by interpreting WSDL, through this service, dissimilar Tele-WoundTM application can communicate through SOA based architecture by using web services. All above services are the part of Tele-WoundTM in order to support all those telemedicine applications that are wonder to be connected to different telemedicine application.

Below is the WSDL code that is being generated by our system to avail the different CRUD service.

```
<s:element name="CRUD_InsertData">
     <s:complexType>
     <s:sequence>
             <s:element                              minOccurs="0"maxOccurs="1"
     name="strTableName" type="s:string"/>
             <s:element minOccurs="0" maxOccurs="1" name="strFields"
     type="tns:ArrayOfString"/>
             <s:element    minOccurs="0"    maxOccurs="1"    name="strValues"
type="tns:ArrayOfString"/>
         </s:sequence>
     </s:complexType>
     </s:element>
```

```
- <wsdl:definitions targetNamespace="http://tempuri.org/TeleMedicineCRUD/Service1">
  - <wsdl:types>
    - <s:schema elementFormDefault="qualified" targetNamespace="http://tempuri.org/TeleMedicineCRUD/Service1">
      + <s:element name="LoginToDatabase"></s:element>
      + <s:element name="LoginToDatabaseResponse"></s:element>
      + <s:element name="LogoutFromDatabase"></s:element>
      + <s:element name="LogoutFromDatabaseResponse"></s:element>
      + <s:element name="CRUD_InsertData"></s:element>
      + <s:complexType name="ArrayOfString"></s:complexType>
      + <s:element name="CRUD_InsertDataResponse"></s:element>
      + <s:element name="CRUD_GetDataset"></s:element>
      + <s:element name="CRUD_GetDatasetResponse"></s:element>
      </s:schema>
    </wsdl:types>
  - <wsdl:message name="LoginToDatabaseSoapIn">
      <wsdl:part name="parameters" element="tns:LoginToDatabase"/>
    </wsdl:message>
  - <wsdl:message name="LoginToDatabaseSoapOut">
      <wsdl:part name="parameters" element="tns:LoginToDatabaseResponse"/>
    </wsdl:message>
  - <wsdl:message name="LogoutFromDatabaseSoapIn">
      <wsdl:part name="parameters" element="tns:LogoutFromDatabase"/>
    </wsdl:message>
  - <wsdl:message name="LogoutFromDatabaseSoapOut">
      <wsdl:part name="parameters" element="tns:LogoutFromDatabaseResponse"/>
    </wsdl:message>
```

**Fig. 6.** WSDL file for CRUD [4]

Following is the screen shot of CRUD insert data dump as an example that is being created for CRUD Type CREATE (see Table 1). In the same, we have implemented different services for different functions of Tele-Wound.

**Table 1.** List of CRUD services used in Tele-Wound $^{TM}$ Application [5]

| Services | CRUD Type | Description |
|---|---|---|
| CRUD_DeleteRecords(intID) | Delete | Delete the patients records |
| Form1_Load() | Read | Load many parameters and variables when form load (i.e pwd of email accounts and user and pwd of database server) |
| CRUD_InsertData (strEmailID, strEmailTxt, strImagePath) | Create | Take email/MMS data from email server and then parse them and insert it into database |
| CRUD_InsertDataResponse (strEmailID) | Create | Insert the unique email id assigned by email server into database. |
| PopulateGrid(sql) | Read | Populate the present the patient's records against the patient's personal number in webpage. |
| CRUD_GetDataset() | Read | Read email from server in a assigned time interval |
| CRUD_GetDataResponse() | Read | |
| CRUD_UpdateRecords(intID) | Update | Update patients records |

## 4   Evaluation of Proposed Architecture

Following are few advantages of proposed architecture:

### 4.1  Technologies Used in Tele-Wound$^{TM}$

Being that architecture as services oriented architecture; the application will communicate with external web services. Web services technologies can be used to create a contract between disparate software systems [3]. We have planned to use

**TeleMedicineCRUDService**

Click here for a complete list of operations.

## CRUD_InsertData

**Test**

The test form is only available for requests from the local machine.

**SOAP**

The following is a sample SOAP request and response. The **placeholders** shown need to be replaced with actual values.

```
POST /test/TeleMedicineCRUD.asmx HTTP/1.1
Host: 80.244.65.131
Content-Type: text/xml; charset=utf-8
Content-Length: length
SOAPAction: "http://tempuri.org/TeleMedicineCRUD/Service1/CRUD_InsertData"

<?xml version="1.0" encoding="utf-8"?>
<soap:Envelope xmlns:xsi="http://www.w3.org/2001/XMLSchema-instance" xmlns:xsd="http://www.w3.org/2001/XMLSchema" xmlns:
  <soap:Body>
    <CRUD_InsertData xmlns="http://tempuri.org/TeleMedicineCRUD/Service1">
      <strTableName>string</strTableName>
      <strFields>
        <string>string</string>
        <string>string</string>
      </strFields>
      <strValues>
        <string>string</string>
        <string>string</string>
      </strValues>
    </CRUD_InsertData>
  </soap:Body>
</soap:Envelope>
```

```
HTTP/1.1 200 OK
Content-Type: text/xml; charset=utf-8
Content-Length: length

<?xml version="1.0" encoding="utf-8"?>
<soap:Envelope xmlns:xsi="http://www.w3.org/2001/XMLSchema-instance" xmlns:xsd="http://www.w3.org/2001/XMLSchema" xmln
  <soap:Body>
    <CRUD_InsertDataResponse xmlns="http://tempuri.org/TeleMedicineCRUD/Service1">
      <CRUD_InsertDataResult>string</CRUD_InsertDataResult>
    </CRUD_InsertDataResponse>
  </soap:Body>
</soap:Envelope>
```

**Fig. 7.** WSDL file for CRUD [5]

different web services. The main advantage of this approach is that it allows organizations to cost-effectively reusable component without adopting expensive technique. Following are the little description and advantages of the applied web services.

By using web services description language (WSDL), the communication protocols and messaging formats have been standardized between external services in the Tele-Wound ™ application. Figure 8 shows that how the WSDL interchange the different platforms share the web services contract and communicate using SOAP, it will help to resolve the problem data interchange from one platform to another, for make it possible both ends need to understand XML to interact data with in different plate form. Web Services- Metadata Exchange (WSME) facilitates to know what the other end of application wants to know which type of information and how to interact between them. By using the above web service, it is possible to resolve data interchange problem and if the two or more different telemedicine application from different vender use these service then the application is able to communicate and

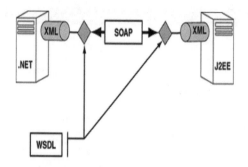

**Fig. 8.** WSDL describe how to process SOAP message [3]

interact their data. By this process the user will not be dependent on a single vendor because if the technologies are changed then user can easily change his telemedicine application according to his new needs.

Now consider the figure 9, which illustrates the fact that when .NET platform and J2EE distributes the web service agreement, they communicate using SOAP messages. Basically these SOAP messages are XML documents [3] because each contributor requires the understanding of XML. The WSDL documents are approved by both parties in order to produce the SOAP messages. For example if the data type incompatibilities are different, then both parties understand and interchange the data to each other because it is understandable to them. That connection shows that both ends should have access to same WSDL definition, so that both SOAP nodes can interchange the messages to the common WSDL definition. This process will certainly resolve the problem.of connectivity and data interchanging in case of same WSDL definition. Tele-Wound $^{TM}$ is consisting of all above service in order to provide a better solution when the application is running on heterogeneous environment.

**Fig. 9.** The WSDL agreement permits dissimilar systems to connect [3]

Hence it's clearly stated that dissimilar Tele-Wound$^{TM}$ application can integrate, connect and communicate through SOA based architecture by using web services. All above services are the part of Tele-Wound$^{TM}$ in order to support all other telemedicine applications that are wonder to be connected to different telemedicine application. Only the difference is the usage of same web service and same WSDL contract in order to go for further communication.

# 5   Conclusion

We have designed and implemented service oriented architecture in our Tele-Wound ™ application in order to avoid the problem of interoperability, vendor lock-in and data interchange. The problems that appear when discussing the interoperability, data interchange and vendor lock-in among telemedicine application, then SOA suggested the better solution. This paper presents the proposed architecture of Tele-Wound ™ application with the implementation of web services based on the experience gained with some previously developed applications. We tried to resolve the problem of interoperability in general; vendor lock-in and data interchange through our Tele-Wound™ architecture, so that our Tele-Wound ™ users will not be dependent on single vendor and can also easily interchange the data between similar and dissimilar application by understanding same WSDL definition.

# References

1. Ganguly, P., Ray, P.: Software Interoperability of Telemedicine Systems, IEEE CNF (July 4-7, 2000), http://ieeexplore.ieee.org/iel5/6934/18631/00857717.pdf? tp=&arnumber=857717&isnumber=18631
2. Craft, R.: Introduction to the Telemedicine System Interoperability Architecture, Sandia National Laboratories, http://telemedicine.sandia.gov/
3. Newcomer, E., Lomow, G.: Understanding SOA with Web Services. Addison Wesley, Reading (2004)
4. Telemedicine CRUD Services,
   http://80.244.65.131/test/TeleMedicineCRUD.asmx
5. Tele-Wound CRUD Web Service Function,
   http://80.244.65.131/test/Testbedforservice.aspx

# Performance Analysis of PDC-OCDMA System with APD Mismatch

Umrani Fahim Aziz[1], T. O'Farrell[1], and Umrani A. Waheed[2]

[1] Institute of Advanced Telecommunications, University of Swansea,
Wales, United Kingdom
[2] Institute of Communication Technologies, Mehran University,
76062 Jamshoro, Pakistan
fahim_umrani@yahoo.com

**Abstract.** This paper presents the practical analysis of the Optical Code-Division Multiple-Access (O-CDMA) systems based on Perfect Difference Codes. The work carried out shows how the mismatch in the gains of the APDs used in the information- and interference-bearing branches of the receiver can degrade the BER performance of the system. The system performance, with consideration of shot noise, thermal noise, avalanche photodiode (APD), bulk and surface leakage currents, and APD gain mismatch, is also investigated.

**Keywords:** OCDMA, Perfect Difference Codes, APD Mismatch.

## 1 Introduction

There has been a recent upsurge of interest in applying Code-Division Multiple-Access (CDMA) techniques to optical networks. Optical CDMA offers an interesting alternative for LANs as compared to traditional LAN multiple-accessing techniques, namely TDMA and WDMA, because neither time nor frequency management of all nodes is necessary [1]. Optical CDMA results in very low latencies because it liberates the network from synchronization, and as dedicated time or wavelength slots do not have to be allocated, so statistical multiplexing gains can be high. CDMA also allows flexible network design because the bit error rate (BER) depends on the number of active users (i.e., soft-limited).

Evaluation of error probability based on the assumption that the gains of APDs qused in the upper and lower branch of the desired user's receiver are matched provides an approximation which may in practice be an overestimate or underestimate of the actual probability of error [2-4]. We examine how the mismatch in APD affects the asynchronous optical CDMA system performance proposed in [5]. In addition, the system performance, with consideration of shot noise, thermal noise, avalanche photodiode (APD), bulk and surface leakage currents, and APD gain mismatch is also evaluated.

The remainder of this paper is organized as follows. In section 2, the structure of the transmitter and receiver is described. Section 3 presents the analysis of APD mismatch gain. Finally, section 4 discusses the experiments and numerical results obtained for the BER and system capacity.

D.M.A. Hussain et al. (Eds.): IMTIC 2008, CCIS 20, pp. 306–313, 2008.
© Springer-Verlag Berlin Heidelberg 2008

## 2 System Model

In [5], an asynchronous OCDMA system is proposed based on PDCs. The analysis is carried out in ideal conditions and is based on some serious assumptions such as the gain of APDs used in receiver of proposed systems are perfectly matched and that there is no uniformity loss in the splitters used at transmitter and receiver side. In what follows we analyze the system proposed in [5] considering practical environment and attempts to critically analyze the obtained results and also propose some ways to minimize or overcome the losses which results due to the consideration of practical losses.

Each subscriber is assigned a unique code. Each active user transmits a signature sequence of $k$ laser pulses (representing the destination address), over a time frame if mark "1" is transmitted. However, if the data bit is space "0," no pulses are transmitted during the time frame. The system presented in [5] proposes the use of PDC to overcome the limitation of codeword synchronization and power loss incurred in the Spectral Amplitude Coding (SAC)-OCDMA Systems.

The arrayed waveguide grating v×v wavelength multiplexer (AWG MUX) proposed in [6] is used as encoder and decoder since both it and PDC have a cyclic-shifted property. It is assumed that a broadband optical pulse entering one of the input ports of the AWG MUX is split into a v number of spectral components. Each spectral component followed a unique route through the AWG MUX in accordance with its particular wavelength.

The transmitter shown in Fig. 1 comprises a switch, a v×v AWG MUX and a l×1 coupler. In accordance with the employed PDC, l output ports are selected in advance. When data bit '1' is to be transmitted, a broadband optical pulse is sent to one of the v input ports of the multiplexer. The choice of input port is determined by the switch in accordance with the signature sequence of the destined user. The optical broadband pulse entering the multiplexer is split into v spectral components. These components exit the multiplexer through the $l$ predetermined output ports and are then combined into a single pulse by the $l$×1 coupler and transmitted to the destined user. To transmit bit '1' to a different user, the transmitter uses the switch in front of the AWG MUX to change the input port of the broadband optical pulse in accordance with the codeword sequence of the new user. Consequently, a different group of spectral components exits from the $l$ predetermined output ports. When a '0' data bit is to be transmitted, nothing is actually sent. In this study, it is assumed that $w_1, w_2,...,w_l$ are the spectral components which make up the signature sequence of the destined user.

The front of the receiver is implemented by adding a $(v - 1) \times 1$ coupler to the transmitter structure. In accordance with the signature address of the destined user, the received optical pulse is directed to the corresponding input port by the switch in front of the AWG MUX. As described above, the optical pulse is then split into several spectral components and each component follows its own particular route through the AWG MUX. The spectral components, w1, w2,...,wl, exiting from the l predetermined output ports, are collected by the l×1 coupler and combined into a single optical pulse. This pulse is transmitted to an APD, which responds by outputting the corresponding photoelectron count, Y1. Meanwhile, the $(v - k) \times 1$ coupler collects

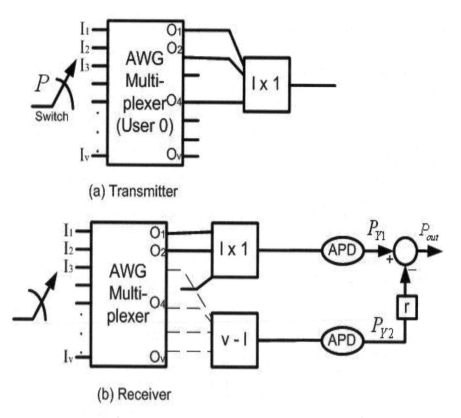

(a) Transmitter

(b) Receiver

**Fig. 1.** System model proposed in [5]

the spectral components which exit through all of the output ports of the AWG MUX other than the l predetermined ports. The output of the $(v - k) \times 1$ coupler, referred to as the filtered multiple-access interference (MAI), is photodetected by a second APD, which outputs the photoelectron count, Y2. The filtered MAI signal is employed to remove the MAI from the spectral components coupled by the l×1 coupler, i.e. the residual MAI.

## 3   APD Mismatch Analysis

We assume that the number of active users is $N$ and that there are $I$ interfering users. Furthermore, without loss of generality, it is assumed that the first user is the desired user and that $b_0$ is the desired bit. The average photon arrival rate $\lambda$ per pulse at the input of the optical correlator in the first branch is given by.

$$\lambda = \eta P / hf \ . \tag{1}$$

where, $P$ is the input power, $h$ is the Planck's constant, $f$ is the optical frequency and $\eta$ is the APD quantum efficiency. The power of each spectral component is:

$$\lambda = \eta P / vhf \ . \tag{2}$$

Each user contributes one spectral component in the desired user's signal and is given as:

$$I = \sum_{k=1}^{K} i_k \ . \tag{3}$$

Given $N = I$ and the desired bit $b_0 = 1$, the mean and variance of output $Y_1$ after the sampler in the first branch can be expressed as [5]:

$$\mu_{y1} = G_1 T_c \left[ (k+1)\lambda + \frac{I_{b1}}{e} \right] + \frac{T_c I_{s1}}{e} \ . \tag{4}$$

$$\sigma_{y1}^2 = G_1^2 F_{e1} T_c \left[ (k+1)\lambda + \frac{I_{b1}}{e} \right] + \frac{T_c I_{s1}}{e} + \sigma_{th}^2 \ . \tag{5}$$

where, $G_1$ is the average APD gain of upper APD, $e$ is the electron charge, $I_b$ is the APD bulk leakage current, $I_s$ is the APD surface leakage current, and $T_c$ is the chip duration, $F_e$ is the excess noise factor, and $\sigma_{th}$ is the variance of thermal noise (the subscript 1 represent the terms of upper branch).

Similarly, given $I$ and $b_0 = 0$, the mean and variance of $Y_1$ is given by (6) and (7), respectively.

$$\mu'_{y1} = G_1 T_c \left[ I\lambda + \frac{I_{b1}}{e} \right] + \frac{T_c I_{s1}}{e} \ . \tag{6}$$

$$\sigma_{y1}^{2'} = G_1^2 F_{e1} T_c \left[ I\lambda + \frac{I_{b1}}{e} \right] + \frac{T_c I_{s1}}{e} + \sigma_{th}^2 \ . \tag{7}$$

Given $I$, the mean and variance of the output $Y_2$ can be expressed by (8) and (9), respectively.

$$\mu_{y2} = G_2 T_c \left[ (k-1)\lambda + \frac{I_{b2}}{e} \right] + \frac{T_c I_{s2}}{e} \ . \tag{8}$$

$$\sigma_{y2}^2 = G_2^2 F_{e2} T_c \left[ (k-1)I\lambda + \frac{I_{b2}}{e} \right] + \frac{T_c I_{s2}}{e} + \sigma_{th}^2 \ . \tag{9}$$

After subtracting $rY_2$ from $Y_1$, we obtain the mean of $Y$ as given by (10) for $b_0 = 1$ and $b_0 = 0$, respectively.

$$E(Y) = \begin{cases} \mu_Y = G_1 T_c \left[ \begin{array}{c} (k+1)\lambda - \\ r\dfrac{G_2}{G_1}(k-1)I\lambda \end{array} \right] + \dfrac{T_c G_1}{e}\left[ I_{b1} - rI_{b2}\dfrac{G_2}{G_1} \right] + \dfrac{T_c}{e}\left[ I_{s1} - rI_{s2} \right] \\[4ex] \mu_Y' = G_1 T_c \left[ \begin{array}{c} I\lambda - r \\ \dfrac{G_2}{G_1}(k-1)I\lambda \end{array} \right] + \dfrac{T_c G_1}{e}\left[ I_{b1} - rI_{b2}\dfrac{G_2}{G_1} \right] + \dfrac{T_c}{e}\left[ I_{s1} - rI_{s2} \right] \end{cases} \quad (10)$$

For the nominal case, that is $G_1 = G_2 = G$, the MAI is cancelled for the condition shown in (11).

$$r = 1/(k-1) . \quad (11)$$

Then variances of $Y$ are given by equations (12).

$$Var(Y) = \begin{cases} \sigma_Y^2 = \sigma_{y1}^2 + r^2 \sigma_{y2}^2 \\ \sigma_Y^{2'} = \sigma_{y1}^{2'} + r^2 \sigma_{y2}^2 \end{cases} . \quad (12)$$

The threshold value based on the maximum *a posteriori* (MAP) theory of the On-Off Keying (OOK) demodulator is given by (13) as defined in [5].

$$\theta = \frac{\sigma_{y1}' \mu_{y1} + \sigma_{y1} \mu_{y1}'}{\sigma_{y1} + \sigma_{y1}'} . \quad (13)$$

The bit error can be derived from (14) to (18)

$$\begin{aligned} \Pr(error \mid N = n) &= \Pr(Y \geq \theta \mid N = n, b_o = 0) \times \Pr(b_o = 0) \\ &+ \Pr(Y < \theta \mid N = n, b_o = 1) \times \Pr(b_o = 1) \\[2ex] &= \frac{1}{2}\sum_{i=o}^{n-1} \Pr(Y \geq \theta \mid I = i, b_o = 0) \times \Pr(I = i \mid N = n, b_o = 0) \\ &+ \frac{1}{2}\sum_{i=o}^{n-1} \Pr(Y < \theta \mid I = i, b_o = 1) \times \Pr(I = i \mid N = n, b_o = 1) \end{aligned} \quad (14)$$

where

$$\Pr(Y \geq \theta \mid I = i, b_o = 0) = \int_\theta^\infty \frac{1}{\sqrt{2\pi\sigma_{Y0}^2}} \exp\left[ -\frac{(y - \mu_{Y0})^2}{2\sigma_{Y0}^2} \right] dy = \frac{1}{2} erfc\left( \frac{\theta - \mu_{Y0}}{\sqrt{2\sigma_{Y0}^2}} \right) . \quad (15)$$

$$Pr(Y < \theta \mid I = i, b_o = 1) = \int_{-\infty}^{\theta} \frac{1}{\sqrt{2\pi\sigma_{Y1}^2}} \exp\left[-\frac{(y - \mu_{Y1})^2}{2\sigma_{Y1}^2}\right] dy = \frac{1}{2} erfc\left(\frac{\mu_{Y1} - \theta}{\sqrt{2\sigma_{Y1}^2}}\right) \cdot \quad (16)$$

$$Pr(I = i \mid N = n, b_o = 0) = Pr(I = i \mid N = n, b_o = 1) = \binom{n-1}{i}\left(\frac{1}{2}\right)^{n-1} \cdot \quad (17)$$

and, $erfc(\cdot)$ stands for the complementary error function, as defined in (18)

$$erfc(x) = \frac{2}{\sqrt{\pi}} \int_{x}^{\infty} \exp\left(-y^2\right) dy \cdot \quad (18)$$

## 4  Experiments and Results

For our experiments, the gain of the upper branch's APD, which is used to extract the information of the desired user, is kept fixed and set to the typical value of 100 [2, 8]. The gain of the lower branch's APD is incremented and decremented in steps of 5%. It is assumed that the power received per pulse is fixed at $P = 140\mu W$. The values of

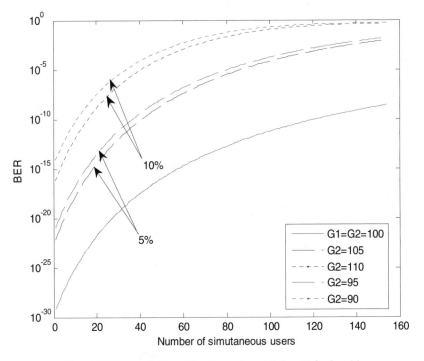

**Fig. 2.** BER probability under Gain mismatch for [5] for k = 14

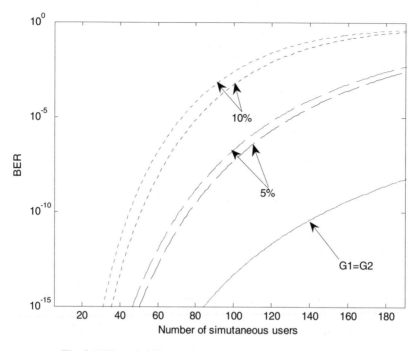

**Fig. 3.** BER probability under Gain mismatch for [5] for k = 17

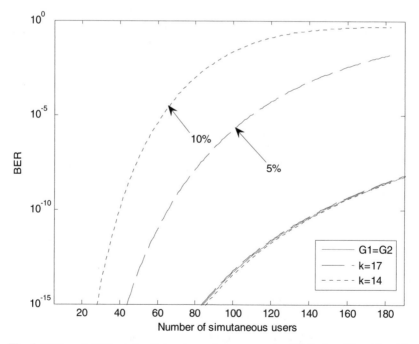

**Fig. 4.** BER probability under Gain mismatch compensation [5] for k = 17 and k = 14

shot noise, thermal noise, and bulk and surface leakage currents, are taken separately for both the APDs. We use the value of $r$ obtained from (11) which assume that the gains of both APDs are perfectly matched. However, in practice this is not the case. As a result, the MAI is not completely cancelled, and consequently only a 5% difference between the gains of the two APDs causes more than 50% reduction in the system capacity, as shown in Fig. 2. Fig. 2 and Fig. 3 shows the mismatch analysis for k = 14 and k = 17, respectively. This problem can be compensated, if the gains of the two APDs used are accurately measured under the prevailing conditions. This condition for the complete cancellation of MAI is defined by $r = G_1(k + I)/G_2(k - 1)I$.

As illustrated in Fig. 4 for the compensated case, the performance of the system is almost similar to the case when the branches are completely matched. The experiments also show that the effect of differences in bulk and surface leakage currents of two APDs are negligible.

**Acknowledgments.** This research is supported by Mehran University of Engineering & Technology, Pakistan.

# References

1. Stok, A., Sargent, E.H.: Lighting the Local Area: Optical Code-Division Multiple Access and Quality of Service Provisioning. IEEE Network (November/December 2000)
2. Weng, X.-S., Wu, J.: Perfect difference codes for synchronous fibre-optic CDMA communication systems. Journal of Lightwave Technology 19(2), 186–194 (2001)
3. Hossam, M., Shalaby, H.: Synchronous Fibre-Optic CDMA Systems with Interference. Estimators 17(11), 2268–2275 (1999)
4. Kwon, H.M.: Optical orthogonal code-division multiple-access system.I. APD noise and thermal noise. IEEE Transactions on Communications 42(7), 2470–2479 (1994)
5. Jhou, J.S., Lin, J.Y., Wen, J.H.: Performance enhancement of perfect difference codes with spectral amplitude coding using sub-optimal decision threshold. European Transactions on Telecommunications (2007)
6. Takahashi, H., Oda, K., Toba, H., Inoue, Y.: Transmission characteristics of arrayed waveguide wavelength multiplexer. IEEE Journal of Lightwave Technology 13, 447–455 (1995)

# Blocking Probabilities in WDM Switching Networks Using Overflow Analysis Method

Arif Ali Rehman, Abid Karim, and Shakeel Ahmed Khoja

Bahria University (Karachi Campus) 13 National Stadium Road 75260 Karachi Pakistan
{rehmanaf, abid, shakeel}@bimcs.edu.pk

**Abstract.** Wavelength Division Multiplexed switching networks are considered as an important candidate for the future transport networks. As the size of network increases, conventional methods used in teletraffic theory to model these networks become computationally difficult to handle, as the state space grows exponentially. In this paper, we have applied overflow analysis to model these networks. Our results show that moment analyses using equivalent random theory (ERT) results in accurate approximations for the modeling of WDM switching networks.

**Keywords:** WDM switching networks, Equivalent Random Theory (ERT).

## 1 Introduction

For communication networks such as wireless or wavelength division multiplexing optical networks with wavelength conversion capabilities, the teletraffic modeling results in a multidimensional systems. Glenstrup *et al.* [1] has demonstrated an exact method for the calculation of the blocking probability of small size systems. Yates *et al.* [2] has developed an approximate method to calculate the blocking probabilities using graph-theoretical approach for the networks with fixed routing with limited conversion degree. Subramanium *et al.* [3] has focused on conversion density using Bernoulli-Poisson-Pascal (BPP) for the evaluation of blocking probability; however for large systems these methods face limitations due to the exponential growth of state space [4].

In this paper, we have shown the use of overflow analysis method for calculating blocking probability of the given network. Our approach is applicable for the WDM switching networks ranging from no conversion to full conversion capability, with any numbers of fibers and wavelengths.

In section 2, a theoretical model is presented to describe the WDM network. Section 3 discusses the analytical approach used for the calculation of blocking probability of the network, whereas discussion of resulted is presented in section 4. The analytical results are shown to agree well with the network simulation.

## 2 Theoretical Model

Consider a WDM switching network as shown in Figure 1. There are $N_1$ incoming fibers and $N_2$ outgoing fiber links to the network. The multiplexing degree which

D.M.A. Hussain et al. (Eds.): IMTIC 2008, CCIS 20, pp. 314–318, 2008.
© Springer-Verlag Berlin Heidelberg 2008

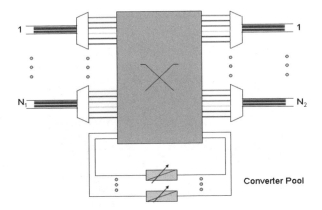

**Fig. 1.** WDM Network Model

corresponds to the number of incoming channels (wavelengths) per fiber is $M$. The network has a provision of a pool of "K" wavelength converters. Each incoming wavelength from the fiber is separated using de-multiplexers.

The internal design of the switching node switches the incoming connection requests of particular wavelengths to switching elements specific for those wavelengths. The switched wavelength is sent to the multiplexer of the respective route. In case if the wavelength on the outgoing wavelength is already occupied then the incoming connection request is internally switched to tunable wavelength converters Each tunable wavelength converter can receive wavelength from any switching element and converts to the wavelength which is free on the outgoing fibers and send the wavelength to the respective switching element.

Considering switching network as primary group of $n_1$ servers and wavelength converters as secondary groups $n_2$ servers respectively (fig. 2) , the above system can be assumed as a overflow traffic model with following assumptions.

- The arriving requests exhibit Poisson distribution with intensity $\lambda$.
- The holding time distribution is negative exponential with parameter $\mu$.
- With parameters $\lambda$ and $\mu$ the intensity of offered traffic is given as $\lambda/\mu$.
- The incoming request is initially offered to primary servers.
- If all primary servers are occupied then the traffic is over flown to the secondary servers forming an interrupted Poison process (IPP) with traffic intensity $\alpha$.
- If all the servers in secondary group are occupied then the incoming request is completely blocked.
- The connection on the converted wavelength will simultaneously keep the possession of wavelength converter as well as outgoing fiber as long as it exists.
- The primary servers have not only to serve the incoming requests but also the converted wavelength requests.
- With each converted wavelength the number of servers is reduced to serve the new incoming requests.

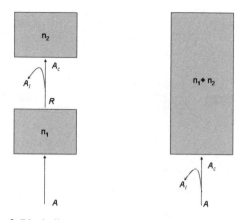

**Fig. 2.** Block diagram representation of overflow system

The primary and secondary groups jointly form loss system where number of servers is equal to $n_1 + n_2$. The division of the capacity into two parts does not affect the overall behavior of the system, as it always forms M/M/n system.

Equivalent random theory (ERT) method provides an approximate method to calculate the blocking probability for non-Poisson traffic. The traffic is defined using mean intensity $R$ and variance $V$ of the occupancy in an infinite system. For the case of several independent contributors the mean intensity and variance is equal to the sum of individual mean intensities and variances respectively. The idea of ERT method is to get a traffic $(R, V)$ from a fictitious channel with offered traffic $A*$ and number of servers $N*$. The values of $A*$ and $N*$ are such calculated that the overflow traffic in the fictitious channel has intensity $R$ and $V$. The values of $A*$ and $N*$ are calculated numerically as described in [5, 6]. Thereafter the blocking in the overflow channel is calculated as:

$$\frac{A^* E\left(N_1 + N^*, A^*\right)}{R}.$$
(1)

and the total blocking of the system is calculated as:

$$\frac{A^* E\left(N_1 + N^*, A^*\right)}{\sum_i A_i}.$$
(2)

## 3   Results and Discussion

For calculations of blocking probability five fibers, fifty wavelengths, while varying the number of wavelength converters used ranging from no conversion to full conversion. Fig. 3 shows the result of numerical analyses as performed using algorithms of one moment (solid lines) and two moment (dotted lines) methods respectively described in [7]. The results indicate that one moment analysis underestimates the

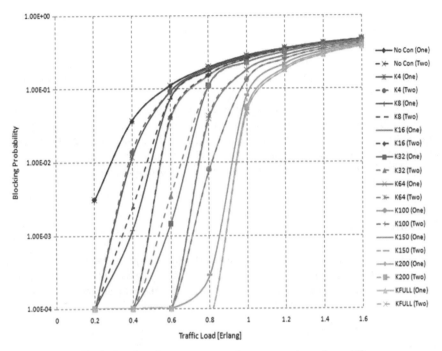

**Fig. 3.** Results of one moment and two moments analyses [7]

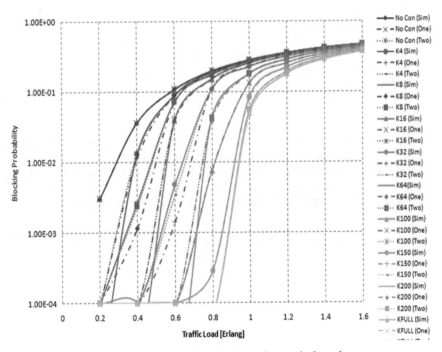

**Fig. 4.** Comparison of simulation and numerical results

blocking probability for small value of offered load; however, for large traffic loads predictions from both methods i.e. one moment and two moments are coincident. The figure also indicates the significant improvement in the blocking characteristics with the presence of wavelength conversion. It is evident from the graph that the presence of limited number of converters results in blocking probabilities that are closer to full conversion case.

Fig. 4 shows the comparison of the results using numerical analysis and simulations. The results of simulation and two moment analysis agree well for small and large values of traffic, whereas there is a significant difference between the results of simulation and one moment analysis for the case of small value of traffic. For large value of traffic load simulation, one moment and two moment analysis result in well agreed model.

# 4  Conclusion

In this paper, we presented the analytical model for analyzing WDM switching networks with limited number of wavelength converters using overflow analysis method. Analytical methods using one moment and two moment analysis techniques of ERT method are presented and comparison of results is presented.

# References

1. Glenstrup, A.J., Iverson, V.B.: Exact Evaluation of Blocking in WDM Networks, COM, Technical University of Denmark (2001)
2. Yates, J.M., Rumsewicz, M.P., Lacey, J.P.R., Everitt, D.E.: Modeling Blocking Probabilities in WDM Networks with Fixed Alternate Routing, Australian Photonics Cooperative Research Centre, University of Melbourne (1997)
3. Subramanium, S., Somani, A.K., Azizoglu, M., Barry, R.A.: The Benefits of Wavelength Conversion in WDM Networks with Non-Poisson Traffic. IEEE Communication Letters 3(3) (1999)
4. Tripathi, T., Sivarajan, K.N.: Computing Approximate Blocking Probabilities in Wavelength Routed All Optical Networks with Limited Range Wavelength Conversion. IEEE Journal on Selected Areas in Communications 18(10), 2123–2129 (2000)
5. Iverson, V.B.: Teletraffic Engineering Handbook. ITC in cooperation with ITU-D SG2, Geneva (2003)
6. Jagerman, D.L.: Methods in Traffic Calculations. AT&T Bell Laboratories Technical Journal 63(7), 1283–1310 (1984)
7. Rehman, A.A., Karim, A.: Modeling and Performance Analysis of WDM Switching Networks with a Limited Number of Wavelength Converters. In: IEEE-INMIC 9th International Multitopic Conference, Karachi, Pakistan, pp. 1–6 (2005)

# Design and Implementation of a Selective Harmonic Elimination SPWM Strategy-Based Inverter

M. Zafarullah Khan, M. Mubashir Hussain, and D.M. Akbar Hussain

Institute of Industrial Control Systems, P.O. Box 1398, Rawalpindi, Pakistan
mmh.office@gmail.com

**Abstract.** After comparison of various Sinusoidal Pulse Width Modulation (SPWM) schemes, the selection of Selective Harmonic Elimination (SHE) SPWM is justified through MATLAB-based simulations. First, the results of MATLAB-based spectral analysis of simulated three-phase SHE SPWM waveforms are described. Next, the three-phase SPWM signals generated using a microcontroller and the SPWM outputs of a three-phase inverter are analyzed using MATLAB and the results compared. Brief descriptions of the algorithm employed and hardware used to implement the SHE SPWM strategy are also given.

**Keywords:** Sinusoidal Pulse Width Modulation (SPWM), Selective Harmonic Elimination SPWM (SHE SPWM), MATLAB, Total Harmonic Distortion (THD), Natural Sampling SPWM, Regular Sampled Asymmetric SPWM, Modified Asymmetric Regular Sampled SPWM, Regular Sampled Symmetric SPWM, PIC Microcontroller, Gate Drive Amplifier (GDA).

## 1 Introduction

Due to pulsating torque, the heating effects and losses incurred because of flow of harmonic currents in motor windings, research is being focused on evolving efficient methods of power conversion and control for motors, which generate minimum amount of harmonics (especially the lower order ones) in the output voltage and current. In addition, extensive efforts are being put in finding easily implementable power conversion techniques that give lower output Total Harmonic Distortion (THD) thereby requiring lesser filtering components and lower switching losses. With the aim to reduce the output THD of inverters, several SPWM strategies were considered and keeping in view simplicity of implementation, fewer switching per cycle and elimination of first few harmonics due to their undesired effects, the Selective Harmonic Elimination SPWM (SHE SPWM) strategy was considered.

## 2 Some SPWM Strategies

Some of the SPWM strategies considered were [1]:

**Natural Sampling Strategy.** This strategy is based on direct comparison of a sinusoidal modulating waveform with a triangular carrier wave to determine the switching

D.M.A. Hussain et al. (Eds.): IMTIC 2008, CCIS 20, pp. 319–331, 2008.

angles and the pulse widths. The pulse width is proportional to the amplitude of modulating waveform at the instant that comparison occurs. The advantage of this strategy is that as the carrier frequency is increased, harmonic content reduces. The disadvantage is that the underlying mathematics make it inappropriate for computer based analysis and digital implementation. It is only suitable for analog implementation.

**Regular Sampled Asymmetric SPWM Strategy.** In this strategy, a Sinusoidal modulating waveform is sampled at both positive and negative peaks of the triangular carrier to generate 'sample and hold' equivalent of the modulating waveform for comparison with the carrier waveform. Its harmonic spectrum is good but requires longer computation time.

**Modified Asymmetric Regular Sampled PWM Strategy.** This strategy uses the sampling height by averaging the two successive regularly sampled points taken at both positive and negative peaks of the carrier triangular wave. This strategy is capable of generating more accurate switching angles and produces lower harmonic distortion.

**Regular Sampled Symmetric PWM Strategy.** In this simple strategy the sinusoidal modulating waveform is sampled at every positive peak of the triangular carrier waveform prior to comparison with the carrier. Both edges of the pulse widths for this strategy are equidistant from the centre point.

## 3 Introduction to Selective Harmonic Elimination SPWM Strategy

The carrier based SPWM strategies described previously have the main advantage of shifting the annoying band of lower order harmonic frequencies closer to the higher carrier frequency, which can then be filtered easily. However, these strategies also generate harmonics of the carrier and tend to increase the harmonic content that can cause oscillations in the output filter. In addition, for a microprocessor/ microcontroller based implementation these strategies require a compromise between the use of lookup tables and on-line computations to generate three-phase outputs. If all timings were computed off-line and programmed into the on-chip memory then expensive and complex devices like advanced microcontrollers with larger on-chip memories would be required. On the other hand, if on-line computation were performed than faster devices like Digital Signal Processors would be required. These SPWM strategies also require large number of switching per cycle thereby increasing switching losses.

The next best option is to use the Selective Harmonic Elimination (SHE) strategy [2], [3]. This strategy eliminates selected harmonics (the annoying lower order ones and their multiples) and doesn't introduce any carrier and its multiples as in strategies described previously. However, the main disadvantage of this strategy is that it requires numerical computations that are impossible to implement within the requisite timing constraints using the existing microcontroller/DSP technology.

In the strategy under discussion, the number of notches (switching) per quarter cycle is selected depending upon the number of lower order harmonics to be eliminated. In a single phase, two level SPWM waveform with odd and half wave symmetries shown in Fig. 1, introduction of say $n$ notches within a quarter cycle gives us $n$

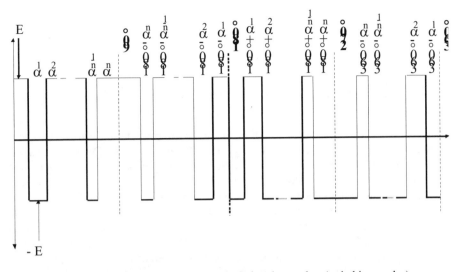

**Fig. 1.** A sample SPWM waveform depicting the notches (switching angles)

degrees of freedom. We then have the choice of either eliminating $n$ harmonics or eliminating $n$-1 harmonics and controlling the amplitude of the fundamental using the remaining degree of freedom. We have used the latter approach.

With $n$ notches, the peak magnitudes of the harmonic components including the fundamental obtained using Fourier analysis are given by the following set of equations:

$$h_1 = \left(\frac{4E}{\pi}\right)\left(1 - 2\cos(\alpha_1) + 2\cos(\alpha_2) - 2\cos(\alpha_3) \cdots\cdots\cdots 2\cos(\alpha_n)\right)$$

$$h_3 = \left(\frac{4E}{3\pi}\right)\left(1 - 2\cos(3\alpha_1) + 2\cos(3\alpha_2) - 2\cos(3\alpha_3) \cdots\cdots\cdots 2\cos(3\alpha_n)\right)$$

$$h_5 = \left(\frac{4E}{5\pi}\right)\left(1 - 2\cos(5\alpha_1) + 2\cos(5\alpha_2) - 2\cos(5\alpha_3) \cdots\cdots\cdots 2\cos(5\alpha_n)\right) \quad (1)$$

$$\cdots\cdots\cdots\cdots\cdots\cdots$$
$$\cdots\cdots\cdots\cdots\cdots\cdots$$

$$h_k = \left(\frac{4E}{k\pi}\right)\left(1 - 2\cos(k\alpha_1) + 2\cos(k\alpha_2) - 2\cos(k\alpha_3) \cdots\cdots\cdots 2\cos(k\alpha_n)\right)$$

where, $h_1$ represents the peak value of fundamental and $h_k$ is the $k^{th}$ harmonic. In addition, $\alpha_j$ is the $j^{th}$ notch or switching angle within quarter cycle of a sinusoidal wave. $E$ is the DC bus voltage. It is pertinent to note that due to the symmetry selected, even harmonics are not generated.

Let us assume that we want to eliminate 3$^{rd}$ and 5$^{th}$ harmonics. For the purpose we will select n = 3, i.e. three notches per quarter cycle and set $h3$ and $h5$ to 0. Depending upon the DC bus voltage $E$ and requisite value of $h_1$ we can then numerically solve the following set of equations for $\alpha_1$, $\alpha_2$ and $\alpha_3$:

$$h_1 = \left(\frac{4E}{\pi}\right)(1 - 2\cos(\alpha_1) + 2\cos(\alpha_2) - 2\cos(\alpha_3))$$

$$0 = \left(\frac{4E}{3\pi}\right)(1 - 2\cos(3\alpha_1) + 2\cos(3\alpha_2) - 2\cos(3\alpha_3)) \qquad (2)$$

$$0 = \left(\frac{4E}{5\pi}\right)(1 - 2\cos(5\alpha_1) + 2\cos(5\alpha_2) - 2\cos(5\alpha_3))$$

By observing the above set of equations it can be inferred that we can proceed with the angle (notch) computation in two ways. First we can keep the DC bus voltage $E$ constant and find the angles by varying $h_1$. Secondly, we can vary E and compute the angles selecting $h_1$ to be a fixed percentage of $E$ each time. The inverter described in this paper uses the latter scheme since it is easier to implement than the former.

As is evident, it is impossible to solve equations like those above numerically and within requisite timing constraints using existing microcontroller technology. This is especially so if more harmonics are to be eliminated. Another disadvantage of this strategy is that the harmonic content increases considerably with slightest error in computed angle values. However, despite these constraints an efficient scheme has been implemented that offers best compromise between offline and on-line computations, and memory requirements.

## 4   Simulation Results

In our implementation of the three-phase SHE SPWM strategy, we have eliminated the 5th, 7th and 11th harmonics and their multiples. For the purpose, first angles $\alpha_1$, $\alpha_2$, $\alpha_3$ and $\alpha_4$ were computed numerically off-line for AC voltage range from 0 to 331 Vrms (82.75% of $E$), where $E$ was selected to be 400 V. The four angles could not be computed reasonably above 331 Vrms. The computed angles are plotted versus the rms voltages in Fig. 2

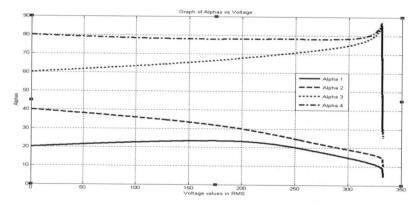

**Fig. 2.** Graph of voltage vs. $\alpha_1$, $\alpha_2$, $\alpha_3$ and $\alpha_4$

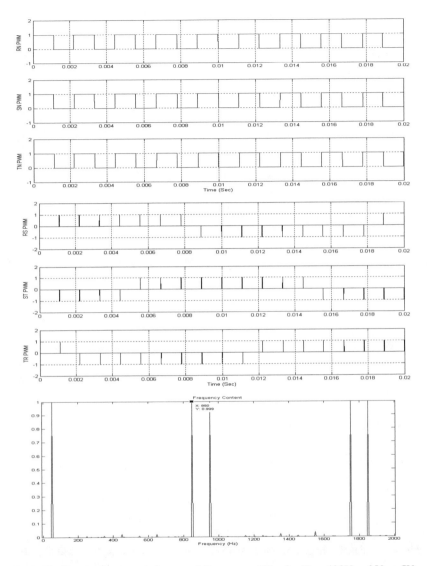

**Fig. 3.** Simulated voltage waveforms and Spectrum of $V_{RS}$ for $V_{DC}$=400V and $V_{AC}$=5V

After the angles $\alpha_1$, $\alpha_2$, $\alpha_3$ and $\alpha_4$ within the range $0 < \alpha_k < 90°$ (Refer to Fig. 1 for finding angles within the range $0 < \alpha < 360°$) were computed, an algorithm was developed in MATLAB to convert angles into time. The following equation was used to convert angles within the range of 0 to 360° to time in seconds:

$$Time = \left( \left( \frac{\alpha}{360} \right) \times \left( \frac{1}{FREQ} \right) \times F_{Ins} \right) \qquad (3)$$

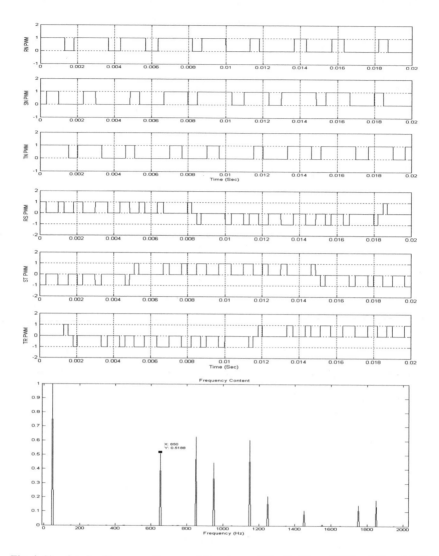

**Fig. 4.** Simulated voltage waveforms and Spectrum of $V_{RS}$ for $V_{DC}$=400V and $V_{AC}$=165V

where:

> $\alpha$: Notch angles (in degrees)
> $FREQ$ : Frequency of operation of Inverter
> $F_{Ins}$ : Instruction Cycle Rate of the microcontroller used

The waveform switching times thus obtained were used to generate a SPWM wave-form corresponding to R-phase. The S-Phase and T-Phase waveforms were obtained just by phase shifting the R-phase waveform. Waveforms corresponding to the three

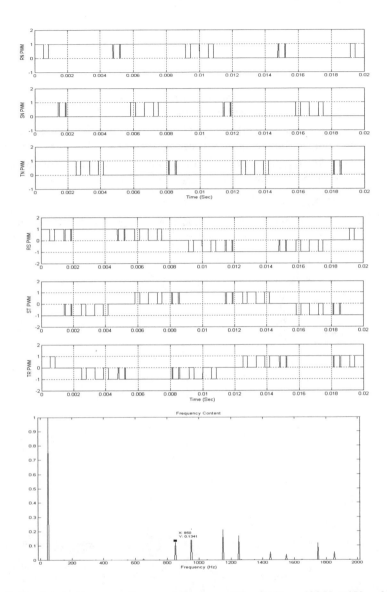

**Fig. 5.** Simulated voltage waveforms and spectrum of $V_{RS}$ for $V_{DC}$= 400 V and $V_{AC}$=331V

phases were implemented in MATLAB and spectral analysis was carried out to determine the harmonic spectra. The simulation results for 5 Vrms, 165 Vrms and 331 Vrms are given in Figures 3, 4 and 5 respectively. Following where the observations:

- If the DC voltage $E$ was assumed to be constant, the switching angles $\alpha$ varied with variation in AC output voltage.

- In case of Phase-to-Neutral voltages $V_{RN}$, $V_{SN}$ and $V_{TN}$, although the 5th, 7th, 11th and their multiple harmonics were eliminated, the 3rd harmonic and its multiples were present.
- In case of Phase-to-Phase voltages $V_{RS}$, $V_{ST}$ and $V_{TR}$, the 3rd harmonic and multiples were missing in addition to 5th, 7th, 11th and their multiple harmonics.
- Even the 13th harmonic and its multiples were missing.
- It was observed that harmonic spectra improved with increasing AC Voltage. If we compare Figures 3 to 5, it can be seen that at lower AC voltages, the magnitudes of higher order harmonics are large in comparison to the fundamental, i.e. 50 Hz. However, at increased AC voltages, the higher order harmonics are attenuated considerably.

After extensively evaluating the pros and cons of the SHE SPWM strategy using MATLAB, the strategy was implemented using a PIC18F452 microcontroller running

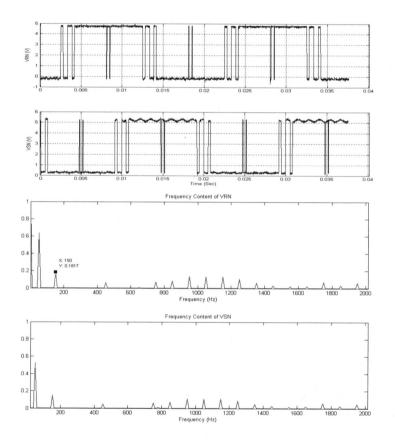

**Fig. 6.** $V_{RN}$ and $V_{SN}$ Voltages at output of microcontroller and their spectrum

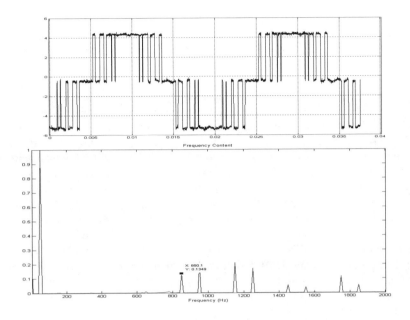

**Fig. 7.** $V_{RS}$ Voltage waveform at the output of microcontroller and its Spectrum

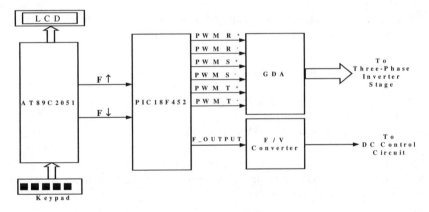

**Fig. 8.** Block Diagram for three phase PWM generation circuit

at 40 MHz. It was observed that the SPWM outputs of the microcontroller matched with the simulated waveforms and results of the spectral analysis were similar, as is evident from comparison of Fig. 3 to Fig. 5 with Fig. 6 and Fig. 8.

## 5   Firmware Implementation

It was observed that two possible schemes are possible for implementing a three-phase SHE SPWM based inverter. In the first approach, DC bus is kept constant and the notch angles are computed offline for every possible output AC Voltage. The main advantage of using this scheme is that the rectifier stage can be implemented using a simpler three-phase AC to DC conversion strategy. On the other hand, the downside is that we would have to compute the timings on-line corresponding to each angle for each output voltage and at each possible frequency. If the timings were computed off-line and stored in on-chip memory of a microcontroller then prohibitively large memory would be required. On the other hand if timings were computed on-line than a faster microcontroller or a DSP would be required.

The other approach is to allow the DC bus to vary. In this approach if the ratio of output AC Voltage to DC Voltage is fixed, e.g., VAC/VDC = 0.8275, then neither the notch angles $\alpha$ nor the harmonics spectrum change with varying VDC and VAC. However, due to the requirement of variable DC voltage output, the rectifier stage must employ a complex AC to DC conversion strategy. On the other hand, the firmware for generating three-phase SPWM is simpler, since in this case we will have to generate only a single set of notch angles, which would be valid from minimum value to maximum value of VDC. Thus as VDC would vary from 0 to 400 V, VAC would vary from 0 to 331 Vrms and the timing corresponding to the notch angles will depend only on the operating frequency. In this particular scheme, the notch angles are computed off-line and their corresponding timings are computed on-line depending upon the frequency of operation.

Due to simpler implementation requirements, the latter scheme was selected for implementing a three-phase inverter. It was observed that due to the symmetry of all of the three SPWM outputs, only the timings corresponding to the first quarter of each SPWM wave was required to be computed and repeated for the rest of the time period, thereby further simplifying the firmware implementation.

## 6  Hardware Implementation

Despite simplifications inherent in the SHE SPWM generation scheme employed, the timing requirements were still stringent and a single PIC18F452 microcontroller capable of operating up to 40 MHz (Minimum timing resolution of 100 nsec) was employed. Another microcontroller was only required to signal increase/decrease of frequency, accepting inputs from the user and for driving a LCD for providing a friendly GUI for the user. The inputs and outputs of the two microcontrollers are depicted in the block diagram of Fig. 8.

## 7  Conclusion

A 5 KVA three-phase Inverter with a frequency range of 0 to 400 Hz based on the SHE SPWM algorithm described in this paper was developed and tested using a resistive load and an Induction Motor. A simplified block diagram of the inverter is shown in Fig. 9. A Gate Drive Amplifier (GDA) for IGBTs was also developed indigenously with all protection features and at low cost.

It was observed that waveforms and harmonic spectrum of three-phase SPWM outputs of the inverter matched with the simulation results as well as the microcontroller output. For 50 Hz operation of inverter, Figs. 10 to 14 depict practical results

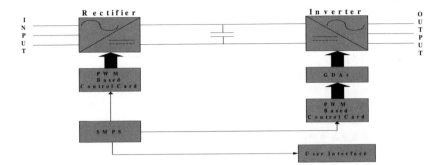

**Fig. 9.** Simplified block diagram of the 5KVA Inverter developed

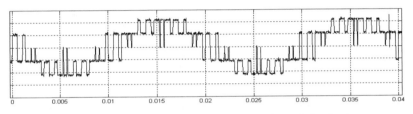

**Fig. 10.** Phase to Neutral voltage waveform for resistive load

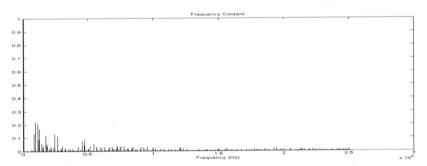

**Fig. 11.** Spectrum of Phase to Neutral voltage for resistive load

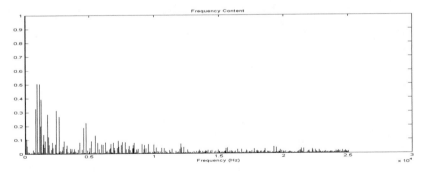

**Fig. 12.** Spectrum of Line Current for resistive load

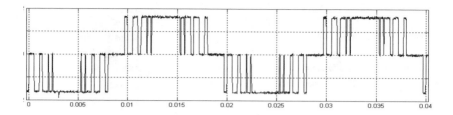

**Fig. 13.** Waveform of Phase to Phase Voltage for resistive load

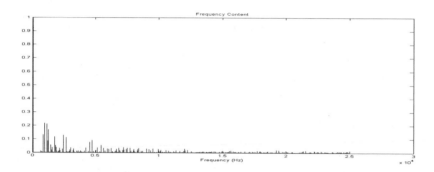

**Fig. 14.** Spectrum of Phase-to-Phase Voltage for resistive load

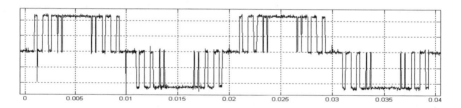

**Fig. 15.** Waveform for Phase-to-Phase Voltage for Inductive load

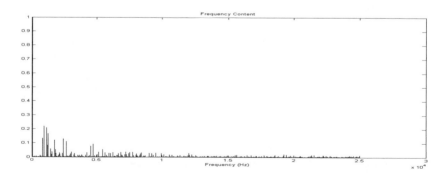

**Fig. 16.** Spectrum of Phase-to-Phase Voltage for Inductive load

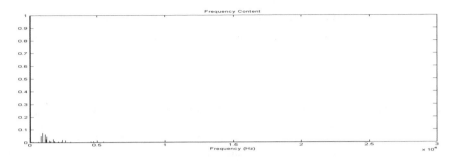

**Fig. 17.** Spectrum of Line Current for Inductive load

for a resistive load and Figs. 15 to 17 depict results for inductive load. As can be observed, the introduction of dead time and delays due to the GDA did not affect the frequency spectrum. In addition, it is evident that the SHE SPWM strategy provides a cleaner frequency spectrum as compared to carrier based SPWM strategies that generate multiple harmonics of the carrier frequency.

## References

1. Khan, M.Z.: Microprocessor Controlled PWM Inverters for UPS Applications, Doctoral Thesis, Loughborough University of Technology, UK (1989)
2. Gole, A.M.: Course notes in 24.437 Power Electronics, PWM Techniques for Harmonic Reduction in VSC, University of Manitoba (2000)
3. Bose, B.K., Sutherland, H.A.: A High-Performance Pulse width Modulator for an Inverter-Fed Drive System using a Microcomputer. IEEE Transactions on Industry Applications IA-19(2), 235–243 (1983)

# Case Study: Investigating the Performance of Interactive Multiple Motion Model Algorithm for a Crossing Target Scenario

D.M. Akbar Hussain[1], Shaiq A. Haq[2], M. Zafar Ullah Khan, and Zaki Ahmed

[1] Department of Software Engineering & Media Technology
Aalborg University, Esbjerg Denmark
akbar@aaue.dk
[2] Department of Mechatronics
Air University, Islamabad Pakistan
shaiq_haq@yahoo.com

**Abstract.** The paper presents a case study on the performance of the interactive multiple motion (IMM) model technique in tracking more than one target, especially when the targets are crossing each other during their motion. We have used a selective approach in choosing multiple motion models, thus providing wider coverage to track straight line as well as manoeuvring targets. Initially, there are two motion models in the system to track each target. The probability of each model being correct is computed through a likelihood function for each. The study presents a simple technique to introduce additional models into the system using deterministic acceleration, which basically defines the dynamics of the system. Therefore, based on this value, more motion models are employed to increase the coverage. Finally, the combined estimate is obtained using *posteriori* probabilities from different filter models. The case study shows that when targets are well separated from each other and may be manoeuvring, the system adequately tracks them. However, when the targets are close to each other or when they cross each other, the system performance decreases and large tracking error occurs.

**Keywords:** Interactive Multiple Model (IMM), Target Tracking, Simulation, Kalman Filter.

## 1 Introduction

Tracking a single straight-line motion target can be adequately achieved using a standard Kalman filter, however, for maneuvering motion model one needs an extended Kalman filter. Most investigations are performed on a single maneuvering target for example using a maneuvering detector for switching the filter motion between a non-maneuvering to a maneuvering model. It should also be noted that large errors typically occur during the switching period and the identification of maneuver/non-maneuver. Also it is difficult to identify the correct time when the target returns to a non-maneuvering mode. The study presented here investigate a complex scenario in

D.M.A. Hussain et al. (Eds.): IMTIC 2008, CCIS 20, pp. 332–342, 2008.

which case more than one target are present in the tracking space and these targets may be maneuvering or non-maneuvering. The tracking technique used in our investigation is based on a pioneering work related to maneuvering target tracking, that is Interacting Multiple Model (IMM) algorithm [1], where more than one model is used to describe the motion of the target [2], [3], [4]. The probability of each model being true is found using likelihood function for the model, and movement between models is taken into account using a transition probability. The combined estimate is then obtained using a weighted sum of estimates from the filters with different models using posteriori probabilities as the weighing factor. IMM algorithm has shown better results than the switching schemes because a smooth transition is achieved from one model to another. On the other hand, IMM algorithm needs more computation power for a fixed number of models. However, when only one of the models matches the target motion good results are obtained, but in a practical situation, it cannot be achieved. Another choice is to increase the number of filter models in order to provide better coverage, one such method has been investigated in reference [5],[6] using a Selected Filter Interacting Multiple Model. Our investigation use a similar approach however, more motion models are employed to track the target. The model employment is achieved through the random bias influencing the dynamics of the system. This bias corresponds to the deterministic acceleration of the target. The investigation here is extended by considering more than one target having multiple motion model trajectories and applying the IMM algorithm to obtained the performance. Section 2 provides the IMM algorithm detail, section 3 is concerned with implementation and results, finally section 4 provides conclusion of our investigations.

## 2   IMM Algorithm

Basically, IMM is a filtering methodology by which a target state estimate is obtained at time $k$, which is a proper combination of $n$ target estimates (tracks) computed in accordance with $n$ target models [7], [8], [9], [10]. IMM algorithm is briefly described in the following paragraph with the assumption of having $m$ filters or models. The transition between models is achieved through transition probability given by Markov chain. The transition probability needed to be initialized (set) at the beginning of the algorithm. The algorithm has five sections: Mode-Match filtering, Mode Probability Update, Mixing Probability, Estimate-Covariance Combining and Interacting as shown in Figure 1.

### 2.1   Mode-Match Filtering

Filters corresponding to models are updated using standard Kalman filtering algorithm the standard Kalman filter equations for estimating the position and velocity of the target motion are:

$$\hat{\underline{x}}_{k+1/k} = \Phi \hat{\underline{x}}_k + \Gamma U_{k+1} \tag{1}$$

$$\hat{\underline{x}}_{k+1} = \hat{\underline{x}}_{k+1/k} + K_{k+1} \underline{V}_{k+1} \tag{2}$$

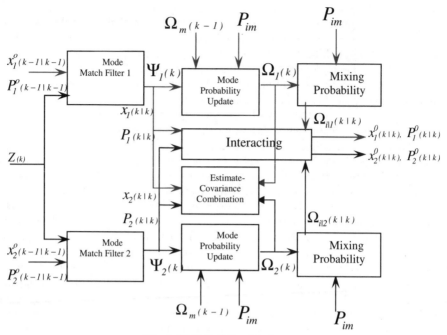

**Fig. 1.** IMM Building Blocks

$$K_{k+1} = P_{k+1/k} H^T B_{k+1}^{-1} \tag{3}$$

$$P_{k+1/k} = \Phi P_k \Phi^T + \Gamma Q_k^F \Gamma^T \tag{4}$$

$$B_{k+1} = R_{k+1} + H P_{k+1/k} H^T \tag{5}$$

$$P_{k+1} = (I - K_{k+1} H) P_{k+1/k} \tag{6}$$

$$\underline{v}_{k+1} = \underline{z}_{k+1} - H \hat{\underline{x}}_{k+1/k} \tag{7}$$

where $\hat{\underline{x}}_{k+1/k}$ , $\hat{\underline{x}}_{k+1}$ , $K_{k+1}$, $P_{k+1/k}$, $B_{k+1}$, and $P_{k+1}$ are the predicted state, estimated state, the Kalman gain matrix, the prediction covariance matrix, the covariance matrix of innovation, and the covariance matrix of estimation respectively. The predicted state includes the deterministic acceleration $U_n$ which is different for each model used here in the investigation. $Q_k^F$ is the covariance of the measurement noise assumed by the filter which is normally taken equal to $Q_k$ . In a practical situation, however, the value of $Q_k$ is not known so the choice of $Q_k^F$ should be such that the filter can adequately track any possible motion of the target. To start the computation an initial value is chosen for $P_0$ .

## 2.2  Mode Probability Update

The mode probability is a posterior probability of a model being correct provided with observation data up to scan **k**. Suppose we have **n** models, the mode probability can be calculated for a model **m** as

$$\Omega_m(k) \equiv P\{M_m(k), Z(k)\} \tag{8}$$

$$\Omega_m(k) = \frac{1}{\xi} \Psi_m(k) \sum_{i=1}^{n} p_{im} \Omega_i(k-1) \tag{9}$$

where $\xi$ is a normalizing constant and ensures that total model probabilities sum to one and $P_{im}$ is a priori probability which is assumed invariant with time, $Z_{(k)}$ is the observation (measurement) data from the sensor (Radar/Sonar). $\Psi_m(k)$ is the likelihood function corresponding to the $m^{th}$ filter. The above equation can be rewritten as

$$\Omega_m(k) = \frac{1}{\xi} \Psi_m(k) \ \overline{\xi}_m \qquad (m=1,....,n) \tag{10}$$

$$\xi = \sum_{m=1}^{n} \Psi_m(k) \ \overline{\xi}_m \tag{11}$$

$$P_{im} \equiv P\{M_{(k)} = M_m | M_{(k-1)} = M_i\} \tag{12}$$

## 2.3  Mixing Probability

Suppose mode probability for a model $M_i$ was in effect corresponding to scan **k** given that $M_m$ is in effect in the successive scan **k+1** conditioned on measurement data $Z_{(k)}$, so mixing mode probability is given by the following relationship.

$$\Omega_{i|m}(k \mid k) \equiv P\{M_i(k) \mid M_m(k+1), Z(k)\}$$

$$= \frac{1}{\overline{\xi}_m} P_{im} \Omega_i(k) \tag{13}$$

where

$$\overline{\xi}_m = \sum_{i=1}^{n} p_{im} \Omega_i(k) \tag{14}$$

## 2.4  Estimate and Covariance Combination

The output is computed by combining the state estimates from each model through the following relationships.

$$X(k \mid k) \quad = \quad \sum_{m=1}^{n} X_m(k \mid k)\Omega_m(k) \tag{15}$$

$$P_{(k \mid k)} \quad = \quad \sum_{m=1}^{n} \Omega_m(k) \tag{16}$$

$$\left\{ P_m(k \mid k) + [x_m(k \mid k) - X(k \mid k)][x_m(k \mid k) - X(k \mid k)]^T \right\}$$

## 2.5  Interacting

The mixed state estimate for the filter model $M_m$ at a scan $k$ is determined using the outputs of models and the corresponding model probability $\Omega_m$ as:

$$x_m^o(k \mid k) \quad = \quad \sum_{m=1}^{n} x_i(k \mid k)\Omega_{i\mid m}(k \mid k) \tag{17}$$

$$(m = 1, ....n)$$

$$P_m^o(k \mid k) \quad = \quad \sum_{m=1}^{n} \Omega_{i\mid m}(k \mid k) \tag{18}$$

$$\left\{ P_i(k \mid k) + [x_i(k \mid k) - x_m^0(k \mid k)][x_i(k \mid k) - x_m^0(k \mid k)]^T \right\}$$

# 3  Implementation

Figure 2 shows a block structure of our system implementation. In the first block observation data can be generated for various scenarios with different random acceleration noise and a data array is used to store the relevant information (position, velocity etc) which is then passed to the first block of the IMM algorithm [11]. The whole system was implemented on a TMS320C6713 DSK board [12]. Most of the blocks are self explanatory; for example the initialization block initiates the target estimates and also some other parameters like probability etc. The last block store the output from Estimate-Covariance Combination and the Interacting blocks (shown by dotted line box) in different arrays for the next scan processing. Both input data (first block 5th item in the Data Array) and output arrays have additional information for example color ids for targets to be used for plotting and simulation purposes. The radar is positioned at the origin of the Cartesian Coordinate Reference system, scan rate is chosen to be 1 second and the probability of detection is 100 % meaning each time we do receive the observation from the target and finally no false or clutter is considered.

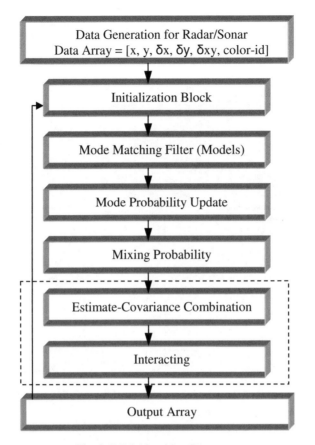

**Fig. 2.** IMM Algorithm Sequence

For the first part of our investigation, we generated data for a single target with various acceleration noise level as shown in figure 3 (a, b, c & d), blue dots are the observations received from the sensor and red-dotted line is the output trajectory estimate from the IMM algorithm. The standard deviation of acceleration processes are ranged around three typical values which corresponds to the minimum value representing straight line motion trajectory 0.1 m/s$^2$, a mid level value representing most situations 1 m/s$^2$ and the large value representing highly maneuvering trajectories 30 m/s$^2$ or more. It can be seen from the figure that the system has adequately followed the target. Table 1 provides the number of models employed by the system and the error observed by the IMM algorithm in each case.

It can be seen that as the random bias increases each time the system was able to increase the number of motion models and hence adapting itself according to the requirement. The mean squared error increases in each case, the reason for such an increase is due to the motion models not exactly matched with the trajectory of the target, which is in fact reasonable in all practical situations. However, due to more motion models the overall coverage is increased and the system is able to track the target.

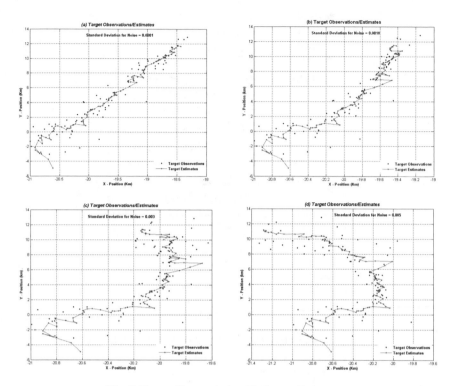

**Fig. 3.** Target Observations with Output Estimates

**Table 1.**

| Figure 3 | No. of Models | Mean Squared Error |
|----------|---------------|--------------------|
| a | 2 | 0.0553 |
| b | 3 | 0.0678 |
| c | 5 | 0.0729 |
| d | 6 | 0.0892 |

For the second part of our investigation we generated data for two targets which cross each other after some time as shown in figure 4. IMM algorithm in itself has no mechanism to deal with observations from multiple targets. Therefore, in such a case system must be provided with separate observations from individual targets before the IMM algorithm is applied for tracking. In our investigation as there are two targets so there will be two observations at each instant (scan), we need to devise a mechanism so that system (IMM algorithm) has separate observations from each target.

We implemented a very simple logic to get the respective observations from each individual targets. The distance between previously received and current observations is calculated and that current observation which is closer to an individual target is

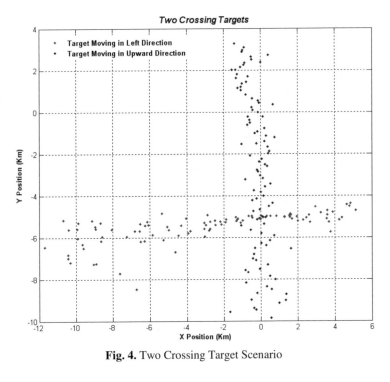

**Fig. 4.** Two Crossing Target Scenario

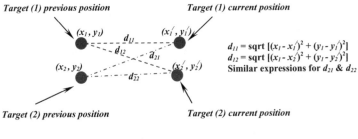

**(a)     Targets are well apart**

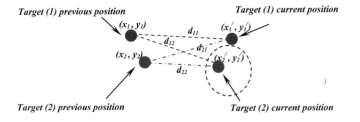

**(b)     Targets are close to each other**

**Fig. 5.** Observation Association for Current Scan

selected and considered to be the logical observation for that target as shown in Fig. 5. As long as the targets are well apart there is essentially no probability of false association (observation from a different target other than the one in question is linked) of observation, however, when target are close to each other there is probability of false association due to the statistical nature of the problem. But it is expected that IMM algorithm would compensate for such false associations by introducing more tracking models in the system. In Fig. 5 (a), where targets are well apart $d_{11} \leq d_{12}$ for observation $(1)'$, therefore, this observation is considered to be the true observation for target $(1)$. Similarly, for observation $(2)'$, $d_{22} \leq d_{21}$, so it is considered as the true observation for target $(2)$. However, situation is quite different in Fig. 5 (b) where targets are close to each other, in this case $d_{12} \leq d_{11}$ & $d_{22} \leq d_{21}$. Therefore, observation $(2)'$ is considered as the true observation for both targets $(1)$ and $(2)$, in fact which has false association for target $(1)$.

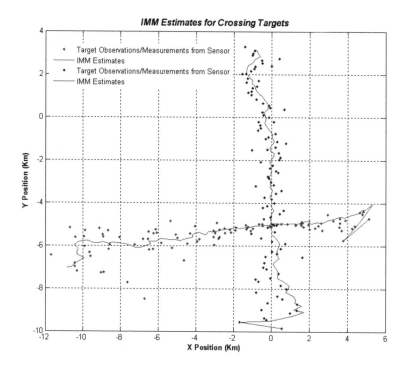

**Fig. 6.** IMM Estimates for Two Crossing Target Scenario

**Table 2.**

| Figure 4 | No. of Models | Mean Squared Error |
|---|---|---|
| For Target Moving Up | 9 | 0.1033 |
| For Target Moving Left | 8 | 0.1208 |

**Table 3.**

| | Scan No. | No. of Models For Target Moving Up | No. of Models For Target Moving Left |
|---|---|---|---|
| | 26 | 2 | 2 |
| | 27 | 4 | 4 |
| | 28-29 | 5 | 7 |
| | 30 | 8 | 8 |
| **Figure 4** | 31 | 9 | 7 |
| | 32 | 8 | 9 |
| | 33 | 9 | 8 |
| | 34 | 8 | 8 |
| | 35-36 | 7 | 8 |
| | 37 | 5 | 5 |
| | 38-41 | 3 | 3 |
| | 42 | 2 | 2 |

Fig. 6 shows the observations (dots) as well as the IMM estimates (solid line) for the said target scenario. It can be seen that system managed to track the targets reasonably well. Table 2 provides the total number of models employed by the system for the whole tracking period 100 Scan (second) and the mean squared error for each individual target. It can be seen that the total number of models introduce by the system and the tracking error is in similar range. In order to see more closely the behavior of IMM in such a scenario Table 3 gives the number of models employed by the system just before the target gets closer (cross) to each other and at the time of crossing. It can be seen that the number of models introduced by the system at crossing time increased due to possible false associations and the system assume a possible maneuver so it has increased the number of models to keep on tracking successfully.

# 4  Conclusions

The main contribution in our investigation here is that we have extended the utilization of IMM for a scenario where more than one target exist, especially, when they are close to each other. One of the difficulty in utilizing IMM algorithm is that it has no build in mechanism to deal multiple targets. We used a very simple approach to deal such a case and provided individual observations to respective targets through a simple calculation of normalized distance. However, it is reasonable to assume that even if a false observation is associated the built in mechanism of IMM algorithm using more motion models compensate for such false associations. The system is extendable for more targets however, more computational power and memory may be required. The implemented system has also provided detailed look at various stages of tracking which helps in understanding and future development of such systems. In future we would like to extend this investigation by introducing maneuvering multiple target in a close proximity. The approach provides a simulation study to design a tracking system for a multiple target tracking using IMM algorithm. The implemented algorithm also shows its adaptability by utilizing the random bias for a decision in employing additional motion models. The results shows a feasible approach to track multiple targets, however, the errors are not quite in the range one might be expecting which is mainly due to the difference in getting an exact model match with the trajectory.

# References

1. Bar-Shalom, Y.: Tracking a maneuvering target using input estimation versus the Interacting Multiple Model algorithm. IEEE Transaction Aerospace and Electronic Systems, AES-25, 296–300 (1989)
2. Bar-Shalom, Y., Li, X.R.: Estimation and tracking: Principles, Techniques and Software. Artech House (1993)
3. Bar-Shalom, Y., Fortmann, T.E.: Tracking and Data Association. Academic Press, Inc., London (1988)
4. Blackman, S.S.: Multiple-Target Tracking with Radar Applications. Artech House, Inc. (1986)
5. Lin, H.-Y., Atherton, D.P.: Investigation of IMM tracking algorithm for maneuvering target tracking. In: First IEEE Regional Conference on Aerospace Control Systems, pp. 113–117 (1993)
6. Hussain, D.M.A.: Motion Model Employment using interacting Motion Model Algorithm. In: Proceedings of International Control Conference (ICC2006), Glasgow, Scotland (2006)
7. Lee, B.J., Park, J.B., Joo, Y.H.: IMM Algorithm Using Intelligent Input Estimation for Maneuvering Target Tracking. IEICE Transactions on Fundamentals of Electronics, Communications and Computer Sciences E88-A(5), 1320–1327 (2005)
8. Simeonova, I., Semerdjiev, T.: About the Specifics of the IMM Algorithm Design. In: Dimov, I.T., Lirkov, I., Margenov, S., Zlatev, Z. (eds.) NMA 2002. LNCS, vol. 2542, pp. 333–341. Springer, Heidelberg (2003)
9. Turkmen, I.: IMM fuzzy probabilistic data association algorithm for tracking maneuvering target. Expert Systems with Applications: An International Journal 34(2), 1243–1249 (2008)
10. Hong, L., Ding, Z.: A distributed multi-rate IMM algorithm for multiplatform tracking. Mathematical and Computer Modeling 32(10), 1095–1116 (2000)
11. Hussain, D.M.A.: Tracking multiple objects using modified track-observation assignment weight approach for data association. In: INMIC-2003, International Multi-topic Conference, Islamabad, Pakistan (2003)
12. Hussain, D.M.A., Durrant, M., Dionne, J.: Exploiting the Computational Resources of a Programmable DSP Micro-processor (Micro Signal Architecture MSA) in the field of Multiple Target Tracking. In: SHARC International DSP Conference 2001, North Eastern University, Boston, USA (2001)

# A Novel Approach for the Control of Dual-Active Bridge DC-DC Converter

Abdul Sattar Larik[1], Mohammad Rafiq Abro[1], Mukhtiar Ali Unar[2], Mukhtiar Ahmed Mahar[1]

[1] Department of Electrical Engineering,
[2] Department of Computer Systems Engineering
Mehran University of Engineering and Technology, Jamshoro, Sindh, Pakistan
sattarlarik@yahoo.com, rafiq_abro@hotmail.com,
mukhtiar_unar@yahoo.com

**Abstract.** During these days high frequency dc-dc converters are frequently used in high Power Electronic applications. It is well known fact that the size of the passive components, such as inductors, capacitors and power transformers decreases as the frequency increases. It is well known that high-voltage converters in the power range of 1-10MW have excessive switching losses at the switching frequencies higher than 4 kHz. The favorable DC-DC converter is realized to be the Dual-Active Bridge when a bi-directional power flow is demanded; however the most important problem is to design an appropriate controller for this type of converter. Switch-mode power supplies represent a particular class of variable structure systems (VSS) and have advantage of non-linear control techniques developed for this class of systems. The sliding mode control can handle large power supply and load variations and provides good dynamic response and simple implementation. The sliding mode controller design has become famous choice for non-linear dynamic systems having class of uncertainties. Moreover, it is maximally robust to all unmatched uncertainties. In this paper the sliding mode control is reviewed and its applications to dc-dc converters are discussed.

**Keywords:** DC-DC converters, High-power, Dual-Active Bridge, Sliding Mode Control.

## 1 Introduction

The Switching of Power Converters have gained tremendous attraction since last decade in consumer products as well as in industrial, medical, uninterrupted power supplies, electric vehicles, battery chargers and aerospace equipment due to their high efficiency, low volume and weight, fast dynamic response and low cost [1], [2], [3]. Even though extensive efforts and resources are continuously devoted to new component technologies and the development of versatile modeling techniques suitable for general switching power converters is still lagging.

High frequency DC-DC converters are gaining popularity in high Power Electronic applications because of the reduction in the size of the passive components. However,

D.M.A. Hussain et al. (Eds.): IMTIC 2008, CCIS 20, pp. 343–349, 2008.
© Springer-Verlag Berlin Heidelberg 2008

the operation of these converters at higher frequencies with conventional hard-switching topologies, the transistor switching losses increases at both the turn-on and turn-off. High-voltage converters in the power range of 1-10 MW will therefore have excessive switching losses if the switching frequency is higher than 4 kHz [4]. Further, the switching of these converters at high frequencies will affect the power quality because of the generation of inrush pulsating current phenomenon with excessive harmonics and high voltage distortion. Power quality problems usually involve a variation in the electric service frequency voltage or current, such as voltage dips and fluctuations, momentary interruptions, harmonics and oscillatory transients causing failure, or malfunctioning of power service equipment [5]. In order to get a high frequency operation of these converters with minimum switching losses, many soft-switching topologies have been worked out and utilized by various researchers [4].

The dual active bridge converter is believed to have attractive features for high-power applications and the favorable DC-DC converter that is realized is a Dual-Active Bridge in particular when a bi-directional power flow is demanded [6], [7], [8], [9], [10].

Various controllers were used to handle the Dual-Active Bridge dc-dc Converter topology and the recent research work though have been done in the proposed area but control strategy design is fraught with some problems and is not giving desired results. Various researchers [7], [9], [11], [12] have also pointed out that the behavior of the conventional controller is not appropriate when the dual-active bridge topology is operating in step-up mode and this is due to an increase of the inductive current resulting in excessive conduction loss and resulting to decrease the efficiency of the converter. This requires that control development should be done in order to obtain desirable characteristics.

In previous work the PI controller was commonly used [7], [9], [11], [12]. The controllers (P, PI) are generally based on linear functions and they often fail to perform satisfactorily under large parameters or load variations and these controllers are also very sensitive to variations of the system parameters. Further, these are optimal at some fixed operating conditions only but their performance may not be optimal under varying operating conditions [10], [13], [14 ]. This means that there is a dire need to investigate some advanced strategies for the applications. The sliding mode controller is a non-linear and advanced controller and hence can be readily used to overcome the weaknesses discussed above.

## 2   Critical Analysis of Existing Control System

The circuit diagram of the simulated controller is given in Figure 1. In order to investigate the performance of PI controller under varying operating conditions and ambiguous situations an already suggested model of Dual-Active Bridge dc-dc converter of Georgios.D. Demetriades[11] as shown in figure 1 has been critically analyzed with some modifications. The results obtained so far are discussed as follows:

Figure 2 shows the output voltage, inductor current and the step in the reference voltage. When the step is applied to the reference voltage, as given in figure 1 of the controller, oscillations are promptly produced. The peak value of the oscillation is observed approximately 0-5% of the output voltage and is stable in after ten periods.

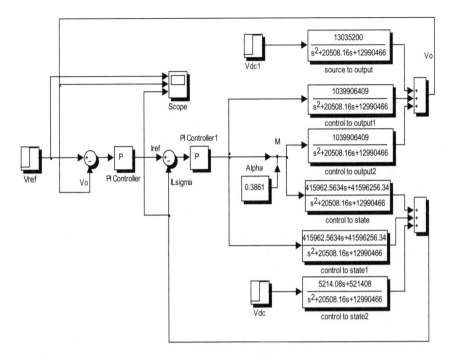

**Fig. 1.** Small-signal controller for DAB converter employing new control strategy

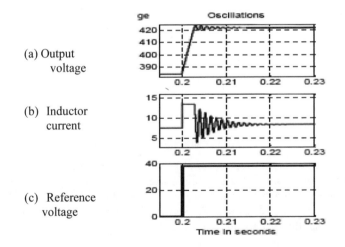

(a) Output
    voltage

(b) Inductor
    current

(c) Reference
    voltage

**Fig. 2.** Controller performance, simulated waveforms. Reference-voltage step.

The oscillations produced in the inductor current, as plotted in figure 2(b), leads in the generation of high currents which affects the efficiency of the converter.

The above results clearly indicate that the PI controller can not successfully handle nonlinear situations. Various researchers [17-21] however have used the techniques of sliding mode controller for tackling the nonlinear cases of conventional type of converters. Therefore, it is deemed that sliding mode controller can readily handle the suggested dual-active bridge DC-DC converter under different uncertainties, in particular, the nonlinear situations.

## 3  Proposed Controller for the DAB Converter

The sliding mode controller is gaining popularity in solving the discrete and dynamics systems. Sliding mode control offers several advantages: stability even for large supply and load variations, robustness, good dynamic response and simple implementation [15], [19]. Moreover, it provides constant stability frequency in steady state, allowing synchronization to external triggers and number of steady state error in the output. Sliding mode controller is non-linear controller [15], [17].

The feasibility of sliding mode controller in handling ambiguous situations is fully justified in this section by utilizing the controller in a general scheme, as discussed fully in the following subsection.

### 3.1  Sliding Mode Control to DC-DC Converters

The P. Mattavelli *et al.* [21] worked on a sliding mode controller for general scheme of dc-dc converters, as shown in Fig-3, is discussed as below:

Let $U_i$ and $uC_N$ be input and output voltages, respectively, $iL_i$ and $uC_j$ ($i = 1-r$, $j = r+1-N-1$) the internal state variables of the converter (inductor currents and capacitor voltages), and $N$ is the system order (state variables). According to the theory, all state variables are sensed, and the corresponding errors $x_i$ (defined by difference to the steady-state values) are multiplied by proper gains $K_i$ and added together to form the sliding function $\psi$. Then, hysteretic block HC controls the switch so as to maintain function $\psi$ near to zero, thus can be written as:

$$\psi = \sum_{1}^{N} K_i \cdot x_i = K^T x = 0 \tag{1}$$

where

$K = [K_1, K_2,.., K_N]^T$ is the vector of sliding coefficients.

The above equation (1) represents a hyperplane in the state error space, passing through the origin. Each of the two regions separated by this plane is associated by block HC to one converter substructure. If it is assumed (**existence condition** of the SM) that the state trajectories near the surface, in both regions, are directed towards the sliding plane, the system state can be enforced to remain near (lie on) the sliding plane by proper operation of the converter switch(es).

Sliding-mode control offers several benefits[20], [21] due to its property of acting on all system state variables simultaneously and are given as follows:

1. Let $N$ be the system order, system response has order $N-1$. In fact, under sliding mode only $N-1$ state variables are independent, the $N$th being constrained by (1).
2. System dynamic is very fast, since all control loops act concurrently.

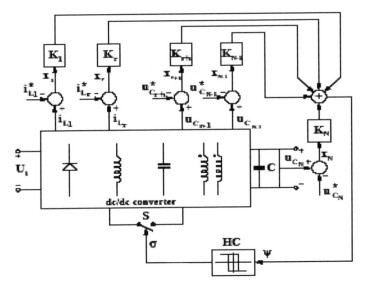

**Fig. 3.** Sliding mode control of dc-dc converters: principle scheme

3. Stability (even for large input and output variations) and robustness (against load, supply and parameter variations) are excellent, as for any other hysteretic control.

4. System response depends only slightly on actual converter parameters.

It is clearly observed from the above discussions, the sliding mode controller can handle the situation easily. Present work is carried out only on the basic converters, i.e. buck, boost, buck-boost, buck and septic converters.

The above justifications reveal that this controller is robust non-linear controller and can be used for the suggested converter that can handle problems which are observed in the previous controllers. At present, we are working on the design of this controller and the results will be presented at a later stage.

## 4 Conclusions

In this paper some serious problems associated with high frequency dual-active bridge DC-DC converter are highlighted. Prior to workout for an appropriate controller for a dual-active bridge DC-DC converter to operate successfully under any ambiguous situation the existing conventional PI controller simulated with the help of Matlab software and the problems associated with it are examined and fully discussed. The results obtained so far clearly reflects that the existing PI controllers does not perform satisfactorily under non-linear situations and does not give desired results when operating in step up mode and this also badly affects the efficiency. In order to improve the performance of the said converters under any uncertain situations the sliding mode controller is suggested to be the powerful tool, which can further ensures the system stability and good dynamic response.

# References

1. Xu, D., Zhao, C., Fan, H.: A PWM Plus Phase shift control Bi-directional DC-DC Converter. IEEE Transactions on Power Electronics 19(3) (2004)
2. Reimann, T., Szeponik, S., Berger, G., Petzoldt, J.: A Novel control principle of bidirectional dc-dc power conversion. In: Proc. of PESC 1997, pp. 978–984 (1997)
3. Jain, M., Jain, P.K., Daniele, M.: Analysis of a bi-directional DC-DC converter topology for low power applications. In: Proc of Canadian Conference on Electrical and Computer Engineering (CCECE 1997), vol. 2, pp. 548–551 (1997)
4. Demetriades, G.D.: Evaluation of different topologies for high power dc-dc converter, Technical Licentiate Thesis, TRITA-EME-0103, Royal Institute of Technology, Sweden (2001)
5. Dash, P.K., Panigrahi, B.K., Panda, G.: Power Quality Analysis Using S-Transform. IEEE Transactions on Power Delivery 18(2) (2003)
6. Kheraluwala, M.H., Gascoigne, R.W., Divan, M.D., Baumann, E.D.: Performance Characterization of a High-Power Dual-Active Bridge dc-dc Converter. IEEE Transactions on Industry applications 28(6) (1992)
7. Zhang, J.M., Xu, D.M., Qian, Z.: An Improved Dual Active Bridge DC/DC Converter. IEEE PESC 1, 232–236 (2001)
8. Kutkut, N.H.: A new dual-bridge soft switching DC-to-DC power converter for high power applications. In: Proceedings of The 25th Annual Conference of the IEEE Industrial Electronics Society (IECON 1999), vol. 1, pp. 474–479 (1999)
9. Li, H., Peng, F.Z.: Modeling of a New ZVS Bi-directional DC-DC Converter. IEEE Transactions on Aerospace and Electronic systems 40(1), 272–283 (2004)
10. Li, H., Peng, F.Z., Lawler, J.: Modeling, Simulation, and Experimental Verification of Soft-switched Bi-directional dc-dc Converters. In: Sixteenth Annual IEEE Applied Power Electronics Conference and Exposition (APEC 2001), vol. 2, pp. 736–742 (2001)
11. Demetriades, G.D.: Analysis and Control of the Single and Dual Active Bridge Topologies, PhD Thesis. TRITA-ETS-2005, Royal Institute of Technology, Sweden (2005)
12. Zang, J., Zang, F., Xie, X., Jiao, D., Qian, Z.: A novel ZVS DC/DC converter for high-power applications. IEEE Transc. on power electronics 19(2) (2004)
13. Raviraj, V.S.C., Sen, P.C.: Comparative Study of Proportional-Integral, sliding mode, and fuzzy logic controllers for power converters. IEEE Transaction on Industry Application 33(2), 518–524 (1997)
14. Tan, S.-C., Lai, Y.M., Tse, C.K., Cheung, M.K.H.: An Adaptive Sliding Mode Controller for Buck Converter in Continuous Conduction Mode. In: Nineteenth Annual IEEE Applied Power Electronics Conference and Exposition (APEC 2004), vol. 3, pp. 1395–1400 (2004)
15. Mahdavi, J., Nasiri, M.R., Agah, A., Emadi, A.: Application of Neural Networks and State-Space Averaging to DC/DC PWM Converters in Sliding-Mode Operation. IEEE/ASME Transaction on Mechatronics 10(1) (2005)
16. Carrasco, J.M., Quero, J.M., Ridao, F.P., Perales, M.A., Franquelo, L.G.: Sliding Mode Control of a DC/DC PWM Converter with PFC Implemented by Neural Networks. IEEE Transactions on Circuit and Systems 1, 44(8), 743–749 (1997)
17. Tan, S.-C., Lai, Y.M., Tse, C.K., Cheung, M.K.H.: Adaptive Feedforward and Feedback Control Schemes for Sliding Mode Controlled Power Converters. IEEE Transaction on Power Electronics 21(1), 182–192 (2006)

18. Ianneli, L., Vasca, F.: Dithering for Sliding Mode Control of DC/DC Converter. In: IEEE 35th Annual IEEE Power Electronics Specialists Conference (PESC 2004), vol. 2, pp. 1616–1620 (2004)
19. Castilla, M., Garcia_de_Vicuna, L., Guerrero, J.M., Matas, J., Miret, J.: Sliding-mode control of quantum series-parallel resonant converters via input-output linearization Industrial Electronics. IEEE Transactions on Industrial Electronics 52(2), 566–575 (2005)
20. Mattavelli, P., Rossetto, L., Spiazzi, G., Tenti, P.: General-purpose sliding-mode controller for DC/DC converter applications. In: Proc. of Power Electronics Specialists Conf. (PESC), Seattle, pp. 609–615 (1993)
21. Mattavelli, P., Rossetto, L., Spiazzi, G., Tenti, P.: Sliding mode control of SEPIC converters. In: Proc. of European Space Power Conf. (ESPC), Graz, pp. 173–178 (1993)

# Importance of Data Center Philosophy, Architecture and Its Technologies

Javed Ali Memon, Nazeer Hussain, and Abdul Qadeer Khan Rajput

IT Division Higher Education Commission, Sector H9 Islamabad, Pakistan
{jmemon, nhussain}@hec.gov.pk
Mehran University of Engineering Technology, Jamshoro, Pakistan
vc@muet.edu.pk

**Abstract.** Currently heavy e-applications/databases are needed for any financial organization and educational institutions for its effective and efficient business processes. In order to run these applications viz. ERP, live video conferencing, streaming services, communication services, other online and real time services seamlessly, there is great need of better underlying networking structuring and power requirement for higher availability and reliability. Therefore data center requires proper and easy maintenance, cooling systems, power system, structured cabling and proper placement of equipment. In this paper a need of planning and designing of critical data centers has been explored and discussed. In this regard some related international standards of planning and designing critical data centers have been reviewed. The designers and implementers of such organization and institutions may utilize this knowledge and work to design and implement critical data centers as per their requirements. This paper presents the architecture followed for building the data center at higher education commission where minimum Tier-II level design recommendation was planned as per the requirements.

**Keywords:** critical and tiers levels, reliability, availability, power and cooling systems.

## 1  Background

In today's competitive telecommunications landscape as well as the business requirement of organizations and educational institutions for research and development activities in Pakistan require reliable, secure and fault tolerant ICT infrastructure. This need has been fuelled in Pakistan by unprecedented growth of financial sector, induction of global market and adoption of e-services for effective processes including e-mail, e-commerce, e-banking, e-payments, e-ticketing etc.

This has certainly demanded reliable ICT facilities for the services to be provided to consumer or customers and to employees for making efficient internal processes of organization. The state of the art environment is essential by establishing the data centers which offer hosting and outsourcing services to corporate or public sector with 99.99% services availability, reliability, integrity, security and Fault Tolerant.

D.M.A. Hussain et al. (Eds.): IMTIC 2008, CCIS 20, pp. 350–356, 2008.

## 2  Design Philosophy

While establishing any tier level data center, it requires key considerations in the design, i.e. availability, connectivity, security, scalability, flexibility and manageability. Data center size and budget normally based on space as well as power/cooling requirements. Selection of site which may vulnerable to threats and size is not planned according to requirement end up with near disaster.

Data center normally planed to be designed according to the requirements and criticality of e-applications including IT services which are offered within organization or outsourced. Selections and purchase of equipment on which these critical applications run are not under the scope of this paper. Only the environment and the requirements of the IT equipment for achieving higher availability, reliability, flexibility to run critical e-applications are discussed in this paper.

Data center may be designed in such a way that it may be easily scalable keeping in consideration the future requirements, i.e. space, power backup, HVAC, etc. Moreover, the design is recommended to be simple with proper tagging/labeling to information outlets, cables etc. which provide ease of management. In case of large setup, the data centers are also established in units for easy maintenance. The disaster recovery site is also established for providing maximum availability and redundancy.

## 3  Planning

Planning for designing data centers involves preliminary study of business processes that are required to be achieved and then needs concentration in its size, site and other recommendation for tier performance. Secondly, the networking availability, reliability, performance and ease of maintenance are also considered.

While planning the design of data center, mainly attributes/standards from three major sources are given consideration. These are

- Uptime institute performance standards tiers
- EIA/TIA 942
- Syska Hennessy critical levels [1]

These common standards define Data Center criticalness or categorize its tier levels [2]. These help mangers to plan accordingly for Data Center availability, reliability for example Tier I categorize basic data center design with single path of power and HVAC and provide 99.671% availability. Tier II level data center lies maintains the redundant components providing 99.741% availability. Tier III data center infrastructure concurrently maintains 99.982% availability. Finally Tier IV level data center design is fully fault-tolerant with multiple active paths of power, cooling etc and providing 99.995% availability.

## 4  Architecture

The challenge arise after site availability for building tier level data center [3] with facilities like power with multiple active paths, cooling, cables management, raise floor, racks setup, fire detectors, etc. Each equipment or component that form part of data center, requires specialized selection in purchase, designing and installation/fixing. Some of critical components, which are essential to build the data center, are discussed below.

## 4.1 Power Efficient Unit

Power unit requires a special consideration while designing the data center. Power units are recommended to be dedicated for data centers. The computing and hardware load must be calculated before purchasing of circuit breakers, UPS(s), generators etc for power units.

Tier IV level data center requires specialize 2 (N+ 1) S+ S power designs which includes two power paths from separate utility feeders each supported by N+1 genera-tors, N+1 UPS and isolation transformers and proper grounding. Placed below is generalized electricity flow diagram, which may be followed for setup of power unit [4]. Moreover, the redundancy components of any tier level may be added:

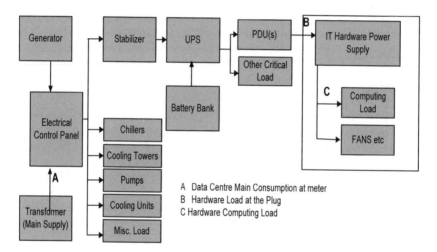

**Fig. 1.** Data Center Power Unit Flow Diagram

## 4.2 Precision Cooling System and HVAC

Proper cooling systems plays critical role in the design of data center. To maintain the cooling system in hot and cold aisle (depends on the tier level design of data center). The heat generated in a data centre (hot aisles) may reach millions of BTU/hr, and the proper cooling system design not only includes precision systems that maintains tem-perature at 22°C ± 1°C and relative humidity between 45% and 50% precisely, but a very critical factor of the design is to manage the air flow to avoid heat zones and short cycling of hot air [hot aisles]. Only a properly designed HVAC system can would ensure removal of the millions of BTU/hr heat generated by the high processing equip-ment in the data centre.

## 4.3  Fire Detection and Suppression System

Automatic fire suppression and detection systems are very critical components of data centers as fire is the leading cause of data centers downtimes after power and HVAC failures multiple sensors with multiple sensitivity levels are required to monitor the systems.

## 4.4  Water Detectors and Sensors

As far as fire suppression and detection systems are vital, water detectors and sensors are also essential components of data centers as water mostly drown from air conditioning may cause of crashing cabling systems etc.

## 4.5  SNMP Based Temperature and Humidity Sensors

Temperature monitoring is also essential in critical data center setups where high temperature can affect the seamless running. Further it can be monitored through some NMS.

## 4.6  Data Structuring Cabling

In order to have ease of management for data centre then structural cabling is required with high focus. The availability and maintainability of data center not only require higher speed cable technology and network speed, but in the data centre proper structuring of the cable plant is a very critical and plays part for the successful implementation of data centers. Further cables with a cross section of 4-10 mm² and above are used in the data center area having shielding to avoid magnetic interference. Also these cables must have good quality of PVC copper to sustain the extreme temperature environment. The laying of cables horizontally and vertically in the data center plan are also defined in EIA/TIA specifications and also discussed at later stage.

## 4.7  Attributes of EIA-TIA 942

EIA-TIA 942 defines tier level attributes that are necessary while establishing the data center. These attributes also defines the various level of availability of data center infrastructure.

## 4.8  Physical Security, Surveillance and Access Control

Physical security is also important in data center where the accessibility of every human must be restricted and this can be obtained through state of art surveillance and access control system installed centrally or independently.

## 4.9  Seismic Protection

Data centers require special building design and specifications, which are not normally given serious consideration while establishment of building of data center. The specialized design requires creation of isolated zones, mantraps, fire proof/fire rated materials/construction, water proof and various other special considerations which are not factors in normal construction.

## 4.10  Disaster Recovery Site

Typically web services, applications services, databases, storage devices organized in multi-tier architecture (N-Tier application architecture). Site to site recovery is the ability to recover the total site from the backup or disaster recovery Site after ensuring the failure of primary or main site. More and more organizations are moving towards the creation of disaster recovery site for providing more reliability, availability and access of information. This also achieves the application redundancy and other aims of business continuity strategy.

This section has really become important particularly after the unfortunate 8 October earthquake incident in Pakistan and currently disaster recovery sites are kept in consideration while establishing data centers.

## 4.11  Access Floors

Raise floors system are laid within the server room of data center to define the under floor chamber in which various conduit and wiring are led through the floor mounted cable trays. EIA-TIA defines the attributes for lying of raise floors. Access floor provides easy management of cables by removing the floor panels.

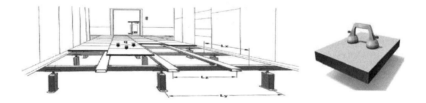

**Fig. 2.** Access floors [Bergvik Flooring]

## 4.12  Miscellaneous Consideration

In addition to above, various other minor components play vital role in the design of data center. These components includes but not limited to, cable racks, fall ceiling electrical lights, cabinet systems, cable trays, cabinet dressing, cable tagging, cable organizers, emergency power off(EPO) etc. Data center should be considered in design that each component of the physical and ICT infrastructure provide the monitoring capability which may provide alerts and alarms to NOC on 24x7 basis. Further all data center lights are connected on standby generator, some of these lights are recommended to be equipped with UPS(s) so that they can continue to function without any interruption.

**Fig. 3.** Cable Tray, Cabinet Rack, Cable Tagging [Google images]

## 5 Infrastructure Standard

These EIA/TIA telecommunication standards [3] are necessary to discuss here which define attributes and guidelines for the design and installation of server room which include cabling structure, network design etc.

These standards include the cabling structure which is basic element of designing any data center. It defines how to organize cables with horizontal and backbone cabling structure and cables for information inlets/outlets. Further, recognize media for backbone and horizontal cabling must meet the applications or network traffic requirement for example fiber and Ethernet cable, patch cords, jumpers etc. The most important standards to design the aisles are discussed below:

### 5.1 Hot and Cold Aisles

In the data center design cabinets and racks placement is much important as all rack mountable communication/data equipment, routers, switches, servers, SAN etc are placed in racks. EIA/TIA standards provide rack pattern in alternative where front of cabinets create "hot" and "cold" aisles. If there is an access floor, power distribution cables should be installed under the access floor on the slab. Hot aisles are behind

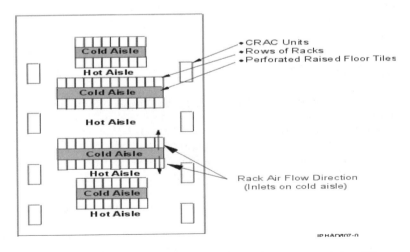

**Fig. 4.** Placement of racks in server room EIA/TIA recommendations [3]

racks and cabinets. If there is an access floor, cable trays for telecommunications cabling should be located under the access floor in the hot aisles.

## 6  New Data Center Technologies - Virtualization

At the end it is very essential to discuss the hot and latest data center technology evaluation i.e. virtualization. It has enabled data center managers to reduce huge investment made for the purchase of IT hardware and great spending on their maintenance.

In the real world environment, through virtualization costs for space, power, cooling etc. are reduced. Operating Systems (OS) and applications from multiple underutilize hardware carried on a single server which maintains isolations between application and OS despite of sharing resources and computing.

In Pakistan, organizations and educational institutions design their data centers keeping the virtualization concept at hardware and operations system level to save the actual costs occurs during the purchase of hardware and such large installation of power and cooling systems.

## 7  Conclusion

The proposed philosophy and architecture will help organizations and educational institutions to plan and design the data center as per their business requirement. Especially in Pakistan where e-culture is being flourish in Corporate as well as government organizations. These data center technologies are required to be understood and considered while building the data center as per requirement. This paper can be taken as initial document to plan and design the strategy for planning and building critical data center with required availability and reliability.

The architecture discussed in the paper is followed for building the data center at higher education commission where minimum tier-II level design recommendations are planned as per requirement. Moreover the paper has also covered the tier-I to tier IV level design recommendations as per the business process of organizations and education institutions.

## References

1. Burkhardt, J., Dennis, R.W., Hennessy, S.: Assessing Criticality Levels in the Data Center (2006), http://searchdatacenter.techtarget.com/tip/0,289483, sid80_gci1218937,00.html
2. Turner, W.P.T., Seader, J.H., Brill, K.G.: Industry Standard Tier Classification Define Site Infrastructure Performance, Uptime Institute white paper, http://www.datacenterdynamics.com/Media/DocumentLibrary/Tier_Classification.pdf
3. ANSI/TIA-942: Telecommunications Industry Association (TIA) Telecommunications Infrastructure Standard for Data Centers (2005)
4. Stanley, J.R., Brill, K.G., Koomey, J.: Four Metrics Define Data Center Greenness, Uptime Institute Inc. (2006), http://uptimeinstitute.org/wp/_pdf/TUI3009F/FourMetricsDefineDataCenter.pdf

# Keynote Address: Is It Right and Is It Legal?

Quintin Gee

Learning Societies Lab, Electronics and Computer Science,
University of Southampton, United Kingdom
qg2@ecs.soton.ac.uk

**Abstract.** Should the rules for IT practitioners be the same as are self-imposed by society in general, or something extra? Engineers, and in particular IT professionals, are being called on to take more and more consideration of non-technical factors when designing their systems. Every system they design is meant to be used (ultimately) by people, and yet people are very diverse and work in unexpected ways. We see this in the manager's role in dealing with his employees, where he now has to cater for their safety, recruitment terms, and what facilities they can and cannot use within the business, for example private e-mail. Information technology itself provides new problems, such as zombies, spam and identity theft. Most importantly, businesses now accumulate vast quantities of digital personal data on their customers. Who is to say how they may or may not use this? Does the IT professional not have a say in what is right, even what is legal? The solutions to this have to be worldwide, and yet most of us can only create a local effect. The best place for an academic to start is in training the future engineers. The paper outlines the background thinking to a Professional & Legal Issues course, given to all second year students of Information Systems, Electronics, Electrical Engineering, and Computer Science, at the University of Southampton, UK.

## 1 Introduction

By what rules should IT practitioners be bound? Should these rules be the same as are self-imposed by society in general, or something extra? What extra should there be?

Let's take some examples. A Prime Minister decides to promote an MP to the Cabinet. A journalist goes into the MP's local video store and asks for the list of the last 10 videos that his household has borrowed. The computer system has been programmed to generate such lists on demand. What is the role of the system designer, system programmer, and counter clerk in this scenario? Can the IT practitioners blithely state "Oh, well. Anything we do can be misused by its users, and therefore we cannot be expected to take account of its uses."

A railway infrastructure operator sub-contracts track maintenance. The maintenance company has an IT system that records all work done, and helps schedule future work. But, it does not prompt the engineers inspecting the track on likely problem areas. That is left to their competency, which is a function of how much training they were given, their own experience, and even how they felt that particular day. Should the IT system be extended to cater for problem areas, if so how and by whom? Or does including this take away skills from a supposedly trained person?

D.M.A. Hussain et al. (Eds.): IMTIC 2008, CCIS 20, pp. 357–365, 2008.
© Springer-Verlag Berlin Heidelberg 2008

Compare that with a reputable car dealer whose service workshop has a checklist for all major and minor services, as well as the Roadworthy test scheme that must be followed to the letter. Even so, the mechanic is still expected to exercise his/her professional judgement if he/she does not believe that the vehicle is safe, and tell the owner so, in order that the workshop may be seen to do a quality job, in their own eyes.

In other words, the remit of the individual always goes further than the role attached to them by their job specification, and we are all expected to "work as a team", even though it might be out of our own sphere of operation, but within our area of competency.

You will have heard that the Hippocratic Oath which doctors take contains the sentence "first, do no harm."[1] (It is more correctly attributed to the Roman physician, Galen). How do we ensure that this applies to our IT systems?

## 2  The Manager's Role

The manager's purpose is the allocation of resources and priorities. She is free to do this as the organisation permits, given her responsibilities and delegated authority.

Other than the distribution and monitoring of finance, the manager has to work almost exclusively with people. That is where a manager's strengths (and sometimes weaknesses) lie.

For our purposes here, we look at three areas: staff, IT, and customers.

### 2.1  People

#### 2.1.1  Recruitment
Before the worker starts with the organisation, they must be recruited. This means that the advertisement for the job, whether in IT or any other discipline, must correctly describe the work to be done, without unnecessary restrictions. Thus, a job that requires "*a refreshment trolley to be pushed around the department from which tea and coffee etc. is distributed*", must not call for a "*tea lady*" since this discriminates against some who could apply. By contrast, and air traffic controller might be required to have "*full-colour vision*", despite the fact that this discriminates against 20% of the male population, since it is an essential characteristic of that job.

Selection between candidates is a tricky task, but as long as the process is clear to all participants, and suitable minutes/records kept, then choosing the 'best' person should be straightforward. Accessibility issues must be addressed; it is wrong to have the interviews on the 3rd floor of a building if wheelchair applicants are expected.

#### 2.1.2  Employment
One of the manager's tasks is to fit the person to the job, and *vice versa*. This may mean some adaptation, and nearly always means training. This may be provided in-house, by attending external courses, or by self-improvement. For the more 'technical' grades of

---

[1] Hippocrates (350 B.C.) *Epidemics*, Bk. I, Sect. XI. One translation reads: "Declare the past, diagnose the present, foretell the future; practice these acts. As to diseases, make a habit of two things—to help, or at least to do no harm."

work, it is termed Continuing Professional Development. The UK Government would have us all indulge in it under the name Life-Long Learning.

### 2.1.3  Termination

Why do people leave their employment? Some move to get another position, some to move to a new location. But if they are forced out, either through action by the employer, or by the employer's negligence, then we are into a legal 'minefield'. The best defence against this is to have formal grievance procedures and dismissal policies clearly stated to the employees, and in full operation.

Thus an air traffic controller whose eyesight subsequently changed to colour blindness, would no longer fit the job description for which they were recruited, and could be removed from that post. They could be moved to a job not requiring such a stringent characteristic, or dismissed with an appropriate and agreed gratuity.

Summary dismissal, for actual theft or racism for example, may no longer be possible. The appropriate procedures must be followed, although the individual may be denied access to the premises, except for the purposes of following the procedures.

Harassment at work may cause a resignation, and this is termed "constructive dismissal" in the UK. Most of these cases are well handled and do not make the newspapers. The unfortunate ones make us all aware of how tribunals operate.

### 2.1.4  Safety

It is the manager's job to promote a safe working environment for the employees, indeed, it is everyone's responsibility. If you see water on the staircase, or a desk blocking an exit route, you are duty bound to report this to the relevant departmental authority.

In the UK, a Safety Officer must be appointed for all businesses, and his/her job is to support managers achieve the correct environment. For example, bookcases must be screwed to the wall, all electrical sockets must be tested regularly as well as all computer power cables, adequate lighting must be in place, the right seats to enable correct posture for computer users must be provided, as must specialists to lift and move computers, and to connect them up.

Suppose the premises have a staff club, incorporating a gymnasium. Who is responsible for safety in the gym: the company or the staff club (to which most staff contribute financially)? Either or both should arrange regular inspections of all equipment, make a qualified first-aider available, and possibly a qualified trainer who can intervene when a particular individual gets too vigorous or misuses equipment.

## 2.2  Information Technology

There are some special requirements for the manager of the IT, brought about by the concentration of data that the organisation collects.

### 2.2.1  Privacy

It is easy to lose files once assembled from data, so their secure retention is high on the list. IT practitioners are always very concerned with the 'cleanliness' of the data, i.e. eliminating duplicate records and inaccuracies. But a deeper problem is always present: who can have access to the data, how is this managed and are there any time restrictions? This leads us not just into caring for the data, but securing it as well, and monitoring that security.

One of the commonest methods of theft is by departing employees taking a list of the customers away when they leave. This is also a wider question since we may want our salesmen to have access to the same customer lists for use on a laptop when travelling on business. What restrictions should we place on them?

Suppose a graduate rings the university and enquires after her marks to show to her employer. Should we give them to her; in what form; how do we check that the graduate is who she claims she is; how can we be sure they are not misused?

### 2.2.2  Ownership

Generally it is clear that the organisation owns the computer hardware, telephones, vehicles, buildings and furniture. (Though not always: the water cooler, the photocopier, the vending machines are normally rented.) But does the organisation own the software it uses? Yes, for all home-grown software; yes, for software that it commissioned and paid for from external parties; possibly yes, for the Windows operating system and utilities. But, with a move to rental as a new mode of financing, this last is tricky since the original software is held and owned by Microsoft, while copies are licensed for use by its customers. Herein lies a problem that needs to be addressed elsewhere.

Generally, it is also clear that the organisation owns the data it has assembled over the years. For example, there are companies that specialise in assembling address lists from which they can select using certain attributes and sell the resulting lists. This data may have been costly to assemble and maintain, and is a critical financial asset of such a company.

Who owns the e-mails? Does this apply to outgoing mails only, or incoming as well? Suppose they are of a personal nature, or even salacious or racist; does that not make them a private communication subject to the Telecommunications Acts? Most companies now have a written policy on these questions, and they ensure that employees are made aware of the policies and the disciplinary procedures arising from them.

Likewise, for phone calls; who owns them? Does this apply to outgoing calls only, or incoming as well? Can the company record this information? How long can they keep the recording? Most companies do not have a written policy on these questions yet, but the advent of VOIP will precipitate a change.

### 2.2.3  Intellectual Property

This term covers much material that cannot be classed as hardware. The organisation owns rights in its designs, its ideas, any software it creates, music, both the original and its performance, logotypes, trademarks, and even brands. These creations can be sold, licensed, and subject of legal ownership disputes. There have been disputes over intellectual property throughout history: the calculus, the sewing machine, the photocopy principle, the cyclone vacuum cleaner, the Fibonacci sort, the name for a food enterprise called McDonald's.

It is worth investigating the difference between freeware, shareware, and open source software.

### 2.2.4 Customers

When organisations, and particularly their managers, operate with customers, they should be seen as partners. Each is likely to hold data on the other and this can be misused by both parties. For example, if the company is taken over, the new holding company may start direct mailing to the extended customer base, quite in contradiction to any previous arrangement that was agreed.

This is why reputable companies usually ask permission of customers to pass on their information, even to other departments within their own enterprise. Those who start abusing their customers, eventually are called to account for it. UK customers have a real weapon now with the Data Protection Act 1998, which enables the individual to get the power of the independent Data Commissioner behind him, even against government institutions. In this way, much spurious mailing can be removed, and incorrect data held by credit checking agencies, for example, can be rectified.

## 3  The IT as a Profession

What is a professional? We mean to encompass doctors, dentists, lawyers, architects, etc. All of these disciplines have the following in common:

- substantial education and training
- members control entry and syllabus for education
- one or more professional bodies
- standards of conduct, and discipline

Information Technology satisfies three of these points, the glaring omission being that of entry. Members do control entry to their professional bodies, but not to working as an IT practitioner. The ratio of members of the British Computer Society to practitioners is about 1:20.

More worrying is that few companies make professional membership/qualification a requirement for working. When a system fails to work to its expected performance, the end-users just "blame the computer", rather than pointing fingers at the designer, since that person or team does not have to put their name to the work, nor do they have to account for it. So they should rather blame the company for having such a lax recruitment policy.

Let us take an extreme example. In any medical IT system on which life depends, who do we get to perform the programming? A software engineering graduate, the professor who taught her, a medical technician trained in programming, a doctor? And what training do we ensure that they have passed in order to be assessed "competent" to undertake such an important role? How do we know that that competency is up-to-date? Who is qualified to test that competency? It is all very problematical. The Medical Information Technologist may soon be the first of the IT practitioner roles in the UK to require registration to practise.

In summary, we can say that an IT professional exhibits behaviour that demonstrates their competency to resolve problems in the best interests of the user, then of the customer, then of society, and finally of himself.

# 4  Legal Environment

Chapter 1 of Bott (2005) gives a fair overview of the environment we all live in. This differs from country to country, though, and we must be sensitive to different cultures and organisations when comparing what can and cannot be undertaken in different systems.

Thus within **society** we select a **government** which enacts **laws** which impose limits to how we as **individuals or groups** can behave. This behaviour is monitored by the **police** and tested by the **judiciary** against existing statute and precedent law. The members of the government, the police and the judiciary are all individuals from within society, and thus are reasonably reflective of it. However, the government and the police are rather more closely intertwined than some people would like, whereas the judiciary is very independent, and removing a judge is quite difficult.

Note that the UK has no national police force, but a combination of 56 or so local police forces. This makes fighting "cyber crime", which is inherently trans-national, difficult. There are moves afoot to create larger units, and the UK may well end up with a system similar to the American model where there are at least three levels of policing.

We also note that H.M. Revenue & Customs has the power to enter premises without a warrant in search of contraband, pirated goods and other items in their jurisdiction, some of which fall in the IT area.

The main areas we are concerned with are theft of cash, intellectual property, and lately identity. The last has become not only a matter for individuals, but recent episodes of company identity theft via inadequate, slow, manual operations at Companies House have been highlighted in the media. It is worth investigating why the system is designed to allow this to happen.

Fraud is deception deliberately practiced in order to secure unfair advantage or unlawful gain. The recent case (2005) of the NHS Trust Chief Executive (Neil Taylor), who had bogus qualifications, highlights the need to be vigilant. He was a chief executive for over 10 years.

## 4.1  Defamation

If a person knowingly makes statements that damage someone's reputation in print (libel) or verbally (slander), it becomes a civil tort. Defamatory statements on web pages, in e-mail and blogs are considered libel. But who does the injured party make the complaint against? Is it just the author, or the site hosting the web page as well as their ISP for propagating the libel? The host can argue that they are only providing an infrastructure, and as long as they take reasonable steps to remove offending material, cannot be considered as the publisher. The ISP's position is unclear.

## 4.2  Harassment

Under this heading we must include deluging phones with texts, calls, and e-mails. This must be the electronic equivalent of stalking. These methods are very intimidating for the individual, and have led to some actions in court.

For companies, the situation is just as grim. It is possible to bring a website to a standstill by bombarding it with e-mail. This is called a Denial of Service (DOS) attack. It can result in a breakdown of the servers, and certainly denies legitimate customers access to the company. This is the equivalent of civil disobedience, such as practised by a *flash mob* who may clutter up a clothing store selling real fur garments so that legitimate customers cannot enter the store, or have a bad shopping experience. The situation has deteriorated over the last two years, with the rise of Distributed DOS (DDoS) attacks mounted by apparently innocent, but insecure, personal computers (called 'zombies', since they under the control of a remote source). These networks of zombies can be marshalled by those with malicious intent into 'botnets' which can then be traded by criminal gangs.

### 4.3  Pornography

The definition of this and its acceptability varies widely between countries. In the UK possession and viewing obscene material is not an offence, but publishing it or possessing it with intent to trade in it is an offence.

What can be done about pornographic web sites? The UK view of banning all obscene sites is seen as many as too narrow, although it has been successful in chasing the problem off-shore. The US view is that such sites will exist, but it is their accessibility to children that worries parents. The USA has been promoting a voluntary code of page tagging, by those who wish to host such material, for filtering software to check. This may be behind the move to create a new web domain .xxx and to get all such sexually explicit material moved this domain. But this page may be considered harmful or harmless in different countries: www.page3.com.

The possession of indecent material involving children is a crime, as well as its viewing and trading.

The UK Government consulted the public in 2005 on whether to make possession of extreme pornography a crime, and Southampton students contributed a joint paper to that discussion.

## 5  Professional and Legal Issues Course

At the University of Southampton, all second year students of information systems, electronics, electrical engineering, computer science, are required to attend a course covering professional and legal issues.

The learning outcomes are set to ensure students:

- understand the legal responsibilities of managers
- understand the characteristics of a professional
- understand the legal context within which organisations operate

Thus, one third of the nominal 100 hours is devoted to each of Management Issues, Professional Issues, and Legal Issues. The study material is provided by part theory and part case study research. The latter results in a group presentation each week

## 5.1  Syllabus

IT Contract restraints
Constraints on Managers of a legal nature
Compromises in systems planning
Copyright and patent
Trade secrets and registered design
Professional societies
Career structures
Codes of conduct and practice
Anti-discrimination policy and Accessibility
Ethics
Spam, Obscene publications, pornography, defamation
Distributed Denial of Service
Contractual obligations in software
Data Protection Act, Privacy, Freedom of Information Act
Computer Misuse Act, Computer-generated evidence

## 5.2  Case Studies

These change from year to year, and the latest used are as follows.

| | |
|---|---|
| Manager's role | When do traditional IT perks become conflicts of interest? |
| The Medical challenge | An Investigation of the Therac-25 Accidents |
| Privacy | Feds consider lowering passenger data requirements, EU privacy law poses compliance challenge for IT, US-EU airline data accord reached |
| Accessibility | Web Accessibility and the Disability Discrimination Act 1995 |
| National ID DB | 60 000 false matches on UK identity database, www.consumer.gov/idtheft/, www.identity-theft.org.uk/ |
| Hacking | Millions hit by Monster site hack |
| Racial stereotyping | Can computers cope with human races? |
| Rights of the Citizen | Computers don't argue |
| Becoming a professional | Which professional organisation should I join and why? BCS or IEE or IPM or IEEE or ACM. |
| Academic ethics | University policy on plagiarism, UK War dossier a sham say experts |

In each case, the student groups are expected to read the source material, research the background, and answer, in a presentation, some basic questions such as:

- What should you have done as a user/operator?
- What should you have done as a design engineer?
- What can Society do to ensure that such scenarios do not happen again?

# 6  Conclusions

As IT becomes more pervasive, the ethical questions it raises become more widespread. Its very flexibility is our undoing. Society can tackle this in several ways, and open debate and full disclosure is our most powerful weapon.

As far as individual IT professionals, and their trainers, are concerned, we need to ensure that engineers have an established habit of ethical thinking in all that they do. We can best inculcate this during their training, and by making it part of their ongoing record of Continuous Professional Development.

It would be nice to see, but we can only enforce it at our training institutions, all project specifications have a section including

- Ethical considerations
- Legal considerations
- Environmental considerations

thus forcing the engineer and IT practitioner to at least give these important topics some thought.

# 7  Future Work

One of the challenges of this course is that daily changes in the law and the web require constant vigilance to keep the legal part of the syllabus up to date. For example, a UK High Court judgement on 23 March 2008 created a precedent, which resolved to allow software patenting. This goes against current government policy, and it will be instructive to follow the resulting actions.

# References

Bott, F.: Professional issues in Information Technology, BCS (2005)

Leveson, N.G., Turner, C.S.: An Investigation of the Therac-25 Accidents. IEEE Computer 26(7), 18–41 (1993)

Tavani, H.T.: Ethics and Technology. John Wiley & Sons, Chichester (2003)

The Risks Digest, Forum On Risks To The Public In Computers And Related Systems, is a source for computer-related disasters compiled by the ACM, and hosted by the University of Newcastle, http://catless.ncl.ac.uk/Risks

# Practical Optimal Caching Using Multiple Virtual Caches in Multiple Query Optimization

Anand Bhaskar Das, Siddharth Goyal, and Anand Gupta

NSIT,
Sector 3, Dwarka, New Delhi, India
{siddharth.goyal,omaranand,dasanand}@nsitonline.in
http://www.nsitonline.in

**Abstract.** Databases today are increasing in both complexity in size and have gained an unprecedented level of importance in commercial and business applications. Of particular interest are the databases based on the client-server model where a centralized database services multiple clients over a network. Clearly, this means that a large number of queries are fired at the central database in a short period of time. Realistically speaking, it is not possible to process the queries in conventional methods and some optimization techniques are used to speed up the entire process. One way to do this optimization is to cache common results of sub-queries to avoid re-computation and redundancy. In the following paper, we demonstrate the use of multiple virtual caches, a novel concept which allows us to practically emulate the optimal caching algorithm with the limited knowledge of the future known by the optimizer. The approach presented significantly speeds up the entire process of query processing and improves both the utilization of the cache and throughput of the CPU. Towards the end of the paper, we will show the results of our performance evaluation, drawing comparisons with the existing approaches that are currently employed.

**Keywords:** Advanced Databases, Multi-Query Optimisation, Multiple Virtual Caches, Optimal Caching Algorithm.

## 1 Introduction

The purpose of a multi query optimizer is to process common sub expressions in multiple queries which are stored in cache, only once so as to minimize the response time of the queries.

Common results among queries are specified based on a query graph (materialized view) [3]. As already proven in [7], identifying common sub-expressions is an NP-Hard problem so heuristic approaches are usually employed[8]. These include the

1)AND/OR Graph
2)Multi-graph Approaches: We will be using this approach to identify and isolate various common sub-expressions.

D.M.A. Hussain et al. (Eds.): IMTIC 2008, CCIS 20, pp. 366–377, 2008.

The process of detecting a common sub expression then becomes a simple process of finding common edges in the multi-graph as explained in [1]. Keeping this in mind we propose a novel way to approach the problem using multiple virtual caches to optimize the evaluation of queries. Unlike in [5], this approach takes into account the fact that the cache size is limited. Moreover we update the results stored in cache after every query evaluation as distinguished from [5]. Keeping cache size in mind, the results are put in the cache after due calculation of the benefit of putting them in the cache which shall be explained in the subsequent sections. In addition, an algorithm to delete the results that will no longer be used again will also be elucidated [Section 4].

### 1.1   Assumptions

In general in most MQO algorithms, the following assumptions are true:

1) The query list is of n sub-queries.
2) The cache space is limited to m results of sub-queries.The cache space is usually expensive and hence limited. Since we are not considering projection cases (condition 1), space occupied by each tuple should be, on an average the same.
3) The size of result of each query, on an average, is p such that $p*n>>m$. As already mentioned, the cache is expensive and thus offers limited space. This space will usually be insufficient to hold all the results of the entire query list or even the results of a single query.
4) The query list is known to us ie we know all queries in the query list. The query list is needed for the calculation of the optimal query plan for any optimizer. However even if the query list is not available, our algorithm reduces to a variation of the demand driven algorithm explained in [6], which can be considered to be a specific case of our proposed algorithm. However the query list is required for complete evaluation of our proposed algorithm.

### 1.2   Important Terms

**The cache or the main cache.** This refers to the cache in which the results will actually get stored and from where the results will actually be retrieved.

**Virtual Cache.** In our simulated approach, we create only one data set which we call "the cache" or the main cache. The results stored in this "cache" shall be operated upon using a set of functions in such a manner that the cache will ultimately get divided into several parts in accordance with certain parameters which we shall be describing in the subsequent sections. We will henceforth refer to all these "parts" of the cache as the virtual caches or the VCs.

Most algorithms like [5] usually consider a single level of cache storage in the main memory. While this technique does exhibit certain advantages it does not offer the versatility of the two level cache storage proposed in this paper which enables us to update the cache, at an extreme case, after and during each query evaluation at a minimal cost. This cost is kept minimal since the updating of the cache takes place at a point where the processor is itself relatively free, that is during the database access itself. What is proposed is to insert valuable results in the main cache as defined above but not allow access to the said main cache directly. This access can take place through the virtual cache which contains the address of the relevant result of the main cache. Since there may be a number of virtual caches for each result, deletion of the result can occur only when each and every one of its entries in the virtual caches has been deleted, which is kept in account by an associated variable stored in the main cache for each result. Thus the Virtual Cache can be defined as simply a path from the main cache to the query.The virtual cache gets created as and when a query is processed. In our approach, it should be assumed that one virtual cache is being created for one query, so that every query "knows" precisely from which virtual cache it has to retrieve results.

**Storing result in the virtual cache.** We will be regularly using this phrase from now on just for the sake of convenience. What this phrase actually means is that the result will be stored in the main cache and its address will be stored in the virtual cache associated with some query.

**Locality of a query.** By this we mean the query as well the queries that occur close to it in the query list. For example, in a query list of 20 queries, by locality of query number 8, we would mean query numbers 7,8, and 9.

The rest of this paper is organized as follows: In section 2 we present the motivation behind dealing with the problem of multiple query optimization. In section 3, we present the related work. Section 4 describes our approach in detail which is compared with the previous approaches and evaluated in section 5. Section 6 is dedicated to conclusion and future work.

## 2   Motivation

Of late Multiple Query Optimization has gained a lot of importance because of the increasing use of complex decision support queries on large sets of data, particularly in data warehouses. Such optimization techniques help in the elimination of redundancies due to re computation of repeated and shared subexpressions which in turn helps in the saving of both time and space.

Clearly, it is preferable to retrieve as many results from the cache as possible as retrieving results from the cache memory is faster by many orders than retrieving results from any other memory. However, owing to cost constraints, cache size is limited and hence only limited number of results can be stored in the cache. Thus a mechanism needs to be developed to find out how beneficial it would be to store a particular result in the cache in as far as retrieving the maximum number of results from the cache is concerned.

## 3   Related Work

In [1], the procedure to find out the common sub-expressions using the multi graph approach has been given in detail which is used in [5] and shall be used in our approach as well. The basic idea elucidated in [1] is that the node of the graph contains the results and the edges connecting these nodes perform certain operations on these nodes thereby, continuously reducing the number of results stored in a particular node which now stores only the common sub expressions.

### 3.1   Optimal Caching

The optimal caching algorithm can be stated as "Cache the result which will be used in the shortest time in the future." This algorithm guarantees a minimum number of database accesses however it is difficult to implement in real time due to the lack of knowledge of the future. This problem has been addressed by us in the following way, if the optimizer possesses the query list then it can be said to have a limited knowledge of the future. The usefulness of each result for caching can then be determined while the next relevant result of the query under consideration is being retrieved. Since secondary storage is much slower than the CPU, the processing occurs in a window where the processor is not being utilized, which means that this processing
    will never be noticed by the user. It will increase the CPU utilization but this utilization should not prove to be a bottleneck providing care is taken while making the query batches.

### 3.2   Multiple Query Optimization in Mobile Databases

In [4], a Multi Query Optimization technique has been explained which does not involve the caching of common sub expressions. This technique can be employed only in cases of Mobile Databases which are bandwidth dependent. [4] diverges from the usual pull and push architectures in answering queries sent by mobile systems. Instead it proposes to use a view channel instead of a downlink channel and only times of broadcasting the results for different queries is sent on the downlink. In effect [4] reduces the redundancy in mobile database systems. The basic idea as elucidated in [4], involves an analysis of a large number of queries (or a query list) to find out the query whose result set contains the results of all the queries.In the simulated approach, the times when the different queries are fired is also known. Let the query set Q be $(q_1,q_2,q_3,q_4,......,q_n)$ and the times at which these queries arrive be (t1,t2,t3,...,tn). Let the query which contains the result superset be qs which arrives at time ts. Then first of all, this query qs is evaluated and its results are stored. Then the results of the query q1 are displayed at time t1, of q2 at t2 and so on. This approach stores the results for a very brief period of time and after displaying the relevant results, no tuples are stored in the cache. However it will be a comparatively weak algorithm if the number of queries is large but the intersection is small between individual results.

### 3.3   Demand Driven Caching

In [6] only the query which comes after the query currently being executed is analyzed. If the current query contains intersecting results with the next query, then these results are simply used again without any formal caching. However this means that if there is even a single query which does not contain the most common result, then the result under consideration will have to be retrieved all over again despite the fact that it is being used throughout the query list.

### 3.4   Caching Intermediate Results

While on one hand we have the multiple query optimization techniques as described above, on the other, we have the techniques which involve the use of cache or the cache memory which as mentioned before is the fastest memory. These strategies involve devising various techniques to intelligently put the most beneficial results in the cache memory. In this approach, the entire query list is analyzed once and the most beneficial results are stored in the cache from which they are retrieved as and when queries in the query list get executed.

## 4   The Method

We first present our system architecture which will be followed by a rigorous elucidation of the algorithms employed:

### 4.1   System Architecture

The system architecture may be described as follows: A batch of n number of queries is sent to the system which are then processed in order. Firstly, the cache is checked if some results of a query can be obtained from it. Those results are then displayed from the corresponding virtual cache which contains results relevant to that particular query. The same results are then removed from the corresponding virtual cache and if necessary from the main cache too(ie if the result is not likely to appear in the next set of j or so queries) The query is then modified and sent to the database from which the remaining results may now be fetched. At the same time, modified query is also sent to the result analyzer which calculates the cost of caching that particular result and hence decides if it would be a good idea to cache the result or not. If it decides to cache the result, the result's address is stored in all the virtual caches associated with the different queries.

### 4.2   Cache Manipulations

The criteria for caching a particular result of a query is fairly straight forward.

1) If a particular result lies in any future query in the query list, store it in the main cache and save its address in the virtual cache relevant to that particular query.
2) In case the main cache becomes full, delete the result whose address is stored only in the last virtual cache from the main cache and replace it with the current result.

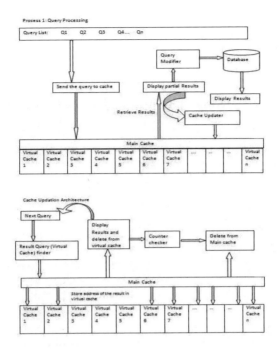

**Fig. 1.** System Architecture

3)Delete the result from the main cache, once its address has been removed from all the virtual caches.

Hence mathematically speaking, if Xi is the cost of caching a result of the ith query in the main cache and Ki is the smallest virtual cache number(associated with query i) where the address of results of queries from a to m are stored, and Np be the total number of virtual caches in which the result(rather the address) can be saved then

Xi is directly proportional to Ki, ie, the higher the number of the virtual cache, more is the cost of putting the result in cache. Hence

1)When Kj<Ki, then X(j)<X(i) even if N(j)<N(x) and result of the jth query can be cached at the expense of the result of the ith query.
2)When Kj=Ki, if N(j)<N(i), then X(j)<X(i) and result of the jth query can be cached at the expense of the result of the ith query.
3)When Kj=Ki, if N(j)=N(i), then X(j)=X(i) and we can choose to cache either result.

Thus it is obvious from the above that it is prudent to cache a result if it is coming earlier in the query list and to remove it from the cache (or the virtual cache) once it is executed (just as in the case of "optimal caching" algorithm). In case of competition for cache space, the frequency of occurrence also becomes an important criteria for deciding whether to cache the result or not.

Since the cache (and the virtual caches) is updated after every query evaluation, it is obvious that the cache hit ratio will be very high in this case as
compared to those algorithms which tend to create one comprehensive cache for
a query list.

## 4.3   Time Based Caching of Multiple Queries Using Multiple Caches

*Conditions*:
1) We are not considering projections and restrictions
2) The query list contains common identifiable sub-expressions.

*Procedure*:
The algorithm can be informally stated as the following process:

Check the cache to find out if there is any usable cached result which will
satisfy the query. Since this is the first query itself, no VCs exist and thus no
cached results are available. After this check, the query is executed and the
results are retrieved from the database. While each result is being retrieved,
the previous sub-result is analyzed for its cacheability. This is done by checking
common edges for the retrieved result with the next queries in order of their
appearance in the query batch. Thus the evaluation will 1st be done for q2, then
q2 and so on. For each time a common sub-result is obtained, if it does not exist
in the main cache, then it is cached in the main cache, and the address of the
main cache is stored in the VC associated with the query found to be using the
result. If the result has already been stored once, its address is stored in the
VC associated with the query found to be using the result. If the VC is null at
the time of declaration, it is created when the common result is obtained whose
address is to be stored in the VC. A frequency variable associated with the main
cache keeps a count of the number of times the result is used. Each time the
result is used once, this variable is decremented by 1 and once it reaches 0, the
result is deleted. Once the query associated with a VC has been evaluated, the
VC is deleted. This reduces useless caching in a large way, since every time a
VC is deleted, the frequency variable is decremented once, since obviously the
query using the result has been executed. Thus as soon as the frequency variable
is reduced to 0, the result is deleted from the main cache. Other than that, if
the cache is full, a common result (say R1) can also be deleted if it is to be
used much later in the query list and the result under analysis (say R3) is to be
used before that result. This can be done simply by checking the last few VCs in
decreasing order (i.e. if the batch is of 15 queries, then q15 first) for the value of
the frequency variable of the result stored by the cache whose address is stored in
the VC (i.e. if VC for q15 stores address of result R1 and the frequency counter
for R1 is 1, it can be deleted.). This is because the low value of the frequency
counter implies that only the VC being analyzed for deletion needs the result
(i.e. R1) and thus the result is needlessly occupying space which would be better
utilized by the result (i.e. R3) being analyzed.

More formally, the algorithm may be divided into the following parts:

---

**Algorithm 1.** The main

Input: *Query Strings in order*

Output: *Result Set*

---

1: **while** number of queries!=0 **do**
2:    tempModQuery=CacheManipulator(A[i], i) {CacheManipulator is the function which displays the initial results from the cache and returns the modified query. tempModQuery is the modified query after few results have been fetched from the cache}
3:    DatabaseFetch(tempModQuery) { Retrieve the rest of the results from the database}
4:    number of queries−
5: **end while**
6: Return 0

---

**Algorithm 2.** Delete Result from cache or virtual cache or both

Input: *VC[j], Result A[i]*

Output: *Modified Cache*

---

    For k=0 to numRemainingResults-1
    **if** ( frequencyofResultA[i]=1 at cache at VC[j][k]) **then**
    deleteCacheResult( )
4: **else**
    Frequency = frequency - 1
    **end if**
    Delete VC[j]

---

**Algorithm 3.** Cache Manipulator

Input: *Query String A[i], Query Number i*

Output: *Modified Query*

---

    Int j=VCfind(i) {find out if a virtual cache for a query exists}
2: k=0
    **while** k<n **do**
    {a fixed constant n is equal to the maximum number of results that can be contained in a particular virtual cache} Result A[i] = VC[j][k]Display(Result(A[i]) VCDelete(VC[j], Result A[i]) {this function deletes the result from the virtual cache} A[i]=Modify(A[i]) {modify the query}
6: **end while**
8: Return A[i]
    ResultAnalyzer(A[i]) {send the modified query to ResultAnalyzer}

---

**Algorithm 4.** Result Analyzer
Input:*Query String A[i],i*
Output:*Modified Cache*

---

    For j=i to numQueries-1
    Int frequency =0
    For k=0 to numRemainingResults-1
3: **If**$(A[i]_Result[k]isinResult(A[j]))$**then**
Int z=VCcreate(j)
**If**$(A[i]_Result[k]incache)$**then**
Frequency = frequency +1
Else
StoreInCache$(A[i]_Result[k], Frequency)$
  **end if**
StoreAddofTuple$(A[i]_Result[k], z)$
**end if**

---

### 4.4   A Concrete Example

Consider the following set of queries:

Select * from lineitem where 1) lorderkey <561 and lorderkey >316 2) lorderkey <2455 and lorderkey >596 3) lorderkey <1470 and lorderkey >742 4) lorderkey <2365 and lorderkey >631 5) lorderkey <1618 and lorderkey >307

   6) lorderkey <1479 and lorderkey >256 7) lorderkey <2128 and lorderkey >679 8) lorderkey <1902 and lorderkey >889 9) lorderkey <1174 and lorderkey >556 10) lorderkey <2037 and lorderkey >697

1. The first query is executed and the first result (lorderkey=317) is fetched from the database. Then the next query is analyzed and as this result does not occur in the query, we move on to the next query. This is repeated until we reach q5. The result is then put in the cache and the address of this location is sent to the virtual cache(VC) associated with the 5th query. This same address is similarly sent to the Virtual Caches associated with q6 and q13 too.
2. Similarly all the results between 316 and 416 are put in the cache and their addresses are sent to the relevant Virtual Caches.At this point the cache is full.
3. From 416 to 561, the results are put in the cache if their benefit exceeds the benefit associated with any result already existing in the cache.
4. When the second query is encountered, the results stored by the caching operation initially are deleted and the new results obtained by evaluation of this query (i.e. q2), viz lorderkey from 889 to 989, which can be used in q3 to q10
5. Since at no point are any of the queries stored in the result cache not needed, hence deletion of queries does not take place until the end of q10, which is when the entire main cache is deleted.

## 4.5   A Special Case

If we increase the size of the batch to a very large number, the CPU may prove to be a bottleneck. However since increasing the batch to a very large number in itself implies extensive knowledge of the future, which is not really very likely in commercial databases, this should not be a very significant problem and should prove to be of theoretical interest only. On the other hand, if the query list is unknown, then although there will be a significant reduction in efficiency, our algorithm reduces to the algorithm proposed in [6] and hence our results will still be optimized and superior to both those obtained through no caching and those through [5].

If the query list is unknown to the system, then clearly no feasible evaluations can be dont to check for common sub-expressions or results. In this case, however, a different approach is adopted by our algorithm, the very next query is analyzed. If the next query is unavailable until completion of the query currently being evaluated, then no optimization techniques are possible. If however the next query is available before the completion of the current one, then the results currently under consideration are checked to see if they satisfy the next query. If they do, then they are cached in a trivial VC. While in this case the VC structure does mean that there will be two cache memory accesses instead of one as in [6], there will only be a limited reduction in efficiency as compared to [6]. In the case that the query list is not known to us, [6] does give a better efficiency however the versatility of clearly superior an use of our algorithm is preferable in cases where there is a high probability of the query list being known.

## 5   Performance Evaluation

The proposed multiple-query optimization using multiple virtual caches method is simulated and evaluated based on TPC-H benchmark [3]. Response time is measured for these 3 possible cases: a) without applying MQO b) the method proposed in [5] and c) the proposed MQO method. The assumptions made for

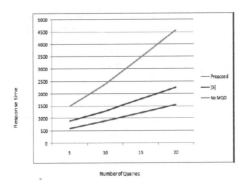

**Fig. 2.** The proposed MQO method response time vs. other methods using sliding windows

the sake of simplicity have already been mentioned before. The figure shows how the response time varies as we increase the number of queries in the query list in each of the 3 cases viz.a) without applying MQO b) the method proposed in [5] and c) the proposed MQO method. ¿From the figure it is quite clear that as the number of queries increases, the average response time increases as well. However the increase in the average response time is less in case of the proposed approach as compared to the current approaches being employed.

# 6    Conclusion and Future Work

From the above performance evaluation, it is clear that the cache hit ratio significantly increases when we put the results in the cache after every query evaluation. What is also significant is the fact that while caching results, the cost of caching a particular result depends not only on its frequency of occurrence in the entire query list but also on the time at which it first arrives. Our devised technique also proves quite conclusively that it makes sense to update the cache after every query run so that those results which "will not occur" in the future at all can be safely removed from the cache once the query( or the queries) of which they are a part, has(have) been run.

However, although this technique improves the cache hit ratio by several factors (as was proven previously), it may also be seen that considerable amount of time is wasted in the preprocessing of the query list which is not really desirable. This is due to the fact that we analyze the query list (which keeps getting shorter and shorter as we run all the queries) every time after a particular query is run and then update the cache accordingly after calculating the benefit of each and every result whose benefit changes after each query run. Hence our future work shall involve devising techniques to save this preprocessing time by a considerable amount simply by doing the preprocessing (and subsequent update of the cache) only a limited number of times when there is a change in the general concentration of the results in their locality which is usually the case.

Our current technique is high on precision but as in it does the preprocessing and updating of cache after each query run which will ultimately be not really required. The cache will then be updated not every time but only on those occasions when it will be required.

# References

1. Chen, F., et al.: Decomposition and Common Subexpression Processing in Multiple-Query Processing. Southern Methodist Univ. Technical Report 94-CSE-30 (August 1994)
2. Sellis, T.K.: Multiple-Query Optimization. ACM Transactions on Database Systems 13(1), 23–52 (1998)
3. Transaction Processing Performance Council (TPC), TPC Benchmark-H, http://www.tpc.org

4. Malladi, R., Davis, K.C.: Applying Multiplequery Optimization in Mobile Databases. In: Proceedings of the 36th Hawaii International Conference on System Sciences, vol. 9(1), pp. 294–303. IEEE, Los Alamitos (2003)
5. Safaeei, A.-A., Kamali, M., Haghjoo, M.S., Izadi, K.: Caching Intermediate Results for Multiple-Query Optimization by Computer. In: Systems and Applications, AICCSA 2007, IEEE/ACS International Conference (2007)
6. Goh, S.-T., Ooi, B.C., Tan, K.-L., et al.: Demand-Driven Caching in Multiuser Environment. IEEE Computer Society, Los Alamitos
7. Diwan, A.A., Sudarshan, S., Thomas, D.: Scheduling and Caching in Multi-Query Optimization, www.cse.iitb.ac.in/comad/2006/proceedings/150.pdf
8. Jarke, M.: Common Subexpression Isolation in Multiplequery Optimization. Query Processing in Database Systems 1(1), 191–205 (1985)
9. Roy, P., Seshadri, S., Sudarshan, S., Bhobe, S.: Efficient and Extensible Algorithms for Multi Query Optimization. In: SIGMOD 2000 (2000)

# A Design and Chronological Survey of Decision Feedback Equalizer for Single Carrier Transmission Compared with OFDM

T.J.S. Khanzada[1], A.R. Ali[1], A.Q.K. Rajput[2], and A.S. Omar[1]

[1] Chair of Microwave and Communication Engineering, University of Magdeburg,
P.O. BOX 4120, D-39106, Germany
[2] Department of Computer Systems & Software Engineering,
Mehran University of Engineering & Technology, Jamshoro, Pakistan
Khanzada@ovgu.de

**Abstract.** Single Carrier Transmission (SCT) is a competing technique for Orthogonal Frequency Division Multiplexing (OFDM) in Broadband Wireless Systems (BWS). Recent developments in Frequency Domain Equalization (FDE) using Decision Feedback Equalization (DFE) have greatly improved the system based on the SC technique, this caused SC-DFE to be selected as a candidate technique for the forthcoming BWS. This paper concentrates the growth of DFE design from scratch till its present form used in SCT and thus restricts our survey to the development of DFE for SCT. A new SCT-DFE modified model is also presented and compared with OFDM, both simulated for WLAN system. This presented equalizer structure performs better than OFDM and improves its performance when the number of iterations is increased. SC-LE and SC-DFE both supersede the performance of OFDM.

## 1 Introduction

Modern Broadband Wireless Communication Systems, which utilize bit rates of tens of megabits per second or more, have variety of applications like Digital Audio Broadcasting (DAB), Digital Video Broadcasting (DVB), Digital Radio Mondiale (DRM), Worldwide Interoperability for Microwave Access (WIMAX), and Wireless Local Area Networks (WLAN) [1][2][3][4][5] in the present era. In the last couple of decades Orthogonal Frequency Division Multiplexing (OFDM) was found to be the optimal candidate technique [6][7]for such systems because of its high spectral efficiency achieved by the orthogonal carriers [8] and ability to resist multipath fading channels [6][7][9][10].

However OFDM systems suffer from high Peak-to-Average-Power Ratio (PAPR) and the Inter-Carrier Interference (ICI) in time-varying channels that characterize, e.g. terrestrial and mobile digital broadcasting and WLANs) [8]. Channel estimation techniques [11][12][13][14][15] are usually applied to overcome these problems. In broadband systems like ETSI Broadband Radio Access Networks (BRAN) and HiperMAN multipath effect can cause severe destruction.

D.M.A. Hussain et al. (Eds.): IMTIC 2008, CCIS 20, pp. 378–390, 2008.

To provide low cost solutions, several variations of OFDM have been proposed as effective anti–multipath techniques [16][17][18][19]. OFDM provides in fact noticeable benefits in performance over SC-TDE and combats multipath fading and the long multipath spread. [16][17].

Conventionally, TD equalizer processing in SCT alleviates power backoff and phase noise sensitivity problems. An equalized SC system offers the same anti–multipath and anti–noise capability as an adaptive OFDM [20][21] system does. A Decision Feedback Equalizer (DFE) yields better performance than a linear equalizer (LE) for radio channels [22]. However, in contrast with OFDM, which uses hundreds or thousands of subcarriers, SC uses a single carrier which encompasses PAPR to be small enough to have less peak power backoff. This enables SC technique to use a cheaper power amplifier than that is used in OFDM systems[18].

We can find a large number of literature explaining the benefits of OFDM over traditional SC systems [6][7][9][10]. Since early 90s [19][20][23] SC-FDE emerged as a competing technique for OFDM, especially in severe multipath dispersive channels.

The rest of the paper is organized as in the following way, section 2 presents the chronological survey of SCT, keeping only DFE developments in view, section 3 discusses conventional OFDM scheme and then presents an iterative DFE for SCT. Section 4 presents the simulation results and section 5 concludes the discussion.

## 2   A Chronological Survey of SCT

As we have already mentioned in Section 1 that SCT technique has been used conventionally for decades with TDE [24][25].

It was Hikmet Sari at el of SAT, Division Telecommunications, France, who initially described the FDE for SCT [19] and then proposed it for terrestrial broadcast channels [20][26] in 1994. In the next year the same author discovered that TDE can not handle ISI on channels with very long impulse responses and may fall on single frequency network. He found a strong analogy between OFDM and SC systems with FDE and proposed it for digital terrestrial TV Broadcasting systems to handle the same type of channel impulse responses as of OFDM systems [23]. This opened a new perspective for research on a technique equivalent to OFDM in modern communication systems. In [19][20][23] the same authors indicated that SC-FDE has essentially the same low complexity as OFDM systems, when combined with FFT processing and the use of a cyclic prefix.

In 1996, a reduced-complexity TD adaptive DFE for long impulse responses was developed by Ariyavisitakul [27] in which he investigated two basic techniques DFE and delayed DF sequence estimation (DDFSE) for highly dispersive channels to minimize the complexity.

Karstten, 1997 [28] elaborated the duality of MC spread spectrum and SCT in 1997. Czylwik [29] compared MC (OFDM) and SC modulation schemes by

simulating different time-variant transfer functions and found that SCT performance was significantly better than that of OFDM with fixed modulation
schemes.

Gusmao, 2000 [30] has emphasized on impact of channel coding and amplification for MC and SCT using FDE and found the equivalent effect. In the same
year, Wang [31] described the MC transmitting with CP and Zero Padding (ZP)
technique with slight increased complexity to eliminate Multi User Interference
(MUI) and ISI. CP technique has the feature of simple equalization while ZP
technique guarantees successful symbol recovery at the receiver end. He also described the Block Equalization types e.g. Pre, Post and Balanced equalizations,
to be used for complexity reduction and efficiency improvement of MC systems.
In the same article he introduced the concept of Generalized MultiCarrier (GMC)
CDMA for efficient MU transmission.

In year 2001, Al-Dhahir [32] proposed a low-complexity scheme for combining
Space Time Block Coding (STBC) with SC-FDE to achieve significant diversity gains. Tubbax, 2001 [33] studied front-end non-idealities of OFDM and
SC-CP WLAN modems and showed by his simulations that for the same data
rate, bandwidth and transmit power constraints (SC-CP) allows the design of a
more power efficient modem than OFDM and is therefore a better candidate for
portable wireless terminals.

Falconer 2002, [18] presented a convertible SC-OFDM receiver, which could al-
ter the receiver processing depending upon the received signal between OFDM and
SCT as shown in fig.1. [18] also proposed employing OFDM in the downlink and
SC-FDE in the uplink for reducing subscriber unit cost and complexity, as shown
in fig.2. In [34] the compatibilities of SC with OFDM, and extensions via DFE &
Overlap-Save Processing of both techniques were discussed by the same author.

A year later, the same author introduced a simpler hybrid time-frequency domain DFE approach [18] to avoid feedback delay problem, by using FD filtering
only for the forward filter part and using conventional transversal filtering for the
feedback part of the DFE . [18] also concluded that the PAPR for SCT signals is

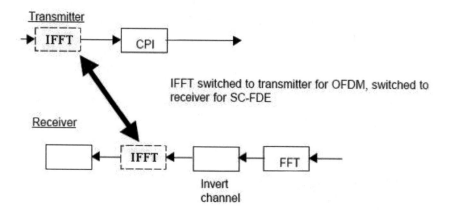

**Fig. 1.** Convertible FD receiver of OFDM and SCT by [35]

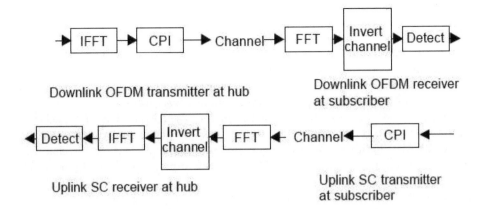

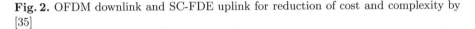

**Fig. 2.** OFDM downlink and SC-FDE uplink for reduction of cost and complexity by [35]

smaller than OFDM, hence it required a smaller linear range to support a given average power or equivalently required less peak power backoff, therefore, this enabled the use of cheaper power amplifier than a comparable OFDM system. In the same year Benvenuto [36] presented a similar FD-DFE for SCT which was simulated for HIPERLAN-2 scenario and achieved the similar conclusions as of [18]. Furthermore, a reduced complexity technique FD-DFE was developed which had similar computational complexity as that of OFDM. Benvenuto also proposed an iterative block DFE (IB-DFE) for SCT which was used to cancel precursors and postcursors of ISI to get better performance [37].

In 2003 Zhou [38] studied STBC for SCBT and proposed novel transmission schemes that achieved a maximum diversity of order $N_t N_r (L + 1)^1$ in rich scattering environments and developed transmissions enabling Maximum Likelihood (ML) optimal decoding based on Viterbi algorithm, as well as turbo decoding for achieving high capacity. It was proposed that single receive and two transmit antennas should be used to have no capacity loss. It was also concluded that the joint exploitation of space-multipath diversity leads to significantly improved performance in the presence of frequency-selective fading channels.

Gusmao [39][40] also verified the performance advantage of the SC-FDE in space diversity within block transmission schemes when compared with MC modulation schemes.

Schniter, 2003 [41] presented a low complexity two-stage receiver for SC-CP to optimally truncate the effective Doppler response and to perform soft interference cancellation in these stages respectively.

In 2004, Zhengdao Wang [42] compared OFDM with SC-ZP block transmissions in aspects of PAPR, BER performance, system throughput, uncoded system performance & complexity and coded system performance & complexity.

---

[1] $N_t(N_r)$ is the number of transmit (receive) antennas, and is the order of the finite impulse response (FIR) channels.

It was confirmed by the simulations that SC-ZP has considerable edge in terms of PAPR, robustness to carrier frequency offset, and un-coded performance at the price of slightly increased complexity. Tran, 2004 [43] proposed SC systems for STBC coded and concatenation of a FEC code with a STBC over ISI fading channels based on successive interference cancelation technique to show that STBC system outperforms the OFDMar receiver under the power constraint scenario and also when multiple receiver antenna diversity is applied.

Zhang, 2004 [44] proposed channel-estimate-based FDE (CE-FDE) scheme using LMS or RLS algorithms. It was proposed that the diversity combining for channels with high-Doppler frequency and tap-selection strategy for channels with sparse multipath propagation improve system performance and reduce the complexity simultaneously.

Dinis, 2004 [45] presented an iterative Layered Space-Time (LST) receiver structure for SC transmission which combined LST principles with iterative block DFE (IB-DFE) techniques to get performance closer to MF Bound (MFB) in few iterations.

Agathe, 2005 [46] improved the iterative FD-DFE with reduced complexity and fast convergent with respect to linear MMSE and ZF equalizers.

Gusmao, 2006 [47] worked on CP assistance methods in SC transmission and presented an algorithm for a Decision-Directed Correction (DDC) of the FDE inputs for insufficient CP and CP-free conditions to improve performance with slight increased complexity.

In the same year, Martin, 2006 [48] proposed a channel shortening equalizer that attempts to directly minimize the BER of the wireless multicarrier systems. In [49] Agathe updated the iterative FD-DFE to improve the BER performance by assuming the perfect equalization at the first iteration with minimized sum of noise and decision error power at threshold detector input. In our previous work, Khanzada, 2006 [50], we have also verified the supremacy of SCT over OFDM for fast varying channels.

Tomeba and Takeda, 2006 in [51][52] and [53] further analyzed the performance of SC-FDE with spaced time delays and using Tomlinson-Harashima precoding.

Most recently Tang, 2007 [54] proposed an extended data model for a low complexity FDE, which was incorporated using a receiver window to enforce the banded channel matrix and inter-block interference reduction. Martin, 2007 [55][56] presented the receiver design for SC transmission and the channel shortening technique to improve its performance in time varying channels for CP systems.

# 3    System Model

In this section we describe the system model for SC-FDE transmission. FDE is computationally simpler than the corresponding TDE. It has been pointed out in [19] that when combined with FFT processing and use of a CP is made, a SC system with FDE (SC-FDE) has essentially the same performance and low complexity than that of OFDM. In Figure. 3 transceiver block diagrams of combined OFDM and SC-FDE are shown . The coding, interleaving, S/P, P/S blocks are skipped in the figure for simplicity. Figure 4 shows the LE for SC transmission. A DFE model is shown in Fig. 5. Both OFDM and SC systems involve one Discrete Fourier Transform (DFT) and one Inverse DFT (IDFT) block. The only difference is that SC-FDE utilizes both blocks at the receiver side [46]. An IFFT operation is located between equalization and decision. This inverse FFT operation spreads the noise contributions of all the individual subcarriers on all the samples in time domain [29].

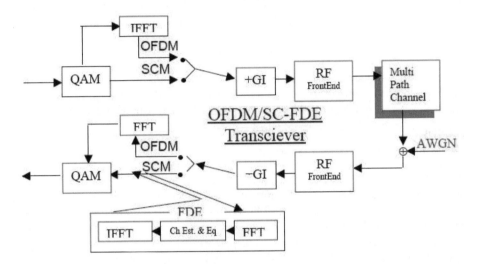

**Fig. 3.** Transmit and receive block diagrams of OFDM and SCT

In order to describe the SC-DFE model, we start with the OFDM conventional model. In an OFDM scheme, the information is coded in symbols of duration $T_s$. The available channel bandwidth $B$ is uniformly subdivided into a number $N$ of sub-bands (sub-channels), whose mid-frequencies characterize orthogonal subcarriers. The data block mapped onto a symbol is subdivided into the same number $N$ of sub-blocks; each corresponds to one of these sub-bands and then $N$-point IDFT is taken for the resulting sub-blocks, the output is serialized and Guard Interval (GI) is added at the transmitter, after that further RF modulation operations are applied. At the receiver, the GI is removed and $N$-point DFT is taken for the resulting signal. If $n$ is the carrier frequency index, $k$ is the OFDM symbol index, the data symbols are denoted by $a_n(k)$, corresponding additive noise by

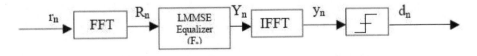

**Fig. 4.** Block diagram of a FD-LE for SCT

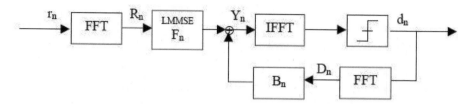

**Fig. 5.** Block diagram of FD-DFE for SCT

$w_n(k)$ and the channel transfer function by $H_n(k)$ then the $N$ signal samples at the DFT output during the $k^{th}$ OFDM symbol can be written as [9]

$$R_n(k) = H_n(k)a_n(k) + w_n(k), \qquad n = 1, 2, \ldots, N \qquad (1)$$

From (1), it is clear that OFDM needs only a complex multiplier bank at the DFT output to completely eliminate channel distortion. Denoting the set of multiplier bank coefficients by $(C_1, C_2, \ldots, C_N)$, their values, which invert the channel transfer function are given by

$$C_n = \frac{1}{H_n}, \qquad n = 1, 2, \ldots, N. \qquad (2)$$

Equation (2) describes the optimum solution which aims the canceling of ISI, regardless of the noise level [10] and known as Zero Forcing (ZF) equalizer criterion. Since the noise contributions of highly attenuated subcarrier can be rather large, a ZF equalizer shows a poor noise performance [29] therefore, a Minimum Mean Square Error (MMSE) equalizer is used for the SC systems. The MMSE criterion [10] can be calculated by (3)

$$C_n = \frac{H_n^*}{|H_n|^2 + \sigma_\omega{}^2} \qquad (3)$$

where $\sigma_\omega^2$ is the variance of the additive noise. Since symbols in SC-FDE are transmitted over the entire channel bandwidth, this technique can operate without channel coding and only channel equalization and ISI compensation are needed. In order to improve the performance of SC-FDE, it was proposed by [18] and [36] to use a DFE with time-domain feedback. Another DFE scheme was proposed in [37] for SCT, where both the Feed Forward (FF) and the feedback parts were implemented in the FD. Furthermore, the DFE was made iterative by using the decision block of the previous iteration to compute the better equalizer output. [46] Most recently, [49] updated the iterative FD-DFE to improve the BER performance by assuming the perfect equalization at the first iteration and with minimize sum of noise and decision error power at threshold detector input.

We can define the generic SC-DFE structure used and updated by [49] as under;

If a symbol block is denoted by $(a_1, a_2, \ldots, a_N)$ and the corresponding received signal block by $(r_1, r_2, \ldots, r_N)$ then $(R_1, R_2, \ldots, R_N)$ would be the DFT output block. This output block is multiplied by FF coefficients $(F_1, F_2, \ldots, F_N)$ of the equalizer and the resulting signal block enters an inverse DFT, which yields the output block $(y_1, y_2, \ldots, y_N)$ on which the threshold detector bases its first decisions for the transmitted signal block. Once the receiver makes a first set of decisions, the decision block is fed to a feedback filter with coefficients $(B_1, B_2, \ldots, B_N)$, and an iterative DFE is implemented. The iterative DFE of [37] is optimized under the MMSE criterion. At the $l^{th}$ iteration, we can calculate the output of DFE by (4)

$$Y_n(l) = F_n(l)R_n + B_n(l)D_n(l-1), \qquad n = 1, 2, \ldots, N \qquad (4)$$

where $F_n(l)$ and $B_n(l)$ are the coefficient sets of the FF and feedback filters respectively, at the $l^{th}$ iteration, and $D_n(l-1)$ are the FD decisions at the previous iteration. The initial equalizer decisions are taken place using $\alpha_0 = 1$ that makes it a MMSE equalizer having

$$F_n(0) = \frac{H_n^*}{\sigma_\omega{}^2 + (|H_n|)^2}, \quad n = 1, 2, \ldots, N$$

where $H_n$ is the channel frequency response and

$$B_n(0) = 0$$

then the $\alpha$ parameter is decreased exponentially by using $\alpha_l = 1 - \frac{\sqrt{l}}{\sqrt[10]{L}}$ where $L$ is the total number of iterations. The FF and feedback coefficients for the rest of iterations are calculated by (5) and (6) respectively.

$$F_n(l) = \frac{H_n^*}{\sigma_\omega{}^2 + (1 - \alpha_{l-1}^2)|H_n|^2} \qquad (5)$$

$$B_n(l) = \alpha_{l-1}[H_n(l)F_n(l) - \frac{1}{N}\sum_{n=1}^{N} H_n F_n(l)], \qquad (6)$$

$$for \ n = 1, 2, \ldots, N$$

This equalizer structure performs better than OFDM and improves its performance when the number of iterations increases. We have designed the equalizer to provide the two types of GI, the CP based GI and Zero Padding (ZP) based GI as discussed in [42][31].

## 4   Simulation Results

We have compared the OFDM technique with SC-DFE by simulating WLAN system.The performance of the presented FD iterative DFE was investigated using QPSK modulation with fast fading channel. Simulations were carried out for both OFDM-CP and OFDM-ZP, as discussed in [42] [31] with SC-LE and SC-DFE(CP and ZP). SC was simulated and compared with different number

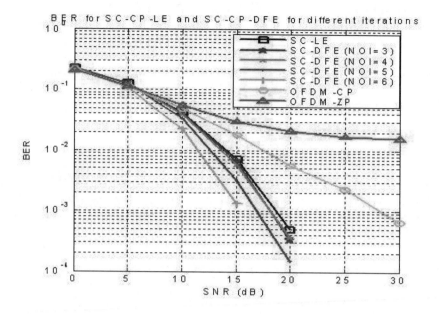

**Fig. 6.** BER Comparisons of SC-CP-LE,SC-CP-DFE with OFDM-CP and OFDM-ZP

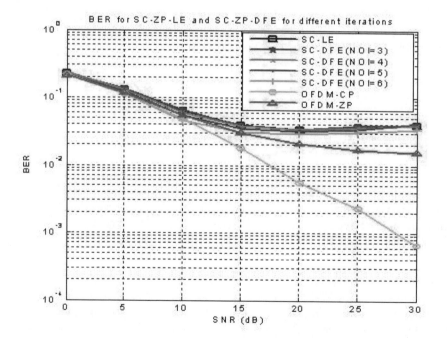

**Fig. 7.** BER Comparisons of SC-ZP-LE,SC-ZP-DFE with OFDM-CP and OFDM-ZP

of iterations for the DFE according to our proposed model. Figure 6 shows the comparisons of the simulation results for BER of SC-CP-LE and SC-ZP-DFE with OFDM-CP and OFDM-ZP. It is clearly shown that OFDM-CP performs better than that of OFDM-ZP and SC-LE gives 10 $dB$ performance gain over OFDM-CP at BER of $10^{-3}$. SC-DFE improves performance further as we increase number of iterations. The results are shown for the number of iterations from 3 to 6, the performance becomes worst as we move further. Figure 7 shows the similar comparisons when SC-DFE is used with ZP instead of CP. We can note that both of the OFDM techniques perform better than SC techniques in this case.

## 5  Conclusion

We have presented a new design for block iterative SC-DFE which is simulated with CP and ZP techniques and compared SC-LE and similar variants of OFDM. It was proved by the simulation results that SC-DFE with CP showed 10 dB performance gain at BER level of $10^{-3}$. A chronological survey of the historical and recent achievements, in order to utilize SC-DFE as promising competitor technique for OFDM in BWS, is also presented.

## Acknowledgment

This research work, along with University of Magdeburg, Germany, is partially supported by Mehran University of Engineering and Technology (MUET), Jamshoro, Pakistan.

## References

1. ETSI. Radio Broadcasting Systems: Digital Audio Broadcasting (DAB) to mobile, portable and fixed receivers, 2 edn. (May 1997)
2. ETSI. Digital Video Broadcasting: Framing structure, channel coding, and modulation of digital terrestrial television
3. ETSI. Digital Radio Mondiale (DRM)-System Specification
4. Smith, C., Meyer, J.: 3G wireless with 802.16 and 802.11. Technical report
5. Bingham, J.: ADSL, VDSL and Multi Carrier Modulation. Wiley, New York (2000)
6. van Nee, R., Prasad, R.: OFDM for Wireless Multimedia Communications. Artech House (2000)
7. Hanzo, L., Mnster, M., Choi, B.J., Keller, T.: OFDM and MCCDMA for Broadband Multi-User Communications, WLANs and Broadcasting. Artech House (September 2003)
8. Weinstein, S.B., Ebert, P.M.: Data transmission by frequency division multiplexing using the discrete fourier transform. In: IEEE Trans. Comm. Tech., vol. 19, pp. 628–634 (October 1971)
9. Prasad, R.: OFDM for Wireless Communications Systems. Artech House (2004)
10. Proakis, J.G.: Digital Communications. McGraw Hill, New York (1995)

11. Ali, A.R., Aassie-Ali, A., Omar, A.S.: A multistage channel estimation and ICI reduction method for OFDM systems in double dispersive channels. In: IEEE Radio and Wireless Symposium (RWS-2006), San Diego,USA (2006)
12. Aassie-Ali, A., Aly, O., Omar, A.S.: High resolution WLAN indoor channel parameter estimation and measurements for communication and positioning applications at 2.4, 5.2 and 5.8 ghz. In: IEEE Radio and Wireless Symposium (RWS-2006), San Diego,USA (2006)
13. Muck, M., de Courville, M., Duhamel, P.: A pseudorandom postfix OFDM modulator-semi-blind channel estimation and equalization. IEEE Transactions On Signal Processing 54(3) (March 2006)
14. Chang, M.X., Su, Y.T.: Blind and semiblind detections of OFDM signals in fading channels. IEEE Trans. Communications 52(5), 744–754 (2004)
15. Aassie-Ali, A., Omar, A.S.: Super resolution matrix pencil algorithm for future fading prediction of mobile radio channels. In: Proc. 8th IEEE International Symposium on Signal Processing and its Application, ISSAP 2005, Sydney, Australia (2005)
16. Cimini. Jr., L.J.: Analysis and simulation of a digital mobile channel using orthogonal frequency division multiplexing. IEEE Trans. Commun. 33(7), 665–675 (1985)
17. McDonnell, J.T.E., Wilkinson, T.A.: Comparison of computational complexity of adaptive equalization and OFDM for indoor wireless networks. In: Proc. PIMRC 1996, Taipei, pp. 1088–1090 (24th/25th September 2003)
18. Falconer, D., Ariyavisitakul, S.L., Benyamin-Seeyar, A., Eidson, B.: Frequency domain equalization for single carrier broadband wireless systems. IEEE Commun. Mag. 40(4), 58–66 (2002)
19. Sari, H., Karam, G., Jeanclaude, I.: Channel equalization and carrier synchronization in OFDM systems. In: Audio and Video Digital Radio Broadcasting Systems and Techniques, Amsterdam,Netherlands, pp. 191–202. Elsevier Science Publishers (September 1993)
20. Sari, H., Karam, G., Jeanclaude, I.: Frequency domain equalization of mobile radio and terrestrial broadcast channels. In: Proc. IEEE Global Telecommun. Conf., vol. 1, pp. 1–5 (November 1994)
21. Aue, V., Fettweis, G.P., Valenzuela, R.: A comparison of the performance of linearly equalized single carrier and coded OFDM over frequency selective fading channels using the random coding technique. In: Proc. ICC 1998 IEEE International Conference on Communications, vol. 2, pp. 753–757 (June 1998)
22. Qureshi, S.U.H.: Adaptive equalization. In: Proc. IEEE, vol. 73, pp. 1349–1387 (September 1985)
23. Sari, H., Karam, G., Jeanclaude, I.: Transmission techniques for digital terrestrial TV broadcasting. IEEE Commun. Mag. 33, 100–109 (1995)
24. Walzman, T., Schwartz, M.: Automatic equalization using the discrete frequency domain. IEEE Trans. Commun. IT-19(1), 59–68 (1973)
25. Ferrara Jr., E.R.: Frequency-Domain Adaptive Filtering. Prentice-Hall, Englewood Cliffs (1985)
26. Sari, H., Karam, G., Jeanclaude, I.: An analysis of orthogonal frequency division multiplexing for mobile radio applications, pp. 1–5 (June 1994)
27. Ariyavisitakul, S., Greenstein, L.J.: Reduced-complexity equalization techniques for broadband wirelesschannels. In: 5th IEEE International Conference on Universal Personal Communications, 1996, vol. 1, pp. 125–130 (September 1996)
28. Karsten Br&#252;ninghaus, ninghaus, and Hermann Rohling. On the duality of multi-carrier spread spectrum and single-carrier transmission, pp. 187–194 (1997)

29. Czylwik, A.: Comparison between adaptive OFDM] and single carrier modulation with frequency domain equalization. In: Proc. IEEE Veh.Technol. Conf., May, vol. 2, pp. 865–869 (1997)

30. Gusmo, A., Dinis, R., Conceio, j., Esteves, N.: Comparison of two modulation choices for broadband wireless communications. In: Proc. IEEE Vehicular Technology Conf., vol. 2, pp. 1300–1305 (2000)

31. Wang, Z., Giannakis, G.B.: Wireless multicarrier communications: Where fourier meets shannon. IEEE Signal Process. Mag. 17(3), 29–48 (2000)

32. Al-Dhahir, N.: Single-carrier frequency-domain equalization for space– time block-coded transmission over frequency-selective fading channels. IEEE Commun. Lett. 5(7), 304–306 (2001)

33. Tubbax, J., Come, B., Van der Perre, L., Deneire, L., Donnay, S., Engels, M.: OFDM versus single carrier with cyclic prefix: a system-based comparison. In: Vehicular Technology Conference

34. Falconer, D., Ariyavisitakul, S.L.: Broadband wireless using single carrier and frequency domain equalization. In: 5th International Symposium on Wireless Personal Multimedia Communications, 27-30 October 2002, vol. 1, pp. 27–36 (2002)

35. Falconer, D., Ariyavisitakul, S.L.: Frequency domain equalization for 211 GHz fixed broadband wireless systems. Tutorial, IEEE 802.16, January 22 (2001)

36. Benvenuto, N., Tomasin, S.: On the comparison between OFDM and single carrier modulation with a dfe using a frequency-domain feedforward filter. IEEE Trans. Commun. 50(6), 947–955 (2002)

37. Benvenuto, N., Tomasin, S.: Block iterative dfe for single carrier modulation. Electron. Lett. 38(19) (12 September)

38. Shengli, Z., Giannakis, G.B.: Single-carrier space-time block-coded transmissions over frequency-selective fading channels. IEEE Transactions on Information Theory, one 49(1), 164–179 (2003)

39. Gusmao, A., Dinis, R., Esteves, N.: On frequency-domain equalization and diversity combining for broadband wireless communications. IEEE Trans. Commun. 51(7), 1029–1033 (2003)

40. Gusmo, A., Dinis, R., Esteves, N.: On broadband block transmission over strongly frequency-selective fading channels. In: Wireless conference 2003, Calgary, Canada, pp. 261–269 (July 2003)

41. Schniter, P., Hong, L.: Iterative equalization for single-carrier cyclic-prefix in doubly-dispersive channels. In: 37th Asilomar Conference on Signals, Systems and Computers, 9–12 November, vol. 1, pp. 502–506 (2003)

42. Wang, Z., Ma, X., Giannakis, G.B.: OFDM or single-carrier block transmissions? IEEE Trans. Commun. 52(3), 380–394 (2004)

43. Tran, T.A., Lai, T.X., Sesay, A.B.: Single-carrier concatenated space-time block coded transmissions over selective-fading channels. Canadian Conference on Electrical and Computer Engineering 3, 1577–1580 (2004)

44. Zhang, Q., Le-Ngoc, T.: Channel-estimate-based frequency-domain equalization (ce-fde) for broadband single-carrier transmission. Wireless Commun. and Mobile Computing 4(4), 449–461 (2004)

45. Dinis, R., Kalbasi, R., Falconer, D., Banihashemi, A.H.: Iterative layered space-time receivers for single-carrier transmission over severe time-dispersive channels. IEEE Communications Letters 8(9), 579–581 (2004)

46. Agathe, F.S., Sari, H.: Single-carrier transmission with iterative frequency-domain decision-feedback equalization. In: 13th european Signal processing conference, Antalya,Turkey (September 2005)

47. Gusmao, A., Torres, P., Dinis, R., Esteves, N.: On SC/FDE block transmission with reduced cyclic prefix assistance. In: ICC 2006. IEEE International Conference on Communications, Istanbul, vol. 11, pp. 5058–5063 (June 2006)

48. Martin, R.K., Vanbleu, K., Ysebaert, G., Klein, A.G.: Bit error rate minimizing channel shortening equalizers for multicarrier systems. In: IEEE 7th Workshop on Signal Processing Advances in Wireless Communications, SPAWC 2006, Cannes, France, pp. 1–5 (July 2006)10.1109/SPAWC.2006.346351

49. Agathe, F.S., Sari, H.: New results in iterative frequency-domain decision-feedback equalization. In: Proc. 14th European Conference on Signal Processing (EUSIPCO 2006), Florence, Italy (September 2006)

50. Khanzada, T.J.S., Ali, A.R., Omar, A.S.: An analytical model for sltdm to reduce the papr and ici in OFDM systems for fast varying channels. In: Proc. 10th IEEE INMIC 2006, Islamabad, Pakistan, pp. 57–61 (December 2006)

51. Tomeba, F.A.H., Takeda, K.: Ber performance of single-carrier transmission in a channel having fractionally spaced time delays. Technical report, IEICE, Tohoku Univ

52. Takeda, F.A.K., Tomeba, H.: Single carrier transmission with frequency-domain equalization using tomlinson-harashima precoding. Technical report, Tohoku Univ. Sendai Japan

53. Takeda, F.A.K., Tomeba, H.: Ber performance of turbo coded single-carrier transmission with joint tomlinson-harashima precoding and frequency-domain equalization. Technical report, Tohoku Univ. Sendai Japan (2007)

54. Tang, Z., Leus, G.: Receiver design for single-carrier transmission over time-varying channels. In: International Conference on Acoustics, Speech and Signal Processing, 2007. ICASSP 2007, Honolulu, HI, USA, April 2007, vol. 3, p. III–129–III–132. IEEE, Los Alamitos (2007)

55. Ysebaert, G., Martin, R., Vanbleu, K.: Bit error rate minimizing channel shortening equalizers for single carrier cyclic prefixed systems. In: International Confrence on Acoustics, Speech and Signal Processing ICASSP 2007, Hawaii,USA (April 2007)

56. Martin, R.K., Ysebaert, G., Vanbleu, K.: Bit error rate minimizing channel shortening equalizers for cyclic prefixed systems. In: IEEE Transactions on Signal Processing, Hawaii,USA, vol. 55, pp. 2605–2616 (June 2007)

# Speaker Verification Based on Information Theoretic Vector Quantization

Sheeraz Memon and Margaret Lech

School of Electrical and Computer Engineering,
Royal Melbourne Institute of Technology,
Melbourne, VIC, 3001, Australia
sheeraz.memon@student.rmit.edu.au, margaret.lech@rmit.edu.au

**Abstract.** This paper explores the application of information theoretic based Vector Quantization algorithm called VQIT for speaker verification. Unlike the K-means and LBG Vector Quantization algorithms, VQIT has a physical interpretation and relies on minimization of quantization error in an efficient way. Vector Quantization based Speaker Verification has proven to be successful; usually a codebook is trained to minimize the quantization error for the data from an individual speaker. In this paper we use a set of 36 speakers from TIMIT database and evaluate MFCC and LPC coefficients of speech samples and later apply it to the K-means Vector Quantization, LBG Vector Quantization and VQIT Vector Quantization and suggest that VQIT performs better than other VQ implementations. We also obtain the results from the GMM classifier for the similar coefficient data and compare it to the VQIT.

**Keywords:** Vector Quantization, VQIT, Information Theory, Speaker Verification.

## 1 Background

Speaker verification is a biometric based identity process where personal identity is verified by the voice of a person. Biometrics based verification has received much attention in the recent times as such characteristics come natural to each individual and they are not required to be memorised, unlike passwords and personal identification numbers. The important thing is the speaker is not bound to speak any restricted phrase to get identified but he is free to utter any sentence. The speaker verification system comprises of three stages in the first stage feature extraction is performed over a database of speakers where vectors representing the speaker distinguishing characteristics are isolated. The second step addresses establishing the speaker model; this translates to finding the distribution of feature vectors. The third step is of decision, which determines the claimed identity of a speaker. Vector Quantization (VQ) based speaker verification has remained a successful method. The basic idea in this approach is to compress a large number of short term spectral vectors into a small set of code vectors. The successful modelling of the underlying acoustic classes allows the

D.M.A. Hussain et al. (Eds.): IMTIC 2008, CCIS 20, pp. 391–399, 2008.

Vector Quantization system to achieve high recognition accuracy even with very short test utterances. Vector Quantization has been applied for speaker recognition systems a number of times, Some of the studies suggest that the use of the LBG Vector Quantization algorithm best fits the speaker verification models [1]; other studies indicate that K-means and LBG can be applied to optimize the means and covariances for a GMM based classifier [2].

## 2    Preliminaries

In this section we shall discuss the K-means, LBG and VQIT algorithms for vector quantization and their application as individual classifiers for speaker verification. Further we will discuss the GMM algorithm and compare its performance to speaker verification using the VQIT classifier.

### 2.1    K-Means Algorithm

It is an algorithm to classify or to group the data-driven features into K number of groups. K is a positive integer number. The grouping is done by minimizing the sum of squares of distances between data and the corresponding cluster centroid. Thus, the purpose of the K-mean clustering is to classify the data. The K-means algorithm [5] was developed for vector quantization codebook generation. It represents each cluster by the mean of the cluster. Assuming that a set of vectors $X = x_1, x_2, x_3, ...., x_T$ is to be divided into M clusters represented by their mean vectors $\mu_1, \mu_2, \mu_3, ..., \mu_M$ the objective of K-means algorithm is to minimize the total distortion given by

$$\textbf{totaldistortion} = \sum_{i=1}^{M} \sum_{t=1}^{T} \|x_t - \mu_i\| \tag{1}$$

The K-means algorithm is an iterative approach, In each successive iteration; it redistributes the vectors in order to minimize the distortion. The procedure is outlined below:

(a) Initialize the randomized centroids as the means of M clusters.
(b) Associate the data points with the nearest centroid.
(c) Move the centroids to the centre of their respective clusters.
(d) Repeat steps b $\varepsilon$ c until a suitable level of convergence has been reached, i.e. the distortion is minimized.

When the distortion is minimized, redistribution does not result in any movement of vectors among the clusters. This could be used as an indicator to terminate the algorithm. The total distortion can also be used as an indicator of convergence of the algorithm. Upon convergence, the total distortion does not change as a result of redistribution. It is to be noted that in each iteration, the K-means algorithm estimates the means of all the M clusters.

## 2.2   LBG (Linde, Buzo and Gray) Algorithm

The LBG algorithm is a finite sequence of steps in which, at every step, a new quantizer, with a total distortion less or equal to the previous one, is produced. We can distinguish two phases, the initialization of the codebook and its optimization. The codebook optimization starts from an initial codebook and, after some iterations, generates a final codebook with a distortion corresponding to a local minimum. The following are the steps for LBG algorithm.

**a. Initialization.** The following values are fixed:

- $N_C$ : number of codewords;
- $\epsilon \geq 0$ :precision of the optimization process;
- $Y_o$ : initial codebook;
- $X = \{x_j; j = 1, ...., N_p\}$

   Additionally, the following assignments are made:

- m = 0; where m is the iteration number.
- $D_{-1} = +\infty$; where D is the minimum quantization error calculated at every $m^{th}$ iteration.

**b. Partition calculation.** Given the codebook $Y_m$, the partition $P(Y_m)$ is calculated according to the nearest neighbour condition, given by

$$S_i = \{x \epsilon X : d(x, y_i) \geq d(x, y_j), j = 1, 2, 3, ...., N_C, j \neq i\} \qquad (2)$$

**c. Termination condition check.** The quantizer distortion $D_m = D(Y_m, P(Y_m))$ is calculated according to following equation,

$$MQE \equiv D(Y, S) = \frac{1}{N_p} \sum_{p=1}^{N_p} d(x_p, q(x_p)) = \frac{1}{N_p} \sum_{i=1}^{N_C} D_i \qquad (3)$$

Where $D_i$ indicates the total distortion of $i^{th}$ cell. if $(| D_{m-1} - D_m |)/D_m \leq \epsilon$ then the optimization ends and $Y_m$ is the final returned codebook.

**d. New codebook calculation.** Given the partition $P(Y_m)$, the new codebook is calculated according to the Centroid condition. In symbols:

$$Y_{m+1} = X(P(Y_m)) \qquad (4)$$

After, the counter m is increased by one and the procedure follows from step b.

## 2.3   VQIT (Information Theoretic Vector Quantization) Algorithm

In Vector Quantization the challenge is to find a way that best represents the data. VQIT [4] uses a new set of concepts from information theory and eliminates the flaws of previous vector quantization algorithms. Unlike LBG and SOM this algorithm addresses a clear physical interpretation of data and relies on minimization of a well defined cost function. In the light of information theory it becomes clear that minimizing distance is actually equivalent to minimizing the divergence between distribution of data and distribution of code vectors. When SOM is converged it is at the minimum of cost function, but this cost function is highly discontinuous and drastically changes if any sample changes its best matching centroid [6]. Now attempts have been made to find a cost function that when minimized gives results similar to the original update rule [7] and information theorists have made attempts to design good vector quantifiers [8], [9] and [10]. Unlikely the [7], [8], [9] and [10] VQIT takes the distribution of data explicitly into account by matching the distribution of the code vectors with the distribution of the data points in a data cluster. This approach leads to the minimization of the well defined cost function.

This algorithm works on the principal of minimizing the divergence between Parzen estimator of the code vectors density distributions and a Parzen estimator of the data distribution. Minimizing the divergence between the Parzen estimates of data points and code vectors means minimizing the dissimilarity, and this is achieved by using Cauchy-Schwartz inequality which is the linear approximation of kullback-leibler divergence. The Parzen density estimator is given by the following equation,

$$p(x) = \frac{1}{N} \sum_{i=1}^{N} K(x - x_i) \tag{5}$$

Where K(.) is the Gaussian Kernel and x is the independent variable for which we seek the estimate and $x_i$ represents the data points. The Parzen estimate of the data has N kernels, where N is the number of data points and the Parzen estimator of the code vectors has the M kernels, where M is the number of code vectors and $M << N$. After evaluating the density estimation the divergence measure is evaluated and this is achieved by using Cauchy-Schwartz inequality,

$$\mid a(x)b(x) \mid \leq \parallel a(x) \parallel \parallel b(x) \parallel \tag{6}$$

The following equation is used to minimize the divergence between a(x) and b(x), where a and b represent the data points and code vectors respectively. Hence Maximizing the following expression is equivalent to minimizing divergence between a and b.

$$\frac{\mid a(x)b(x) \mid}{\parallel a(x) \parallel \parallel b(x) \parallel}$$

In order to minimize the divergence between data points (Say a(x)) and Code Vectors (Say b(x)) the following is minimized,

$$D_{C-S}(a(x), b(x)) = -log\frac{(\int (a(x)b(x))d_x)^2}{\int a^2(x)} \tag{7}$$

The cost function used to evaluate the centroids for code vectors can further be achieved as,

$$J(w) = log \int a^2(x)dx - 2log \int a(x)b(x)dx + \int b^2(x)dx \qquad (8)$$

This cost function is minimized with respect to the location of the centroids (w). When the centroids reach to a location so that a local minimum is achieved then no effective force acts on them, this uses gradient descent method for addressing the local minima. Eq.8 has three terms the first term represent the data points which are stationary thus differentiating Eq.8 with respect to centroids will yield zero for the first term, the middle term called cross information potential and the last term called the entropy of centroids [4] will have non-zero derivatives. Consider the cross information potential term; the Parzen estimator for a(x) and b(x) puts Gaussian kernels on each data point $x_j$ and each centroid $w_i$ respectively, where the variances of the kernels are $\sigma_a^2$ and $\sigma_b^2$. Initially the locations of the centroid are chosen randomly.

$$C = \int a(x)b(x)dx \qquad (9a)$$

$$= \frac{1}{MN} \int \sum_i^M G(x - w_i, \sigma_b^2) \sum_j^N G(x - x_j, \sigma_a^2)dx \qquad (9b)$$

$$= \frac{1}{MN} \sum_i^M \sum_j^N G(w_i - x_j, \sigma_f^2), and \ \sigma_f^2 = \sigma_a^2 + \sigma_b^2. \qquad (9c)$$

Where M represents the number of centroid kernels and N represents the number of data point kernels. The gradient update for the centroid $w_k$ from the cross information potential term then becomes,

$$\frac{d}{dw_k} 2logC = -2\frac{\Delta C}{C} \qquad (10)$$

Where $\Delta C$ denotes the derivative of C w.r.t $w_k$, and $\Delta C$ is calculated as,

$$\Delta C = -\frac{1}{MN} \sum j^N G_f(w_k - x_j, \sigma_f)\sigma_f^{-1}(w_k - x_j) \qquad (11)$$

Similarly for the entropy term we have,

$$V = \int b^2(x)dx = \frac{1}{M^2} \sum_i^M \sum_j^M G(w_i - w_j, \sqrt{2}\sigma_b) \qquad (12a)$$

$$\frac{d}{dw} logV = \frac{\Delta V}{V}. \qquad (12b)$$

With,

$$\Delta V = -\frac{1}{M^2} \sum i^M G(w_k - w_i, \sqrt{2}\sigma_b)\sigma_b^{-1}(w_k - w_i) \qquad (13)$$

The update for point k consist of two terms, cross information potential and entropy of the centroids,

$$w_k(n+1) = w_k(n) - \eta(\frac{\Delta V}{V} - 2\frac{\Delta C}{C})$$ (14)

Where $\eta$ is the step size, the VQIT consists of a loop over all $w_k$.

## 2.4   Gaussian Mixture Models

The probability density function (pdf) drawn from a GMM is a weighted sum of M component densities, as depicted in fig. 1 and it is given by the equation,

$$p(x|\lambda) = \sum_{i=1}^{M} p_i b_i(x)$$ (15)

Where x is a D dimensional random vector, $b_i(x), i = 1, 2, ...., >< DEFANGED.$ $19616M$ are the component densities and $p_i, i = 1, 2, ...., M$ are the mixture weights. Each component density is a D-variate Gaussian function of the form

$$b_i(x) = \frac{1}{2\pi \mid \Sigma_i \mid^{1/2}} exp\{-\frac{1}{2}(x - \mu_i)'\Sigma_i^{-1}(x - \mu_i)\}$$ (16)

Where $\mu_i$ is the mean vector and $\Sigma_i$ is the covariance matrix. The mixture weights satisfy the constraint that $\Sigma_{i=1}^{M} p_i = 1$. The complete Gaussian mixture density is the collection of the mean vectors, covariance matrices and mixture weights from all components densities,

$$\lambda = \{p_i, \mu_i, \Sigma_i\}, i = 1, 2, ....., M$$ (17)

Each speaker is represented by a mixture model and is referred by the speaker model $\lambda$.

The Expectation Maximization (EM) algorithm is a broadly applicable approach to the iterative computation of maximum likelihood estimations (MLE), the estimates of a speaker model are passed through a number of iterations of EM algorithm so that every next iterative estimate yields good parameter estimate values. The basic idea of the EM algorithm is beginning with a speaker model $\lambda$ to estimate a new model $\bar{\lambda}$ such that $p(x \mid \bar{\lambda}) \geq p(x \mid \lambda)$ .

# 3   Results and Discussion

Experiments were conducted on TIMIT databases to investigate the suitability of VQIT algorithm for speaker verification. In this section the results of the experiments are presented.

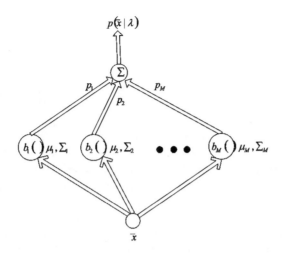

**Fig. 1.** Fig.1. A Gaussian mixture density is a weighted sum of Gaussian densities, where pi, i=1,..,M are the mixture weights and bi(), i = 1,..,M are the component Gaussians

### 3.1   Experimental Setup

**Speaker Corpus.** We have enrolled a set of 36 speakers of the New England dialect from the TIMIT corpus by keeping a half of male and female speakers; we are using two min of speech files after removing the silence, one minute and thirty seconds for training and thirty seconds for testing. The TIMIT corpus of read speech has been designed to provide speaker data. The speech was recorded at Texas instrumentation (TI), transcribed at Massachusetts Institute of Technology (MIT), and has been maintained, verified, and prepared for CD-ROM production by the National Institute of Standards and Technology (NIST).

**Preprocessing and extracting features.** The speech data is preprocessed before extracting the coefficients. We are using a high-pass filter: $v(k) = x(k) - 0.95x(k-1)$ which pre-emphasizes the speech data where $x(k)$ contains the speech sampled data and $v(k)$ returns the pre-emphasized speech data. We are using a logarithmic technique suggested by [11] for separating and segmenting speech from noisy background environments, a set of efficient rules is used to generate speech and noise metrics from the input speech. The rules are derived from the statistical principles about the characteristics of the speech and noise waveform and are based on time-domain processing to have zero-delay decision; at the final stage the algorithm compares the speech and silence metrics using a threshold scheme to control the speech/silence decision. Therefore silence is removed from the speech files before applying the feature extraction algorithms to pre-emphasised speech data. The feature extraction algorithms used to extract the coefficients are linear predictive coefficients (LPC) and mel-frequency Cepstral coefficients (MFCC) which generates the coefficient data ready for training.

## 3.2    Experimental Results

Our principal objective of carrying on this experiment was to test VQIT for the speaker verification data. We carry the experiments with some other implementations of vector ¿¡DEFANGED.19617 quantization such as K-means and LBG to make a comparative analysis and we observed that VQIT behaves as a better vector quantization approach than the other VQ implementations as listed in tables 1 and 2 based on our experiments. Table1 indicates the performance comparison of different VQ implementations when LPC is used as a feature extraction algorithm, and Table2 indicates the performance comparison when MFCC is used as a feature extraction technique. In the second part of our experiments we compare the VQIT with the Gaussian mixture models, Reynolds [12] described and first used the GMM for speaker verification and established the concept of universal background model, we also compare the performance of VQIT with the GMM as listed in Table3 and observe a narrow performance leakage with information theoretic vector quantization VQIT algorithm.

**Table 1.** Performance comparison of different vector quantization techniques when feature extraction algorithm is LPC

| Algorithm | Centroids | | Speech Length | | Rate of verifying an speaker |
|---|---|---|---|---|---|
| | Training | Testing | Training | Testing | |
| K-means | 32 | 32 | 1min.30sec | 30sec | 60.75% |
| LBG | 32 | 32 | 1min.30sec | 30sec | 79.6% |
| VQIT | 32 | 32 | 1min.30sec | 30sec | 89.5% |

**Table 2.** Performance comparison of different vector quantization techniques when feature extraction algorithm is MFCC

| Algorithm | Centroids | | Speech Length | | Rate of verifying an speaker |
|---|---|---|---|---|---|
| | Training | Testing | Training | Testing | |
| K-means | 32 | 32 | 1min.30sec | 30sec | 75.5% |
| LBG | 32 | 32 | 1min.30sec | 30sec | 82.6% |
| VQIT | 32 | 32 | 1min.30sec | 30sec | 98.5% |

**Table 3.** Performance comparison of VQIT and GMM

| Algorithm | Centroids | | Speech Length | | Rate of verifying an speaker | |
|---|---|---|---|---|---|---|
| | Training | Testing | Training | Testing | LPC (%) | MFCC (%) |
| GMM | 32 | 64 | 1min.30sec | 30sec | 87.5 | 98.5 |
| VQIT | 32 | 64 | 1min.30sec | 30sec | 80 | 98 |

# 4    Conclusion

In this paper we evaluate the performance of VQIT algorithm and conclude that it proves better in performance than the other Vector Quantization algorithms, we test it with different feature extraction algorithms like the linear predictive coefficient and mel-frequency Cepstral coefficients and for both of the implementations we observe the performance superiority of VQIT algorithm, when used as a classifier. In the second part of the experiment we compare VQIT to the GMM implementation for the same training data and observe that it establishes approximately similar results to GMM, which can be described as a better indication because the GMM computational complexity is twice that of VQIT, thus using VQIT instead of GMM for text-independent speaker verification we show that they deliver comparable performance. VQIT addresses a significantly reduced computational complexity than GMM, thus making it an ideal choice for low-cost speaker verification applications.

# References

1. Jialong, H., Liu, L., Gunther, P.: A new codebook training algorithm For VQ-based speaker recognition. In: IEEE international conference on acoustics, speech and signal Processing, vol. 2, pp. 1091–1094 (1997)
2. Singh, G., Panda, A., Bhattacharyya, S., Srikanthan, T.: Vector quantization techniques for GMM based speaker verification. In: IEEE international conference on acoustics, speech and signal Processing, vol. 2, pp. 1165–1168 (2003) DE-FANGED.19618
3. Pelecanos, J., Myers, S., Sridharan, S., Chandran, V.: Vector Quantization Based Gaussian Modelling for Speaker Verification. In: International conference on pattern recognition, vol. 3, pp. 294–297 (2000)
4. Tue, L., Anant, H., Deniz, E., Jose, C.: Vector Quantization using information theoretic concepts. Natural Computing: an international journal 4(1), 39–51 (2005)
5. Furui, S.: Digital Speech Processing, Synthesis and Recognition. Marcel Dekker Inc., New York (1989)
6. Erwin, E., Obermayer, K., Schulten, K.: Self organizing maps, ordering, convergence properties and energy functions. Biological Cybernetics 67(1), 47–55 (1991)
7. Heskes, T., Kapen, B.: Error potentials for Self organization. In: IEEE international conference on Neural Networks, vol. 3, pp. 1219–1223 (1993)
8. Heskes, T.: Energy functions for self organizing maps. In: Kohonen Maps, E., Oja, Kaski, S. (eds.) Kohonen Maps, pp. 303–315. Elsevier, Amsterdam (1999)
9. Hulle, M.V.: Kernel based topographic map formation achieved with an information-theoretic approach. Neural Networks 15(8-9), 1029–1039 (2002)
10. Bishop, C.M., Svensen, M., Williams, C.K.I.: GTM: a principled alternative to the self-organizing map. In: International Conference proceedings on Artificial neural networks - ICANN 1996, pp. 165–701 (1996)
11. Lynch Jr., J.J., Crochiere, R.: Speech/Silence segmentation for real-time coding via rule based adaptive endpoint detection. In: IEEE International Conference on Acoustics, Speech, and Signal Processing, vol. 12, pp. 1348–1351 (1987)
12. Douglas, A.R.: Robust Text-Independent Speaker Identification Using Gaussian Mixture Speaker Models. IEEE Transactions on Speech and Audio Processing 3(1), 72–83 (1995)

# Detecting Trojans Using Data Mining Techniques

Muazzam Siddiqui, Morgan C. Wang, and Joohan Lee

University of Central Florida
siddiqui@mail.ucf.edu,
cwang@mail.ucf.edu,
jlee@cs.ucf.edu

**Abstract.** A trojan horse is a program that surreptitiously performs its operation under the guise of a legitimate program. Traditional approaches using signatures to detect these programs pose little danger to new and unseen samples whose signatures are not available. The focus of malware research is shifting from using signature patterns to identifying the malicious behavior displayed by these malwares. This paper presents the novel idea of extracting variable length instruction sequences that can identify trojans from clean programs using data mining techniques. The analysis is facilitated by the program control flow information contained in the instruction sequences. Based on general statistics gathered from these instruction sequences, we formulated the problem as a binary classification problem and built random forest, bagging and support vector machine classifiers. Our approach showed a 94.0% detection rate on novel trojans whose data was not used in the model building process.

**Keywords:** Data Mining, Trojan Detection, Random Forest, Principal Component Analysis, Support Vector Machines, Disassembly.

## 1 Introduction

Computer virus detection has evolved into malware detection since Cohen first formalized the term computer virus in 1983 [13]. Malicious programs, commonly termed as malwares, can be classified into virus, worms, trojans, spywares, adwares and a variety of other classes and subclasses that sometimes overlap and blur the boundaries among these groups [20]. The most common detection method is the signature based detection that makes the core of every commercial anti-virus program. To avoid detection by the traditional signature based algorithms, a number of stealth techniques have been developed by the malware writers. The inability of traditional signature based detection approaches to catch these new breed of malwares has shifted the focus of malware research to find more generalized and scalable features that can identify malicious behavior as a process instead of a single static signature.

The analysis can roughly be divided into static and dynamic analysis. In the static analysis the code of the program is examined without actually running the

D.M.A. Hussain et al. (Eds.): IMTIC 2008, CCIS 20, pp. 400–411, 2008.

program while in dynamic analysis the program is executed in a real or virtual environment. The static analysis, while free from the execution overhead, has its limitation when there is a dynamic decision point in the programs control flow. Dynamic analysis monitors the execution of program to identify behavior that might be deemed malicious. These two approaches are combined also [19] where dynamic analysis is applied only at the decision-making points in the program control flow.

In this paper we present a static analysis method using data mining techniques to automatically extract behavior from trojans and clean programs. We introduce the idea of using sequence of instructions extracted from the disassembly of trojans and clean programs as the primary classification feature. Unlike fixed length instructions or n-grams, the variable length instructions inherently capture the programs control flow information as each sequence reflects a control flow block.

The difference among our approach and other static analysis approaches mentioned in the related research section are as follows.

First, the proposed approach applied data mining as a complete process from data preparation to model building. Although data preparation is a very important step in a data mining process, almost all existing static analysis techniques mentioned in the related research section did not discuss this step in detail except [22]. Second, all features were sequences of instructions extracted by the disassembly instead of using fixed length of bytes such as n-gram. The advantages are:

1. The instruction sequences include program control flow information, not present in n-grams.
2. The instruction sequences capture information from the program at a semantic level rather than syntactic level.
3. These instruction sequences can be traced back to their original location in the program for further analysis of their associated operations.
4. A significant number of sequences that appeared in only clean program or trojans can be eliminated to speed up the modeling process.
5. The classifier obtained can achieve 94% detection rate for new and unseen trojans.
6. Instruction sequences are a domain-independent feature and can be used to detect other malwares e.g. virus, worms, spywares etc.

## 2    Related Research

Data mining has been the focus of many malware researchers in the recent years to detect unknown malwares. A number of classifiers have been built and shown to have very high accuracy rates. Data mining provides the means for analysis and detection of malwares for the categories defined above. Most of these classifiers use n-gram or API calls as their primary feature. An n-gram is a sequence of bytes of a given length extracted from the hexadecimal dump of the file. Besides file dumps, network traffic data and honeypot data is mined for malicious activities.

In a pioneering work [16] used three different types of features and a variety of classifiers to detect malicious programs. Their primary dataset contained 3265 malicious and 1001 clean programs. They applied RIPPER (a rule based system) to the DLL dataset. Strings data was used to fit a Naive Bayes classifier while n-grams were used to train a Multi-Naive Bayes classifier with a voting strategy. No n-gram reduction algorithm was reported to be used. Instead data set partitioning was used and 6 Naive-Bayes classifiers were trained on each partition of the data. They used different features to built different classifiers that do not pose a fair comparison among the classifiers. Naive-Bayes using strings gave the best accuracy in their model.

A similar approach was used by [14], where they built different classifiers including Instance-based Learner, TFIDF, Naive-Bayes, Support vector machines, Decision tree, boosted Naive-Bayes, SVMs and boosted decision tree. Their primary dataset consisted of 1971 clean and 1651 malicious programs. Information gain was used to choose top 500 n-grams as features. Best efficiency was reported using the boosted decision tree J48 algorithm.

[22] developed PEAT (The Portable Executable Analysis Tool) to detect structural anomalies inside a program. PEAT rested on the basic principle that the inserted code in a program disrupts its structural integrity and hence by using statistical attributes and visualization tools this can be detected. The visualization tools plot the probability of finding some specific subject of interest in a particular area of the program. These subjects include sequence of bytes, their ASCII representation, their disassembly representation and memory access via register offsets. Statistical analysis was done on instruction frequencies, instruction patterns, register offsets, jump and call offsets, entropy of opcode values and code and ASCII probabilities. The experimental results were provided for only one malicious program.

[10] used n-grams as features to build multiple neural network classifiers and adopted a voting strategy to predict the final outcome. Their dataset consisted of 53902 clean files and 72 variant sets of different viruses. For clean files, n-grams were extracted from the entire file while only those portions of a virus file are considered that remain constant through different variants of the same virus. A simple threshold pruning algorithm was used to reduce the number of n-grams to use as features. The results they reported are not very promising and even according to them, the procedure is not sufficient to be used as sole criteria in a virus scanner.

[9] used n-grams to build class profiles using KNN algorithm. Their dataset was small with 25 malicious and 40 benign programs. As the dataset is relatively small, no n-gram reduction was reported. They reported 98% accuracy rate on a three-fold cross validation experiment. It would be interesting to see how the algorithm scale as a bigger dataset is used.

In a more recent work [24] applied association rules mining on API execution sequences. They developed an Objective Oriented Association based mining algorithm and reported better performance on polymorphic and novel malwares than traditional classifiers including Naive-Bayes, support vector machines and boosted decision trees.

[18] proposed a signature based method called SAVE (Static Analysis of Vicious Executables) that used behavioral signatures indicating malicious activity. The signatures were represented in the form of API calls and Euclidean distance was used to compare these signatures with sequence of API calls from programs under inspection.

Besides data mining, other popular methods includes activity monitoring and file scanning. [15] proposed a static analysis method to identify the location of system calls within an executable, which can be monitored at runtime to verify that every observed system call is made from a location identified using the static analysis. [23] proposed a system where kernel carry out a check to see if an executable has been tampered with by doing a signature calculation and comparison against a highly secured signature database with a resulting decision to allow or deny the execution based on the result.

All of this work stated above, that does not include data mining as a process, used very few samples to validate their techniques. The security policies needed human experts to devise general characteristics of malicious programs.

Data preparation is a very important step in a data mining process. Except [22], none of the authors presented above have discussed their dataset in detail. Malicious programs used by these researchers are very eclectic in nature exhibiting different program structures and applying the same classifier to every program does not guarantee similar results.

## 3   Data Processing

Our collection of trojans and clean programs consisted of 4722 Windows PE files, of which 3000 were trojans and the 1722 were clean programs. The clean programs were obtained from a PC running Windows XP. These include small Windows applications such as calc, notepad, etc and other application programs running on the machine. A number of clean programs were also downloaded from [1] to get a representation of programs downloaded from the Internet. The trojans were all downloaded from [8]. The dataset was thus consisted of a wide range of programs, created using different compilers and resulting in a sample set of uniform representation. Figure 3 displays the data processing steps.

### 3.1   Malware Analysis

We ran PEiD [5] on our data collection to detect compilers, common packers and cryptors, used to compile and/or modify the programs. Table 1 displays the distribution of different packers and compilers in the collection. Table 2 displays the summary of packed, not packed and unidentified trojans and clean programs.

Before further processing, packed programs were unpacked using specific unpackers such as UPX (with -d switch) [6], and generic unpackers such as Generic Unpacker Win32 [3] and VMUnpacker [7].

**Table 1.** Packers/Compilers analysis details for trojans and clean programs

| Packer/Compiler | Number of Trojans | Number of Clean Programs |
|---|---|---|
| ASPack | 349 | 2 |
| Borland | 431 | 39 |
| FSG | 38 | 1 |
| Microsoft | 1031 | 937 |
| Other Not Packed | 229 | 597 |
| Other Packed | 118 | 24 |
| PECompact | 48 | 2 |
| Unidentified | 174 | 72 |
| UPX | 582 | 48 |

**Table 2.** Packers/Compilers analysis summary for trojans and clean programs

| Type of Program | Not Packed | Packed | Unidentified |
|---|---|---|---|
| Clean | 1573 | 77 | 72 |
| Trojan | 1691 | 1135 | 174 |
| Total | 3264 | 1212 | 246 |

## 3.2   File Size Analysis

Before disassembling the programs to extract instruction sequences, a file size analysis was performed to ensure that the number of instructions extracted from clean programs and trojans is approximately equal. Table 3 displays the file size statistics for trojans and clean programs.

**Table 3.** File size analysis of the program collection

| Statistic | Trojans Size (KB) | Cleans Size (KB) |
|---|---|---|
| Average | 176 | 149 |
| Median | 66 | 51 |
| Minimum | 1 | 3 |
| Maximum | 1951 | 1968 |

To get an even distribution of the number of programs in the collection, we finally chose 1617 trojans and 1544 clean programs after discarding unidentified programs and large trojans that were pulling the average program size a little higher than the clean programs.

## 3.3   Disassembly

Binaries were transformed to a disassembly representation that is parsed to extract features. The disassembly was obtained using Datarescues' IDA Pro [4]. From these disassembled files we extracted sequences of instructions that served

```
mov       dword ptr [ebp-4], 4
lea       eax, [ebp-24h]
mov       [ebp-84h], eax
mov       dword ptr [ebp-8Ch], 4008h
mov       dword ptr [ebp-94h], 8
mov       dword ptr [ebp-9Ch], 3
push      10h
pop       eax
call      __vbaChkstk
lea       esi, [ebp-8Ch]
mov       edi, esp
movsd
movsd
movsd
movsd
push      10h
pop       eax
call      __vbaChkstk
```

**Fig. 1.** Portion of the output of disassembled Win32.Flood.A trojan

as the primary source for the features in our dataset. A sequence is defined as instructions in succession until a conditional or unconditional branch instruction and/or a function boundary is reached. Instruction sequences thus obtained are of various lengths. We only considered the opcode and the operands were discarded from the analysis. Figure 1 shows a portion of the disassembly of the Win32.Flood.A trojan.

### 3.4 Parsing

A parser written in PHP translates the disassembly in figure 1 to instruction sequences. Figure 2 displays the output of the parser. Each row in the parsed output represented a single instruction sequence. The raw disassembly of the trojan and clean programs resulted in 10067320 instruction sequences.

```
mov lea mov mov mov mov push pop call
lea mov movsd movsd movsd movsd push pop call
```

**Fig. 2.** Instruction sequences extracted from the disassembled Win32.Flood.A trojan

### 3.5 Feature Extraction

The parsed output was processed through our Feature Extraction Mechanism. Among the 10067320 instruction sequences, 2962589 unique sequences were identified with different frequencies of occurrence. We removed the sequences that were found in one class only as they will reduce the classifier to a signature detection technique.

### 3.6 Primary Feature Selection

The Feature Selection Mechanism considered frequency of occurrence of each sequence in the entire data to be the primary selection criteria. Sequences with

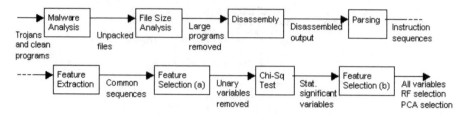

**Fig. 3.** Data preprocessing steps

less than 10% frequency of occurrence were identified as unary features are were not included in the dataset. This removed 97% of the sequences and only 955 sequences were selected. The dataset consisted of frequency of occurrence of each of these sequences in each file. A binary target variable identified each file as trojan or clean.

Using the occurrence frequency as the primary data item in the dataset enabled us to consider the features as count variables.

### 3.7    Independence Test

A Chi-Square test of independence was performed for each feature to determine if a relationship exists between the feature and the target variable. The variables were transformed to their binary representation on a found/not found basis to get a 2-way contingency table. Using a p-value of 0.01 for the test resulted in the removal of the features that did not showed any statistically significant relationship with the target. The resulting number of variables after this step was 877.

### 3.8    Secondary Feature Selection

After performing the primary feature selection and independence test, we applied two more feature selection algorithms to create three different datasets. These algorithms include random forest and principal component analysis. The first dataset retained all the original 877 variables. In the rest of the paper, we will refer to this set as *All variables*.

**Random Forest.** Besides classification, the random forest also gives the important variables used in the model. The importance is calculated as the mean decrease in accuracy or mean decrease in Gini index if the variable is removed from the model. We rejected the variables for which the mean decrease in accuracy was less than 10%. Only 84 variables were selected. This dataset will be referred later in the paper as *RF Selection*.

**Principal Component Analysis.** PCA is a technique used to reduce multidimensional data sets to lower dimensions for analysis. PCA involves the calculation of the eigenvalues that represent the linear combination of original variables such that the lower order eigenvalues explain most of the variance in the data. We kept the 146 variables that explained 95% variance in the dataset and rejected others. This dataset will be referred later in the paper as *PCA Selection*.

# 4 Experiments

The data was partitioned into 70% training and 30% test data. Similar experiments showed best results with tree based models for the count data [17]. We built bagging and random forest models using R [2]. We also experimented with SVMs using R.

## 4.1 Bagging

Bagging or Bootstrap Aggregating is a meta-algorithm to improve classification and regression models in terms of accuracy and stability. Bagging generates multiple versions of a classifier and uses plurality vote to decide for the final class outcome among the versions. The multiple versions are created using bootstrap replications of the original dataset. Bagging can give substantial gains in accuracy by improving on the instability of individual classifiers. [11]

We used classification trees with 100 bootstrap replications in the Bagging model.

## 4.2 Random Forest

Random forest provides a degree of improvement over Bagging by minimizing correlation between classifiers in the ensemble. This is achieved by using bootstraping to generate multiple versions of a classifier as in Bagging but employing only a random subset of the variables to split at each node, instead of all the variables as in Bagging. Using a random selection of features to split each node yields error rates that compare favorably to Adaboost, but are more robust with respect to noise.[12]

We grew 100 classification trees in the Random forest model. The number of variables sampled at each split was ranged from 6 to 43 depending upon the number of variables in the dataset.

## 4.3 Support Vector Machines

SVMs are set of tools for finding the optimal hyperplane that separates the linear or non-linear data into two categories [21]. The separating hyperplane is the hyperplane that maximizes the distance (margin) between the two parallel hyperplanes. The non-linear classification is obtained by applying a kernel function to these maximum-margin hyperplanes.

We used C-Classification with a radial basis kernel function.

## 5   Results

We tested the models using the test data. Confusion matrices were created for each classifier using the actual and predicted responses. The following four estimates define the members of the matrix.

*True Positive (TP)*: Number of correctly identified malicious programs.
*False Positive (FP)*: Number of wrongly identified benign programs.
*True Negative (TN)*: Number of correctly identified benign programs.
*False Negative (FN)*: Number of wrongly identified malicious programs.

The performance of each classifier was evaluated using the detection rate, false alarm rate and overall accuracy that can be defined as follows:

*Detection Rate*: Percentage of correctly identified malicious programs.
$$DetectionRate = \frac{TP}{TP+FN}$$

*False Alarm Rate*: Percentage of wrongly identified benign programs.
$$FalseAlarmRate = \frac{FP}{TN+FP}$$

*Overall Accuracy*: Percentage of correctly identified programs.
$$OverallAccuracy = \frac{TP+TN}{TP+TN+FP+FN}$$

Table 4 displays the experimental results for each classifier over the three test sets.

**Table 4.** Experimental results for the new and unseen trojans

| Classifier | Detection Rate | False Alarm Rate | Overall Accuracy |
|---|---|---|---|
| Random Forest (All variables) | 93.1% | 6.3% | 92.6% |
| Random Forest (RF Selection) | 92.4% | 9.2% | 94.0% |
| Random Forest (PCA Selection) | 89.7% | 10.1% | 89.6% |
| Bagging (All variables) | 91.1% | 7.4% | 89.8% |
| Bagging (RF Selection) | 91.1% | 9.6% | 91.9% |
| Bagging (PCA Selection) | 89.4% | 10.5% | 89.4% |
| SVM (All variables) | 83.6% | 15.0% | 82.6% |
| SVM (RF Selection) | 83.3% | 12.3% | 79.7% |
| SVM (PCA Selection) | 82.5% | 19.8% | 84.7% |

## 6   Discussion

Random forest stood out to be the winner among all the classifiers and feature selection methods in our experiments. The best results for overall accuracy, false positive rate and area under the ROC curve were obtained using random forest classifier on all the variables. The best detection rate was obtained using random forest model with feature selection using a previous run of random forest. This comes with a slight sacrifice in overall accuracy but a much simpler model. Tree

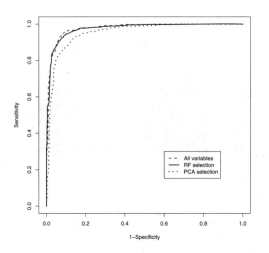

**Fig. 4.** ROC curve comparing random forest test results on datasets with all variables, RF selection and PCA selection

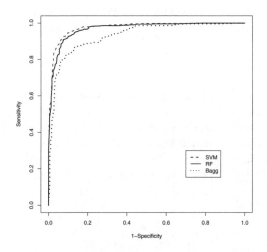

**Fig. 5.** ROC curve comparing random forest, bagging and svm test results on dataset with RF selection

based method performed better on the count data. More specifically random forest performed slightly better than Bagging which is an endorsement of its superiority over Bagging as claimed in [12]. ROC curves are provided to compare results for each classifier and each dataset. For space considerations, we are only providing ROC curves for RF selection dataset for each classifier in figure 4. To compare the results over each dataset figure 5 displays ROC curve for random forest test results for the three datasets.

**Table 5.** Area under the ROC curve for each classifier for each dataset

| | |
|---|---|
| Random Forest (All variables) | 0.9772 |
| Random Forest (RF Selection) | 0.9754 |
| Random Forest (PCA Selection) | 0.9537 |
| Bagging (All variables) | 0.9714 |
| Bagging (RF Selection) | 0.9686 |
| Bagging (PCA Selection) | 0.9549 |
| SVM (All variables) | 0.9295 |
| SVM (RF Selection) | 0.9305 |
| SVM (PCA Selection) | 0.8971 |

## 7   Conclusions

In this paper we presented a data mining framework to detect trojans. The primary feature used for the process was the frequency of occurrence of variable length instruction sequences. The effect of using such a feature set is two fold as the instruction sequences are domain-independent features and same technique can be used to detect other malwares. We used the sequences common to both trojans and clean programs to remove any biases caused by the features that have all their occurrences in one class only. We showed 94% detection rate and a 6.3% false positive rate.

## 8   Future Work

The information included for this analysis was extracted from the executable section of the PE file. To achieve a better detection rate this information will be appended from information from other sections of the file. This will include Import Address Table and the PE header. API calls analysis has proven to be an effective tool in malware detection [18]. Moreover header information has been used in heuristic detection [20]. Our next step is to include this information in our feature set.

## References

1. Download.com, http://www.download.com/
2. The r project for statistical computing, http://www.r-project.org/
3. Generic Unpacker Win32, http://www.exetools.com/unpackers.htm
4. IDA Pro Disassembler, http://www.datarescue.com/idabase/index.htm
5. PEiD, http://peid.has.it/
6. UPX the Ultimate Packer for eXecutables, http://www.exeinfo.go.pl/
7. VMUnpacker, http://dswlab.com/d3.html
8. VX Heavens, http://vx.netlux.org
9. Abou-Assaleh, T., Cercone, N., Keselj, V., Sweidan, R.: N-gram-based detection of new malicious code. In: Proceedings of the 28th Annual International Computer Software and Applications Conference - Workshops and Fast Abstracts - (COMPSAC 2004), vol. 2, pp. 41–42 (2004)

10. Arnold, W., Tesauro, G.: Automatically generated win32 heuristic virus detection. In: Virus Bulletin Conference, pp. 123–132 (2000)
11. Breiman, L.: Bagging predictors. Machine Learning 24(2), 123–140 (1996)
12. Breiman, L.: Random forests. Machine Learning 45(1), 5–32 (2001)
13. Cohen, F.: Computer Viruses. PhD thesis, University of Southern California (1985)
14. Kolter, J.Z., Maloof, M.A.: Learning to detect malicious executables in the wild. In: Proceedings of the 2004 ACM SIGKDD International Conference on Knowledge Discovery and Data Mining (2004)
15. Rabek, J.C., Khazan, R.I., Lewandowski, S.M., Cunningham, R.K.: Detection of injected, dynamically generated, and obfuscated malicious code. In: Proceedings of the 2003 ACM Workshop on Rapid Malcode, pp. 76–82 (2003)
16. Schultz, M.G., Eskin, E., Zadok, E., Stolfo, S.J.: Data mining methods for detection of new malicious executables. In: Proceedings of the IEEE Symposium on Security and Privacy, pp. 38–49 (2001)
17. Siddiqui, M., Wang, M.C., Lee, J.: Data mining methods for malware detection using instruction sequences. In: Proceedings of Artificial Intelligence and Applications, AIA 2008. ACTA Press (2008)
18. Sung, A.H., Xu, J., Chavez, P., Mukkamala, S.: Static analyzer of vicious executables. In: 20th Annual Computer Security Applications Conference, pp. 326–334 (2004)
19. Symantec. Understanding heuristics: Symantec's bloodhound technology. Technical report, Symantec Corporation (1997)
20. Szor, P.: The Art of Computer Virus Research and Defense. Addison Wesley for Symantec Press, New Jersey (2005)
21. Webb, A.: Statisitcal Pattern Recognition. Wiley, Chichester (2005)
22. Weber, M., Schmid, M., Schatz, M., Geyer, D.: A toolkit for detecting and analyzing malicious software. In: Proceedings of the 18th Annual Computer Security Applications Conference, p. 423 (2002)
23. Williams, M.: Anti-trojan and trojan detection with in-kernel digital signature testing of executables. Technical report, NetXSecure NZ Limited (2002)
24. Ye, Y., Wang, D., Li, T., Ye, D.: Imds: intelligent malware detection system. In: KDD 2007: Proceedings of the 13th ACM SIGKDD international conference on Knowledge discovery and data mining, pp. 1043–1047. ACM Press, New York (2007)

# Enabling MPSoC Design Space Exploration on FPGAs

Ahsan Shabbir, Akash Kumar, Bart Mesman, and Henk Corporaal

Eindhoven University of Technology,
5600MB Eindhoven, The Netherlands
{a.shabbir, a.kumar, b.mesman, h.corporaal}@tue.nl
http://www.es.ele.tue.nl/

**Abstract.** Future applications for embedded systems demand chip multiprocessor designs to meet real-time deadlines. These multiprocessors are increasingly becoming heterogeneous for reasons of cost and power. Design space exploration (DSE) of application mapping becomes a major design decision in such systems. The time spent in DSE becomes even greater with multiple applications executing concurrently. Methods have been proposed to automate generation of multiprocessor designs and prototype them on FPGAs. However, only few are able to support heterogeneous platforms. This is because heterogeneous processors require different types of inter-processor communication interfaces. So when we choose a different processor for a particular task, the communication infrastructure of the processor also has to change. In this paper, we present a module that integrates in a multiprocessor design generation flow and allows heterogeneous platform generation. This module is area efficient and fast. The DSE shows that up to 31% FPGA area can be saved when heterogeneous design is used as compared to a homogeneous platform. Moreover, the performance of the application also improves significantly.

**Keywords:** FSL, FPGAs, FIFO, MPSoC.

## 1 Introduction

The overall execution time of an application mapped onto an architecture depends on a number of factors such as memory hierarchy, communication structure etc, however type of processor remains a key contributor. For example, any signal processing application will run faster on a DSP, whereas any control dominated application will not be able to exploit the resources of such processors effectively. Performance can be enhanced if different parts of the application run on different processors which are optimized for those characteristics. Heterogeneous multiprocessor platforms [1] are good candidate for these type of applications.

### 1.1 Synchronous Data Flow Graphs

Synchronous Data Flow Graphs [2](SDFG) are used to model Digital Signal Processing (DSP) and Multimedia applications. Tasks (Actors) are vertices in

D.M.A. Hussain et al. (Eds.): IMTIC 2008, CCIS 20, pp. 412–421, 2008.

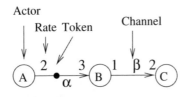

**Fig. 1.** An example of SDF Graph

the graph and the directed edges represent dependencies between the tasks. Tasks also need input data or control information before they can start and they usually produce output data; such information is referred to as *tokens*. Actor execution is also called *firing*. An actor is called *ready* when it has sufficient input tokens on all of its input edges and sufficient buffer space at its output edges;an actor can only fire, when it is ready. Figure 1 shows a simple SDFG consisting of three actors and two channels. Actor B can fire as soon as three tokens are available on channel *alpha*. Its firing results in consumption of three tokens from channel *alpha* and production of one token on channel *beta*.

## 1.2   Problem Description

DSP and multimedia applications are mapped on to multiprocessor platforms by using the SDF graphs [7]. Actors are mapped onto processors and communication between the actors is modeled as First in First Out Channels (FIFOs). FPGA vendors provide Platform [6] FPGAs. These Platform FPGAs are very suitable for prototyping multimedia applications. Heterogeneous platform generation on these FPGAs is difficult because each type of processor has its own communication interface. This slows down the design space exploration as both the processor and the communication infrastructure is changed at every design point. The situation becomes even more complex if accelerator attachment is also a possibility. In this paper we propose to have only one type of communication infrastructure for all types of processors in the MPSoC. We also propose to use the same interface for accelerator attachment. To show validity of our proposal we choose Virtex FPGA by Xilinx. These FPGAs contain up to four hard wired PowerPC-405 [4] cores. Xilinx also provides Microblaze [5] soft cores. Microblaze processors have a FIFO based communication link called Fast Simplex Link [13](FSL). Microblaze processors can be connected to each other through these FSLs. However the PowerPC processors do not have FSLs so we can not directly connect these processors with Microblaze processors. To enable rapid heterogeneous platform generation we have designed an interface, which connects with Processor Local Bus (PLB) of PowerPC and provides a standard FSL interface. We have included the interface into Multi-Applications Multi-Processor Synthesis (MAMPS) design methodology [3]. Our design flow takes in application(s) specifications and generates high level hardware description file (MHS file) for Xilinx FPGAs. This paper does not discuss the MAMPS flow, however interested readers are encouraged to read [3]. This paper describes the interface in detail and presents some results about its performance.

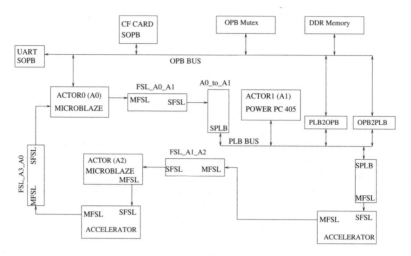

**Fig. 2.** Proposed Architecture

The proposed architecture is shown in figure 2. FSL is used as the basic communication infrastructure in the design. The architecture also supports Shared memory access. "Mutex" modules are used to access the shared sections of the memory.

The paper is organized as follows. Section 2 discusses some similar work. We explain implementation details in section 3. In section 4, we give some experimental results before concluding the paper in section 5.

## 2   Related Work

In ESPAM [8], a communication controller has been designed to interface with the instruction/data bus of Microblaze and PowerPC processors. ESPAM has two configurations for multiprocessor platforms. In the cross-bar based configuration, Every processor is connected with communication controller and the communication controller is further attached to the cross bar and communication memory. In the point to point configuration (Which is similar to our work), Every processor writes to its own communication memory through communication controller. Other processor in Kahn network [9], which has to use this data connects with this communication memory through its own communication controller. As the Communication Controller has been connected to processors using the processor address and data bus. The performance of Communication controller suffers due to this bus sharing [10]. On the other hand, we use the standard FSL bus for communication which means that Microblaze processor does not need additional hardware and has faster interface as compared to ES-PAM. For the PowerPC processor, the Processor Local bus (PLB) is used to connect peripherals which provide master and slave interface to FSL channels.

So our design is area efficient and performs better due to standard interface on Microblaze side.

## 3   Implementation Details

FSL is directly integrated into the pipeline of microblaze processor so it is very efficient and fast interface. The bus contains FIFO buffers. Depth of these buffers is programable. FSL also has a control bit which shows that the location being read contains data or control information. FSL bus can be used synchronously or asynchronously. Every bus has only one Master and Slave so it is dedicated point to point bus. FSL Master and Slave interface signals are shown in figure 3. It is a unidirectional 32-bit bus. Hence to have bidirectional communication, a

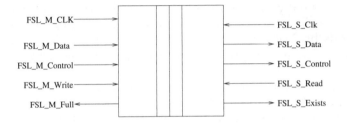

**Fig. 3.** FSL Interface signals

set of master and slave is required on each side as the master only sends data to FSL and Slave only receives. In our approach we have designed one slave PLB peripheral for each direction as shown in figure 4. By doing so we can have the flexibility of having any combination of FSL buses. For example we can have three buses sending data from processor "A" (PowerPC) to processor "B " (Microblaze) and two buses from processor "B" to processor "A". The peripheral, that reads FSL data is named as "Microblaze_to_PowerPC"', and the peripheral that sends data to FSL is named as "PowerPC_to_Microblaze". FIFOs inside the FSL are used in asynchronous mode. The reason for this choice is because PLB and FSL can have different clocks and if we use the FSL bus synchronously then we restrict our designs to use the same clock for PLB and FSL. So our design allows Microblaze and PowerPC processor to run at different frequencies and still communicate over FSL. Microblaze can run up to 100 MHZ, where as the operating frequency of PowerPC processor in Virtex FPGAs can be as high as 400MHz. This gives a large number of Task distribution options among the processors as the PowerPC running at higher frequency can take more of application load than a microblaze.

We implement our interface on Xilinx Virtex-II Pro 2VP30 FPGA, using the Xilinx Embedded Development Kit (EDK8.2i) [12]. The 2VP30 consists of 13,696 slices and up to 2,448 Kbits of on-chip BlockRAM memory. The FPGA contains two PowerPC-405 processors. We recommend to use ISCOM and DSCOM buses

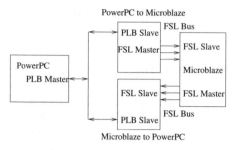

**Fig. 4.** Peripheral IPs are used to connect PowerPC with Microblaze

to connect instruction and data memories with the PowerPC processors as these are dedicated buses. On the other hand, if we use PLB bus for instruction and data along with our FSL interface peripheral, the bus contentions will drop the performance of the system. Sixteen FSLs can be attached to the PLB bus. On the PLB side two registers are designed which are used to read the status of FSL bus and also to enable the sending of a control bit along with the data if required.

The peripherals can be easily integrated into the designs by copying only a "pcores" directory. Software driver files are included in this directory. The whole design space of the application is explored in very short time by mapping different tasks of the application on to different processors and monitoring the execution time of the configuration. A case study of JPEG mapping is presented in the next section.

## 4   Application Mapping

We select JPEG as target application for our MPSOC platform validation. The JPEG Encoder application software is obtained from `http://www.opencores.org/people.cgi/info/quickwayne`. First we map the JPEG application on three Microblaze processors. These processors are connected to each other through FSL as shown in figure 6. The application is divided among three actors. These three actors are

1. File Parser (FP).
2. Color Covesion (CC) and Discrete Cosine Transform (DCT).
3. Variable Length Coding (VLC).

Figure 5 shows the "XML" file snippet used for this design. Three actors defined above are visible in the figure. Number of input/output tokens is also specified in the file. The token size can have the granularity of a macroblock or can be as small as a byte. In the "XML" file we also specify the type of processor as an "attribute". Design space exploration is performed by changing only the processor attributes of the actors. The MAMPS tool takes this file as input and generates the XPS [12] project files which are then synthesized and run to get the

```
<?xml version="1.0"?>
<sdfMapping version="1.0">
 <applicationGraph>
  <sdf name="g" type="G">
   <actor name="FileParser" type="a0">
    <port name="IN" type="in" rate="1"/>
    <port name="OUT" type="out" rate="1"/>
   </actor>
   <actor name="CCDCT" type="a1">
    <port name="IN" type="in" rate="1"/>
    <port name="OUT" type="out" rate="1"/>
   </actor>
   <actor name="VLC" type="a2">
    <port name="IN" type="in" rate="1"/>
    <port name="OUT" type="out" rate="1"/>
   </actor>
```

**Fig. 5.** Snippet of JPEG Encoder application specification

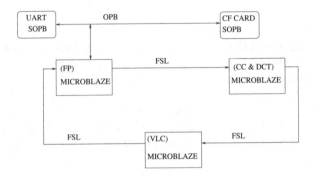

**Fig. 6.** Homogeneous platform, consisting of 3 microblaze processors,connected to each other through FSL

results. Various performance measuring timers are also configured in the XML file to measure actor execution times. Performance results at different design points are compared and best configuration is selected.

Figure 6 shows the homogeneous platform consisting of three microblaze processors. Compact Flash card (CF) and UART are connected to actor0 through on chip Peripheral Bus (OPB). UART is used for debugging and information display where as the CF card contains the input BMP file to be converted into JPEG format. First actor (actor0) opens the BMP file stored in the CF card, and sends the data to next actor (actor1). Actor1 converts the RGB format into YCrCb(4:2:0) format, computes the discrete cosine transform (DCT) and forwards the data to third processor (actor2). The final processor (actor2) performs VLC of the input, and sends the encoded data to actor0. Actor0 writes back the JPEG encoded stream received from Actor2 into the CF card.

For Heterogeneous platform generation we replace one microblaze processor with PowerPC processor as shown in figure 7. We first choose "Actor1" as a candidate for replacement with PowerPC processor and write its XML file. Two PLB slave peripherals are included in the design to have FSL interfaces. The peripheral "A0_to_A1" receives data from actor0 where as peripheral "A1_to_A2"

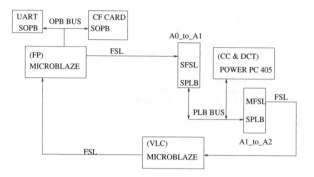

**Fig. 7.** Heterogeneous platform, Two microblaze and one PowerPC processor. PLB slave peripherals provides interface to connect PowerPC processor with the FSL buses.

sends data to actor2 as shown in figure 7. In both Homogeneous and Heterogeneous configurations, the cache of all processors is kept disabled. The input BMP file size is 983,094 bytes and we use Xilinx University Program (XUP) Virtex-2 Pro development board for our experiments.

Now we further investigate the accelerator attachment problem. We profiled the application to accelerate the tasks mapped to "actor1" in the previous configuration. Table 1 shows the profiling results of task "CC" and "DCT". Both microblaze and PowerPC are operating at 100MHZ. Task DCT uses 60% of the execution time and takes more cycles per call as compared to the Color Conversion task. So we decide to map the DCT on hardware. The area utilization results

**Table 1.** Profiling Results for Processors performing CC and DCT

| Function | DCT | CC |
|---|---|---|
| | (60% of execution time) | (40% of execution time) |
| PowerPC | 19964 cycles/call | 3478 cycles/call |
| Microblaze | 20927 cycles/call | 4425 cycles/call |

of implementation are shown in table 2. It is clear that by replacing the microblaze with PowerPC processor, the occupied slices have reduced by 27% and 4 input LUT utilization has reduced by 31%. However, eight additional BRAMS are needed because the PowerPC instruction and data memory interfaces are 64bit wide and require more memory space than Microblaze processor. Columns four and five show the additional area required because of the DCT accelerator. Note that the same accelerator is used for both designs and no change is required in the interface.

Figure 8 shows the performance comparison for homogeneous and heterogeneous platforms. We call the first set of columns in figure 8 as the "software only configuration". Here both platforms are performing "CC" and "DCT" in

**Table 2.** Area utilization of both designs

| Platform Type | Homogeneous | Heterogeneous | Homogeneous with DCT Accelerator | Heterogeneous with DCT Accelerator |
|---|---|---|---|---|
| Occupied Slices | 5067 (36%) | 3701 (27%) | 6746 (49%) | 5712 (41%) |
| 4 Input LUTs | 8589 (31%) | 5881 (21%) | 10787 (39% ) | 8406 (30%) |
| BRAM blocks | 40 (29%) | 48 (35%) | 40 (29%) | 48 (35%) |

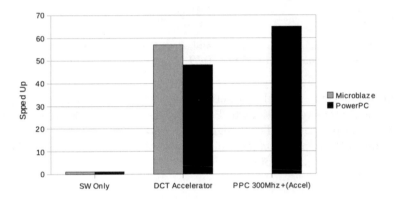

**Fig. 8.** DCT kernel Speed Up comparison for Microblaze and PowerPC processors

software. Table 1 shows the required number of execution cycles for both microblaze and PowerPC processors. Both processors utilize almost same number of execution cycles for both tasks. This is also the base-line performance and the results obtained by "accelerator based configurations" are compared with this configuration. The second set of columns shows the case when Hardware accelerators are used by both processors. Here the performance of homogeneous platform is is slightly better and a DCT kernel speed up of 57 is achieved. The Speed up by PowerPC-accelerator is only 48. This is a lower speed up as compared to microblaze based system. we share the PLB between two peripherals so the traffic on the PLB increases resulting in lower performance. However, as our interface supports asynchronous clock operation we run the PowerPC processor at 300 MHz and the PLB at 150 MHz. Consequently, the DCT kernel speed up of 65 is achieved. Figure 9 shows the best configuration after going through all the design points in the design space.

Similar results are obtained when PowerPC processor is assigned the VLC actor. However the performance deteriorates by a factor 1.5, when we map function FP to PowerPC as compared to the case when we map DCT and CC to PowerPC. This is because of the fact that file parsing is a slow function and involves low speed interface to OPB. The PowerPC connects to OPB through PLB2OPB bridge. This bridge is responsible for loss in performance as the microblaze connects with the OPB directly. These results suggest that most

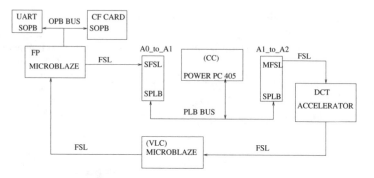

**Fig. 9.** Heterogeneous platform, Two microblaze and one PowerPC processor. PowerPC only performs color conversion and task DCT is mapeded onto hardware accelerator.

performance benefits from the PowerPC can be obtained by mapping the most computation intensive functions to PowerPC and relatively slow I/O functions to Microblaze processors. The designed interface helps in rapid design space exploration of the application mapping and it took only 3 hours to find the most appropriate application mapping.

## 5   Conclusion

In this paper, we present an interface, which enables quick DSE of multimedia applications on Virtex FPGAs. The interface is very easy to use and provides up to 31% savings in hardware resources for the same design mapped to Microblaze based platforms. We use JPEG encoder as a case study and explore the whole design space. We also propose to use FSL as standard communication architecture in the design. We observe that due to sharing of the interface at the PowerPC side, some performance is lost but it can be regained if we run the PowerPC processors at higher frequency. DCT Kernel speed up of 65 is achieved by operating the PowerPC processor at 300 MHZ. The interface has been integrated in MAMPS design flow and Platform configuration can be changed very quickly by only modifying an XML file. Our architecture also supports accelerators with standard FSL interface. These accelerators can be easily included in the design flow and their impact on performance can be observed.

**Acknowledgments.** First author is thankful to National Engineering and Scientific Commission Islamabad, Pakistan for their financial support.

## References

1. Enslow Jr., P.H.: Multiprocessor Organization – A Survey. ACM Computing Surveys 9(1), 103–129 (1977)
2. Lee, E.A., Messerschmitt, D.G.: Statis scheduling of synchronous dataflow programs for digital signal processing. IEEE Transactions on Computers (1987)

3. Kumar, A., Fernando, S., Ha, Y., Mesman, B., Corporaal, H.: Multi-processor System-level Synthesis for Multiple Applications on Platform FPGA. In: Proceedings of Field Programmable Logic (FPL) Conference, Amsterdam, The Netherlands, pp. 92–97 (2007)
4. PowerPC Processor Reference Guide. Xilinx Inc. (2007)
5. Microblaze Processor Reference Guide. Xilinx Inc. (2007)
6. www.xilinx.com.www.altera.com
7. Sriram, S., Bhattacharyya, S.S. (eds.): Embedded Mutiprocessors: Scheduling and Synchronization, Marcel Dekker (2000)
8. Nikolov, H., Stefancov, T., Deprettere, E.: Multi-processor System Design with ESPAM. In: Proc. 4th IEEE/ACM/IFIP Int. Conference on HW/SW Codesign and System Synthesis (CODES-ISSS 2006), Madrid, Spain, pp. 323–328 (2006)
9. Kahn, G.: The Semantics of a Simple Language for Parallel Language. In: Proc. of IFIP Congress (1974)
10. Thompson, M., Nikolov, H., Stefanov, T.: A Frame Work For Rapid System-Level Exploration, Synthesis, and Programming of Multimedia MP-Socs. In: Proc. 5th IEEE/ACM/IFIP Int. Conf. on HW/SW Codesign and System Synthesis (CODES-ISSS 2007), Salzburg, Austria (2007)
11. Wolf, W.: The Future of multiprocessor system on chip. In: Proceedings of 41st Annual Conference on Design Automation (DAC 2004), San Diego, California, pp. 681–685 (2004)
12. Embedded Systems Tools Guide. Xilinx Embedded Development Kit, EDK version 8.2i ed., Xilinx, Inc. (2004)
13. Fast Simplex Link Channel (FSL), Product Specification Xilinx (2004)

# Range Based Real Time Localization in Wireless Sensor Networks

Hemat K. Maheshwari, A.H. Kemp, and Qinghua Zeng

Institute of Integrated Information Systems, School of E & EE,
University of Leeds, Leeds, LS2 9JT, U.K.
{elhkm,A.H.Kemp,eenqz}@leeds.ac.uk

**Abstract.** In today's world, freight containers and shipping play a massive role for the development of international commerce by handling containers worldwide. Due to lack of a reliable positioning and tracking system, the record of containers is not accurate and results in a number of misplaced, unidentified and lost containers which can become overwhelming over time. This indirectly effects the development of international commerce, especially for countries that rely on imported food, resources and raw materials. In this paper, we discuss the paradigm of Wireless Sensor Networks for a real time localization system to track the location and status of freight containers in a port. We have considered the time of arrival (ToA) method and a two-way ranging scheme. The ISM 2.4 GHz band of Zigbee (IEEE 802.15.4 standard) is considered which can be used worldwide for a real time positioning system because of its low cost, communication and low power consumption.

**Keywords:** Wireless Sensor Networks, Real Time Localization, Two-Way Ranging, IEEE 802.15.4, freight containers.

## 1 Introduction

Today, shipping and cargo in any country have a very important responsibility to improve financial conditions. As the number of twenty-foot equivalent units (TEU) is increasing throughout the world, the percentage of containers which are misplaced or delivered to the wrong destination is also increasing. Container terminals and ports with efficient tracking and locating systems which can make a mammoth difference to productivity. For that, multiple positioning systems like differential global positioning system (DGPS) or real time location systems to automate container and terminal equipment positioning or perhaps radio frequency identification (RFID) are in use. The overall effort for an efficient system is to avoid errors that might send a container to the wrong destination in the shipyard, resulting in delayed departures or incomplete shipments. Recently, RFID is one of the technology which has become a method to track the container in a port. According to facts and figures given by world shipping council, the equivalent of about 141 million loaded TEU containers were moved across the oceans in 2007 [1] and this number is expected to increase rapidly.

D.M.A. Hussain et al. (Eds.): IMTIC 2008, CCIS 20, pp. 422–432, 2008.

During the last decade, rapidly growing attention has been focused on large scale Wireless Sensor Networks (WSNs) due to low cost, be free of deployment hurdles, and wrapping a huge variety of different applications nature. Because of this nature, the presences of WSNs have excited the research community. A wireless sensor network, which is composed of sensor nodes, sensor seeds or anchors and can be used by any specific application, has a lot of different challenges in terms of architecture, robust and reliable routing, localization, time synchronization, security and privacy, fault tolerance, dependency, cooperation with nodes and many more ([2] - [10]).

Today, ports are using different technology for real time tracking. The port of singapore [11] currently uses RFID to track the containers on a yard, international terminal solutions has designed G-POS [12], a GPS based system to track and report the position information about the containers, but the loss of LOS satellite signals with GPS may not get the correct position of the containers. Apart from these, some of the ports are based on manual systems to update container information, so there is no record of lost or misplaced containers. It takes long time to manually track the position or status of any container in a port. Some times, it is impossible to realize different level unauthorized attempts to open a container, any sort of leakage or drop, illegal movements etc. [13]. Figure 1 is of the general application where a real time localization system can be more suitable to locate the position of containers in a port, or any other object. In the outdoor environment, one source of primary power can be used perhaps through street lights, which can help anchors / sensors to work on more complex envirnoment.

In figure 1, on a port with the large number of stacks of the containers (ISO Standard size L=20', W=8', H=8'6") are connected with the sensors on the front

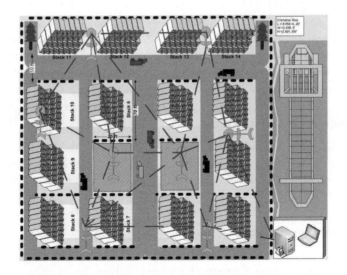

**Fig. 1.** Simple design of Freight containers on a port with real time locating system

of the door. The main reason to connect these to the front door of the container to keep it safe from the other containers. In case of a sensor attached on the roof of the container, it will be difficult to keep it safe from other containers which are supposed to be in a stack. In case of a sensor attached on the side of the containers, horizontally placed containers will create hurdles. Anchor nodes (reference point) will be attached to the electric poles/load carriers, where the electric power can be used to increase the life of anchor nodes.

Localization and retrieval of freight containers in a port is a challenging problem because of the open environment which may be subject to interference signals. It is anticipated that significant multipath propagation will impact the accuracy and reliability of the system. Measurement of the propagation for such a container environment is currently not available. The interference problem in such an open environment can be because of other users in the ISM band and due to the self interference of multipath propagation.

In this paper, we discuss real time localization systems that can be implemented on a yard to locate containers with different status to save time, to save money, to improve the productivity and to improve the international commerce. We discuss the Zigbee based wireless network using the method of time of arrival (ToA) and the concept of two-way ranging. The paper is structured as follows: in the second section, an overview of generalized location and ToA approach is elaborated. In section three, the position in localization is discussed. Section four of the paper deals with the technology and approach. In section five, we discuss system description in terms of traffic model and performance measurement parameters. Conclusions and future work are covered by section six and seven respectively.

## 2   Localization in WSNs

Localization systems are based on different algorithms because of changes in topology, signal transmission representations, energy and power requirement, computational complexity, cost and management etc. All of these factors have forced researchers to divide approaches into two different categories in WSNs: range-based and range-free [2].

### 2.1   Range Free Localization

It is cost effective, free from the need of any special hardware and based on the contents of the messages received through simple operations. These systems are based on the connectivity of the network. Only very coarse localization is achievable with these systems but this may be adequate for many applications.

### 2.2   Range Based Localization

It is based on the absolute position or estimates of the actual distance between nodes using ranging and it is one of the main advantage of range-based localization. Range-based 3D location sensing technology SpotON [19], RADAR [8], cricket indoor location-support system [20] are based on RSSI measurement and ultrasound pulses respectively to achieve the location information. In [21,22] angle of arrival (AOA) scheme is proposed between the sensor nodes to locate the position. Collaborative and iterative trilateration [23] scheme can be used to locate the position of unknown sensor nodes. Once the sensor nodes are estimated, they can behave as anchor nodes.

## 2.3  Time of Arrival (ToA)

Time of Arrival is a localization method which takes advantage of the relationship between the distance and transmission with the help of known propagation speed. ToA has been categorized in different directions. One is based on the single way ranging, which absolutely requires high resolution clocks for time synchronization. Unlike one-way ToA measurement, two -way ToA measurement don't require high resolution clocks since the offset time between transmitter and receiver can be used to calculate the round trip time. The roundtrip time can be measured by adding UpTime and DownTime (represents the time from transmitter to receiver and vice versa) and dividing it by 2 with multiplication of propagation speed, c.

## 2.4  Two-Way Ranging Method

The two-way ranging method [26] is shown diagrammatically in the Figure 2. In the diagram, the master device starts the sequence by transmitting a signal at known time within the master device time scale, but essentially an arbitrary time with respect to the slave device. The slave device must therefore search for the arrival of the signal. The signal transmitted by the master has a propagation time $\Delta t$ seconds between the master and slave device. However, as the slave device is not synchronized to the master device the offset it measures is denoted by $\Delta B$ seconds which will in general be non-equal to $\Delta t$. The slave device sends a return signal to the master exactly one frame after it received the signal from master. In practice it is unlikely to be possible to exactly synchronize the slave transmission to be exactly aligned with the time of reception of the master's signal, given the timing precision required for high accuracy ranging and the physical resolution by which the slave clock is able to be shifted (for example a 1m resolution would require a 300MHz clock within the device, which will be impractical in a low power device). However such high accuracy time shifts are not required; it is only necessary to know the timing error associated with the signal sent by the slave to the master as this can be removed from the final timing measurement made at the master. The signal sent to the master is received by it and the accurate time offset is measured. The offset will be with respect to master time, with t=0 seconds assumed to be the point that is exactly one frame after the signal was sent from the master to the slave. This is of course also based upon the assumption that the master clock remains free running throughout this period. If the signal was sent at exactly the correct time from the slave to

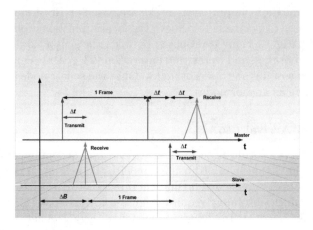

**Fig. 2.** Two-way ranging concept

the master then the offset measured by the master would be equal to the 2-way range, as the signal would have been transmitted from the slave $\Delta t$ seconds after the master begins to look for the signal and the propagation time encountered before receiving the signal is a further $\Delta t$ seconds. If there is any residual timing error at the slave, which could be reported to the master as part of the payload data sent by the slave, which could be reported to the master as part of the payload data sent by the slave, then this is simply subtracted from the offset measured by the master. To realize the accurate ranging function with two way ranging method, slight modification has been realized in the Zigbee module for future research.

## 2.5   Trilateration

Trilateration can be used as a second phase to estimate the location of sensor nodes. Figure 3 shows the simple concept of trilateration where a unknown node is within the range of at least 3 anchor nodes. After network initialization, the fixed anchors or reference points will propagate their coordinates to the unknown sensors. Once an unknown is within the range of at least 3 anchors, the distance between the anchor and sensor will be determined. After retrieving the position of nodes connected with at least 3 anchors, they will serve as anchor nodes for remaining unknown nodes. This process will locate all unknown nodes in a network. Unknown sensors with two or less anchors in range will need special consideration with the help of collaborative trilateration.

In equation 1, $(x, y)$ are the unknown coordinates and $(d_1, d_2, d_3)$ are distances from anchor-1, anchor-2 and anchor-3 to unknown node respectively.

$$2 \begin{bmatrix} X_3 - X_1 & Y_3 - Y_1 \\ X_3 - X_2 & Y_3 - Y_2 \end{bmatrix} \begin{bmatrix} x \\ y \end{bmatrix} = \begin{bmatrix} (d_1^2 - d_3^2) - (X_1^2 - X_3^2) - (Y_1^2 - Y_3^2) \\ (d_2^2 - d_3^2) - (X_2^2 - X_3^2) - (Y_2^2 - Y_3^2) \end{bmatrix} \quad (1)$$

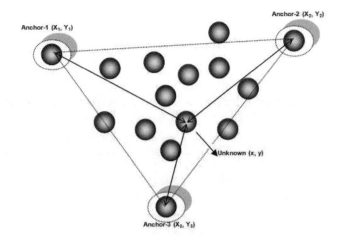

**Fig. 3.** Simple concept of Trilateration

# 3   Localization Positions in WSNs

Based on a specific application, a WSN based positioning system can be based on relative positioning or absolute positioning.

## 3.1   Relative Positions

The advantage of using the relative position in WSNs is that, we are not required to use the sensors with extra power capability or the high density of anchors [14]. In some cases, we only need three fixed, known anchors to move from relative to absolute position.

## 3.2   Absolute Positions

In a general frame of reference, for all objects and embedded positions in Universal Transverse Mercator (UTM) coordinates describe an absolute coordinate system for any possible place on earth. As compared to relative position, the absolute position is expensive in terms of anchors deployment, power and energy consumption. There are many ways for a node to determine its absolute position, including GPS and techniques based on stored maps, landmarks, or beacons [15].

GPS [16] can be used with WSNs absolute for localization but it is expensive in terms of energy, power, cost. It has the problem of requiring a line of sight to four or more satellites which won't be available in many buildings [17], beneath heavy foliage or under water. Based on these expenses, the utilization of GPS within each sensor node is not suitable for cheap and efficient localization. Instead of connecting each sensor node, providing support to nodes in order to estimate their location based on the information transmitted by the nodes who know their own location because of the attached GPS receivers (anchor nodes) [18].

# 4  Technology

## 4.1  IEEE 802.15.4 (Zigbee)

The swift growth of wireless networking in our daily life has forced us to break the
wireless applications into two different standard directions. One of the directions
which are rapidly getting a lot of attention is especially for low-data-rate, low-
power-consumption, and low cost applications. Based on the comparison on the
different wireless standards in table 1, the Zigbee standard has many advantages
for the real time positioning system in terms of battery life, network size, and
ranging.

In [24], Ultra-Wideband (UWB) precision asset location (PAL) has been developed and tested in open cargo space and performed to estimated accuracy of a
few feet. The authors considered the expensive and traditional method by UWB
where UWB reference and active tags were deployed on the grid and expected
as LOS to locate the containers on a yard where system receiver antennas were
mounted on tripods in corners of the cargo.

UWB forces challenges to accurate TOA [25] estimation in practical system
because of an important factor, clock jitter. Since UWB pulses have very short
(nanosecond) duration, clock accuracy and drifts in the target and the reference
nodes affect the TOA estimates. Apart from this, the set of delay positions
that includes TOA is usually very large compared to the chip duration. With a

**Table 1.** The Comparison parameters

| COMPARISON OF WIRELESS STANDARDS | | | |
|---|---|---|---|
| **Parameter** | **Zigbee IEE 802.15.4** | **UWB IEEE 802.15.3** | **Bluetooth 802.15.1** |
| Application Focus | Monitoring and Control | Streaming, home entertainment | Wire Replacement |
| Data Rate | 250, 40 and 20 Kbps | 500Mpbs (2m) 110Mpbs (10m) | 480Kpbs |
| Range | 100-4km location range 100m | Below 10m | 10m |
| Frequency | 2.4 GHz USA AUS 915 MHz Europe 868 MHz | 3.1-10.6GHz | 2.4 GHz |
| Channels | 0 = 868MHz(1) 1-10 = 915 (MHz) (10) 11-26 = 2.4 GHz (16) | Evolving | 23 or 79 |
| Throughput | Low | High | Medium |
| Power | Very Low | Ultra Low | Low |
| Topology | Star,Mesh,P-p, Tree | Peer to Peer | Star |
| Single Network | 64K , 216 Nodes | Evolving | 7 or more |
| Complexity (Device and Application) | Low | Medium | High |

promotion of the special ranging function discussed in 2.4, Zigbee devices can be used for each container on a yard which can be useful to improve the accuracy of the container localization. The table 1 shows the comparison parameters between different wireless standards.

## 5  System Description

### 5.1  Traffic Model

To evaluate and measure a real time localization system in a port to retrieve or locate containers within different and unpredictable environments, we consider a random point mobility model in terms of mobility.

### 5.2  Performance Measures

The important parameter to consider for performance measure of efficient and reliable localization is accuracy or mean error of estimated position. The other factors are, location estimation error ($L_E$), object location time ($S_T$).

**Location Estimation Error ($L_E$).** The location estimation error is the quantity by which an estimate differs from a true value. It is the obvious metric when measuring performance of a system. In equation 2, ($x_o,y_o$) is the actual sensor position and ($x_u,y_u$) is the estimated localization.

$$R_e = \sqrt{(x_o - x_u)^2 + (y_o - y_u)^2} \tag{2}$$

$$R_e = \frac{\sum_{i=1}^{n} \sqrt{(x_o - x_u)^2 + (y_o - y_u)^2}}{n} \tag{3}$$

**Object Location Time ($L_T$).** In localization, if object location time is not accurate or synchronized, we can lose the important positioning information at that time. The information can appear as a wrong position in later stages which will make the estimation error very high.

**Effect of Node Density ($N_D$).** As the sensor density increases, the communication between hops will be more accurate and fast with large connectivity but system will be complex in terms of calculatoion.

**Effect of Anchor Density ($A_D$).** The anchor nodes density can play a sensitive role in object tracking. Fewer number of anchors can make the system less complex in cost and computation but more complex in connectivity with other sensors.

## 6    Analysis

To evaluate the proposed idea in 2D, we made analysis in MATLAB. In our analysis, it is assumed that 100 nodes connected with containers are deployed randomly on a port. The anchor nodes with fixed known position and extra power capability (through electric poles) are deployed. The sensor and anchor nodes can communicate with each other if the distance between them is less or equal to the range. The sensors in the simulation have a radio range of 30 m. It is assumed that there is ideal LOS between sensor and anchors nodes. The factors which may cause significant errors are multipath fading, background interference and irregular signal propagation. Our main focus is on the performance of location estimation. Initially, anchor nodes will be used to track the unknown sensors within the range of 30m. Once all the sensor nodes connected with anchors at the range of 30m are tracked, the unknown nodes will be used as anchors to track other nodes. Later on, if a sensor is connected with four or more sensors then anchors will be selected on the basis of signal to noise ratio or shortest distance between anchor and sensor nodes.

## 7    Conclusion and Future Work

This paper presents a basic idea of localization to track freight containers on a port using the two-way time of arrival ranging method. We discussed the iterative trilateration estimation scheme as a second phase of the localization. The idea is based on the importance of the misplaced, lost and illegal access to containers using practical realization of IEEE 802.15.4 Zigbee standard. We consider the 2D positioning for this proposed idea.

In future our research will be focused on the implementation of the discussed idea on real sensors. Future research will also consider the importance of 3D positioning in localization with different level of real world errors and their impact on routing algorithms.

## References

1. World shipping council: Container shipping Information Service: Facts and Figures (2007), http://www.worldshipping.org/
2. He, T., Huang, C., Blum, B.M., Stankovic, J.A., Abdelzaher, T.: Range-free localization schemes for large scale sensor networks. In: MobiCom 2003: Proceedings of the 9th annual international conference on Mobile computing and networking, pp. 81–95. ACM, New York (2003)
3. Lazos, L., Poovendran, R.: SeRLoc: secure range-independent localization for wireless sensor networks. In: Proceedings of the 3rd ACM Workshop on Wireless Security, WiSe 2004, Philadelphia, PA, USA, October 01 - 01, 2004, pp. 21–30. ACM, New York (2004)
4. Sarma, H.K.D., Kar, A.: Security Threats in Wireless Sensor Networks. In: Carnahan Conferences Security Technology. Proceedings 2006 40th Annual IEEE International, October 2006, pp. 243–251 (2006)

5. Gupta, G., Younis, M.: Fault-tolerant clustering of wireless sensor networks. In: Wireless Communications and Networking, 2003. WCNC 2003, March 20, 2003, vol. 3, pp. 1579–1584. IEEE, Los Alamitos (2003)
6. Courtois, B., Kaminska, B.: 5B: emerging technologies - reliable and fault-tolerant wireless sensor networks. In: 23rd IEEE Proceedings on VLSI Test Symposium, 2005, May 1-5, 2005, p. 173 (2005)
7. Lazos, L., Poovendran, R.: HiRLoc: high-resolution robust localization for wireless sensor networks. IEEE Journal on Selected Areas in Communications 24(2), 233–246 (2006)
8. Bahl, P., Padmanabhan, V.: Radar: An In-building RF-Based user location and tracking system. In: IEEE Infocom 2000, vol. 2, pp. 775–784 (2002)
9. Wu, J., Tong, J.: An Analysis on Information Dependability Measurement of Wireless Sensor Networks. In: The Sixth World Congress on Intelligent Control and Automation, WCICA 2006, vol. 1, pp. 90–93 (2006)
10. Basagni, S., Chlamtac, I., Syrotiuk, V.R., Woodward, B.A.: A distance routing effect algorithm for mobility (DREAM). In: Osborne, W.P., Moghe, D. (eds.) Proceedings of the 4th Annual ACM/IEEE international Conference on Mobile Computing and Networking, MobiCom 1998, Dallas, Texas, United States, October 25 - 30, 1998, pp. 76–84. ACM, New York (1998)
11. DHont, S.: The cutting edge of RFID technology and applications for manufacturing and distribution, FID – Key to Container Positioning System at Port of Singapore, Texas Instruments TIRIS (2002)
12. International Terminal Solutions, Real time data capture for ports and terminals, http://www.portautomation.com/
13. Craddock, R.J., Stansfield, E.V.: Sensor fusion for smart containers, Signal Processing Solutions for Homeland Security, 2005. The IEE Seminar on (Ref. No. 2005/11108), 12 (October 11, 2005)
14. Shang, Y., Rumi, W., Zhang, Y., Fromherz, M.: Localization from connectivity in sensor networks. IEEE Transactions on Parallel and Distributed Systems 15(11), 961–974 (2004)
15. Bulusu, N., Heidemann, J., Estrin, D., Tran, T.: Self-configuring localization systems: Design and Experimental Evaluation. Trans. on Embedded Computing Sys. 3(1), 24–60 (2004)
16. Global Positioning System Standard Positioning Service Specification, 2nd edn. (1995), http://www.navcen.uscg.gov/pubs/gps/sigspec/gpssps1.pdf
17. Brooks, R.R.: Wireless Sensor Networks. Raghavendra, C.S., Sivalingam, K.M., Znati, T. (eds.) International Journal of Distributed Sensor Networks 3(4), 371–371 (2007)
18. Hu, L., Evans, D.: Localization for mobile sensor networks. In: Proceedings of the 10th Annual international Conference on Mobile Computing and Networking, MobiCom 2004, Philadelphia, PA, USA, September 26 - October 01, 2004, pp. 45–57. ACM, New York (2004)
19. Hightower, J., Vakili, C., Borriello, G., Want, R.: Design and Calibration of the SpotON Ad-Hoc Location Sensing System (August 2001) (unpublished)
20. Priyantha, N.B., Chakraborty, A., Balakrishnan, H.: The cricket location-support system. In: MobiCom 2000: Proceedings of the 6th annual international conference on Mobile computing and networking, pp. 32–43. ACM, New York (2000)
21. Niculescu, D., Nath, B.: Ad hoc positioning system (APS) using AoA. In: INFO-COM 2003. Twenty-Second Annual Joint Conference of the IEEE Computer and Communications Societies, March 30 - April 3, vol. 3, pp. 1734–1743. IEEE, Los Alamitos (2003)

22. Rong, P., Sichitiu, M.L.: Angle of Arrival Localization for Wireless Sensor Networks. In: Sensor and Ad Hoc Communications and Networks, 2006. 2006 3rd Annual IEEE Communications Society on SECON 2006, September 28, 2006, vol. 1, pp. 374–382 (2006)
23. Savvides, A., Park, H., Srivastava, M.B.: The bits and flops of the n-hop multilateration primitive for node localization problems. In: Proceedings of the 1st ACM international Workshop on Wireless Sensor Networks and Applications, WSNA 2002, Atlanta, Georgia, USA, September 28, 2002, pp. 112–121. ACM, New York (2002)
24. Fontana, R.J., Gunderson, S.J.: Ultra-wideband precision asset location system. In: IEEE Conference on Ultra Wideband Systems and Technologies Digest of Papers 2002, pp. 147–150 (2002)
25. Gezici, S., Tian, Z., Giannakis, G.B., Kobayashi, H., Molisch, A.F., Poor, H.V., Sahinoglu, Z.: Localization via ultra-wideband radios: a look at positioning aspects for future sensor networks. Signal Processing Magazine 22(4), 70–84 (2005)
26. Bordin, G., Bryant, E.: Intelligent pervasive location and tracking (iplot), wp4: Bluetooth ranging, development and performance prototype bluetooth ranging system. Prepared by Path Track Ltd., UK (November 2006)

# PrISM: Automatic Detection and Prevention from Cyber Attacks

Ahmed Zeeshan, Anwar M. Masood, Zafar M. Faisal,
Azam Kalim, and Naheed Farzana

Informatics Complex (ICCC), H-8/1, Islamabad, Pakistan
{zeeshan, hmfzafar, masood, farzana, kalim}@iccc.org.pk

**Abstract.** Network security is a discipline that focuses on securing networks from unauthorized access. Given the escalating threats of malicious cyber attacks, modern enterprises employ multiple lines of defense. A comprehensive defense strategy against such attacks should include: (1) an attack detection component that determines the fact that a system is compromised, (2) an attack identification and prevention component that identifies attack packets so that one can block such packets in the future and prevent the attack from further propagation. Over the last decade, significant research time has been invested in systems that can detect cyber attacks, either statically at compile time, or dynamically at run time. However, not much effort has been spent on automated attack packet identification or attack prevention. In this paper, we present a unified solution to these problems. We implemented this solution after reverse engineering an Open Source Security Information Management (OSSIM) system, called Preventive Information Security Management (PrISM) system, which correlates input from different sensors so that the resulting product can automatically detect any cyber attack against it, and prevent attack by identifying the actual attack packet(s). PrISM was always able to detect the attacks, identify the attack packets and most often prevent attack by blocking the attacker's IP address to continue normal execution. There is no additional run-time performance overhead for attack prevention.

**Keywords:** Information Security Management System, Network security, Computer Security, Intrusion Detection, Intrusion Prevention.

## 1 Introduction

Security is the process of maintaining an acceptable level of perceived risk. Achieving a security goal in a networked system requires the cooperation of a variety of devices, each device potentially requiring a different configuration. Many information security problems may be solved with appropriate management of these devices and their interactions, giving a systematic way to handle the complexity of real situations [1].

The use of networks is growing continuously, constantly increasing the vulnerability of the computer systems that use them. Current solutions for network security, such as firewalls, cannot support sophisticated trust relationships with external entities and lack a comprehensive approach to security. Research in security has shown the

D.M.A. Hussain et al. (Eds.): IMTIC 2008, CCIS 20, pp. 433–444, 2008.

usefulness of mandatory security mechanisms such as intrusion detection and prevention systems for supporting sophisticated trust relationships and secure end points in addition to secure communication channels. Modern intrusion detection systems are comprised of three basically different approaches, host based, network based, and a third relatively recent addition called procedural based detection.

Computer security is a matter of controlling how data are shared for reading and modifying. The information security manager must establish and maintain a security process that ensures three requirements: the confidentiality, integrity, and availability [2] of the information resources [3]. Confidentiality is the protection of information in the system so that unauthorized persons cannot access it. Hackers, masqueraders, unauthorized user activity, unprotected downloaded files, local area networks (LANs), trojan horses are some of the most commonly encountered threats to information confidentiality [3, 4].

Security solutions are slowly emerging, but interoperability, universally accepted security standards, application programming interfaces (APIs) for security, vendor support and cooperation, and multiplatform security products are still problematic. For the most part, no single vendor or even software/vendor consortium has addressed the overall security problem within "open" systems and public networks. This indicates that the problem is very large [3]. We present an approach, preventive information security management (PrISM), with added benefits adjacent to this strategy which will include a more cost-effective and seamless integration of security policies, security architectures, security control mechanisms, and security management processes to support the whole environment.

Typically, upon detecting an attack, victim application is simply terminated, and another instance if necessary is restarted. While terminating a compromised application helps prevent further propagation of the attack, it may lead to a denial of service attack. For network applications with a substantial number of states such as a DNS sever, it takes some time for them to re-acquire the necessary state at start-up in order to provide the full service. For these applications, abrupt termination is not an acceptable attack recovery strategy. Moreover, because existing attack detection systems cannot prevent the same attacks from taking place again, vulnerable applications may be repeatedly victimized and re-started in the presence of recurring attacks as in the case of worms. In the mean time, these applications cannot render any useful service to their intended users [5].

To address the limitations of existing systems that focus only on detection of attempted attacks, this project aims to develop an information security management system called PrISM that can automatically correlate and transform information from different plug-ins into a form that:

- Can detect a cyber attack when the control-sensitive data structure it tampers with is activated,
- Can identify the network packets that lead to the attack, and send these packets to a front-end content filter (viz. firewall) to prevent the same attack from compromising the application again.

The main contribution of this paper is the development of a unified solution to both problems. Even though on the surface attack detection, identification, and prevention appear to be completely orthogonal functions, a careful examination reveals that they can actually be unified into a single implementation framework.

Working of prevention depends heavily on detection. The major disadvantage of intrusion detection system (IDS) is the generation of false alarms. Taking intrusion prevention system (IPS) actions based on false alarm can lead to very difficult conditions including self Denial of Service (DOS) attack. Therefore, extreme care should be taken for prevention steps based on IDS alarms. One solution is to verify alerts generated by IDS through correlation which is performed in PrISM.

Preventive module enables the system to automatically respond to the security events, thereby reducing burden on the administrator's shoulder. It greatly lessens incident handling time. As a result, the overall impact of the event is minimum.

The rest of this paper is organized as follows. Section 2 reviews related work. Section 3 describes the PrISM with its essential components of intrusion detection and prevention for different attacks. In the end, discussion will be concluded by the final remarks of the authors.

## 2   Related Work

An examination of the potential problems that can arise on a poorly secured system helps in understanding the need for security [6]. As attack tools become more user-friendly and automated, more script kiddies can use them to randomly scan the Internet for victims with unpatched vulnerabilities. Given the escalating threats of malicious cyber-attacks, modern enterprises employ multiple lines of defense to protect themselves. First, content-aware intrusion prevention systems (IPS), including firewalls, try to filter out network packets containing attack payloads, including virus, worms, and spyware/adware. Then, file system scanning tools further eliminate those attack programs that somewhat evade the IPS deployed at the enterprise's network entry point [7].Two basic kinds of malicious behavior are *denial of service* and *disclosure of information*. Denial of service occurs when a hostile entity uses a critical service of the computer system in such a way that no service or severely degraded service is available to others. Denial of service is a difficult attack to detect and protect against, because it is difficult to distinguish when a program is being malicious or is simply greedy. Probably the most serious attack is disclosure of information and one of the possibilities is control-hijacking. If the information taken off a system is important to the success of an organization, it has considerable value to a competitor. Corporate espionage is a real threat, especially from foreign companies, where the legal reprisals are much more difficult to enforce. Limiting user access to the information needed to perform specific jobs increases data security dramatically [2, 3].

Over the last decade, a significant amount of research has been invested in the detection of control-hijacking attacks. Some are based on program analysis techniques [8, 9, 10, 11, 12, 13] that statically determine whether a given program contains buffer overflow vulnerability. Others use program transformation techniques [14, 15, 16, 17, 18, 19] to convert applications into a form that can either detect control-hijacking attacks [15, 16, 17, 19] or prevent control-sensitive data structures from being modified at run time [18]. Still others develop operating system mechanisms that ensure that it is not possible to execute code injected into the victim program [20, 21]. Regardless of their approach, most if not all of these efforts could only determine whether a program is under a control-hijacking attack, but could not actively repair a victim program after it has been compromised [5].

The problem of detecting a buffer overflow attack relies on a mechanism to monitor a particular memory location (such as a return address). A similar problem exists in software debugging in which case a dynamically monitored memory location is called a watch-point. Existing solutions of this problem can be divided into runtime dynamic checking techniques [22, 23] and hardware based techniques [24, 25, 28].

The problem of automatic identification of malicious code became increasingly important in the past few years since worms epidemics started to happen more and more frequently and at higher speeds. Given the speed of propagation of the recent worms, it is hopeless to rely on a human-based method for signature generation as by the time the proper signature is created and distributed among computer users, the worm is likely to infect a significant number of computer systems. Autograph [26] is a system that generates worm signatures automatically by detecting common byte sequences in suspicious network flows.

In this system, a network flow is considered suspicious if it comes from a host that is believed to perform port scanning. Toth and Kruegel [27] propose a system that detects malicious code in packet payloads by performing abstract execution of the payload data. Buttercup [10] is a system aimed at preventing polymorphic worms with known signatures from entering the system. It identifies the ranges of possible return addresses for existing vulnerabilities and checks whether a network packet contains such addresses. Another approach to identifying malicious code is to analyze the execution trace of a compromised program.

Finally, yet another approach to malicious input identification is to use a technique similar to Perl taint mode. The idea is to assign different tags to all user inputs and propagate these tags along through all memory operations. Upon discovering a compromised data structure, one can identify the origin of the malicious data by looking at the tag currently associated with that memory location. System support for rollback and reverse execution is another related area of systems research.

# 3 Preventive Information Security Management (PrISM) System

## 3.1 General Architecture

The PrISM system is a complete network security solution. It includes a wide range of services and functions including Intrusion Detection and Prevention, Integrity Checks, Incident Management, and Managerial Reporting. Our primary interest is to help the national interest organizations to reduce Operational Risk using Comprehensive Security Monitoring. PrISM is based on an open source ISMS i.e. Open Source Security Information Management (OSSIM) system [29] which is only an IDS. There is a major functional shift from intrusion detection to intrusion prevention in PrISM by automating the Incident Handling and other tasks. Functionally of the PrISM package is divided into four components: Server, Framework & Framework-D, Agent and, Databases. The functional interaction among them is illustrated in Fig 1.

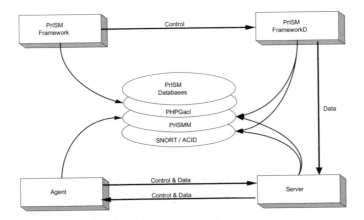

**Fig. 1.** Functional interaction among different components of PrISM

Using its correlation engine via *server*, PrISM screens out a large percentage of false positives, enables to perform a range of tasks from auditing, pattern matching and anomaly detection to forensic analysis in one single platform and offers high level state indicators that allow guiding inspection and measuring the security situation of network.

PrISM integrates a number of powerful open source security tools in a single distribution using an interfacing unit named as *agent* to communicate with server. These tools are:

- Arpwatch [30], used for MAC anomaly detection.
- POf [31], used for passive OS detection and OS change analysis.
- Pads [32], used for service anomaly detection.
- Nessus [33], used for vulnerability assessment and for cross correlation (IDS vs Security Scanner).
- Snort [34], the IDS, also used for cross correlation with Nessus.
- Spade [35], the statistical packet anomaly detection engine. Used to gain knowledge about attacks without signature.
- Tcptrack [36], used for session data information which can grant useful information for attack correlation.
- Ntop [37], which builds an impressive network information database from which we can get aberrant behaviour anomaly detection.
- Nagios [38]. Being fed from the host asset database it monitors host and service availability information.

These tools are linked together in PrISM's console giving the user a single, integrated navigation environment. Besides getting the best out of well known open source tools, PrISM provides a strong correlation engine, detailed low, mid and high level visualization interfaces as well as reporting and incident handling tools, working on a set of defined assets such as hosts, networks, groups and services.

All this information can be limited by network or sensor in order to provide just the needed information to specific users allowing for a fine grained multi-user security environment.

A typical deployment consists of:

- A database host.
- A server which hosts the correlation, qualification and risk assessment engine.
- $N$ agent hosts which do information collection tasks from a number of devices.
- A control daemon which does some maintenance work and ties some parts together. It's called 'FrameworkD'.
- The frontend, called 'Framework', is web based, unifying all the gathered information and providing the ability to control each of the components.

What follows is a brief description of different modules of PrISM. However, intrusion and prevention modules are discussed in detail in the following sections.

*IDS*: PrISM includes Snort, although it is capable of receiving and saving alerts from other IDSes. It is configured and parameterized for maximum performance. We have also included a number of our own alerts, to introduce the preventive measures, which allow PrISM to trounce attacks.

*Anomaly Detection*: PrISM includes three types of anomaly detection:
  i)    Connections that are anomalous in origin or destination,
  ii)   User data that is anomalous in relation to a threshold,
  iii)  Anomalies in data with periodic tendencies learnt using the Holt Winters forecasting algorithm.

*Correlation*: *PrISM* has a powerful correlation engine that can:
  i)    Correlate an alert according to the version of the affected product and operating system.
  ii)   Correlate Snort with Nessus (if there is a possible buffer-overrun and Nessus determines that we are vulnerable, the alert is prioritized)
  iii)  Define logical directives for sequences of events that can correlate: a)alerts, b) anomalies, and c) states by queries to monitors

*Forensic Console*: PrISM utilizes an extension of ACID for its Forensic Console, this console allows to exploit the event database (EDB) collected through the process of normalization. Using ACID, PrISM allows storing and exploiting other events besides those of Snort.

*Risk Monitor*: PrISM includes a monitor of "accumulated risk" called Riskmeter that utilizes a scoring algorithm called CALM [29]. This monitor offers a real time indicator of the security situation of a host, a network, a group of machines, or even the global security situation.

*Auditing*: PrISM integrates Nessus for auditing. Using Nessus, a vulnerability index can be obtained, i.e. the state of network vulnerability, which can be used as an objective or technical assessment of security. Vulnerabilities are stored and correlated to prioritize and discard attacks identified by the IDS.

*Usage Monitor*: PrISM includes Ntop, a monitor that collects all traffic data via passive listening and creates a user profile for each machine. This information is stored in circular databases that enable to save detailed information for a long period of time.

Control Panel: PrISM integrates, summarizes, and links together all of the above tools in a single control panel. Its purpose is to enable the user to analyze and interrelate information from the most abstract to the most concrete. The control panel allows creating reports with information cross-referenced from the various tools that make up PrISM.

## 3.2  PrISM's Operational Plan

The system's processes are performed into two basic stages. The first stage corresponds the pre-processing, carried out by the monitors and detectors and the second stage corresponds to post-processing, executed in a centralized console.

Fig. 2 shows the data flow diagram beginning with the generation of an event gives the basic understanding of process integration. The basic databases used by the system are: **EDB**, the event database, is the largest because it stores every individual event perceived by detectors. **KDB**, the knowledge database, parameterizes the system to be familiar with the network and define the security policy. **UDB,** the profile database, stores all information gathered by the profile monitor.

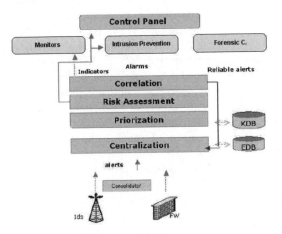

**Fig. 2.** Data Flow Diagram

## 3.3  Intrusion Detection in PrISM

Intrusion detection is the art of detecting inappropriate, incorrect, or anomalous activity [39]. Originally, system administrators performed intrusion detection by sitting in front of a console and monitoring user activities. Although effective enough at the time, this early form of intrusion detection was ad hoc and not scalable [40, 41].

In computer networks, intrusion detection is usually achieved in two ways – at the network level or at the host level. Network Intrusion Detection (NID) means scanning packets in network segments, looking for evidence of known attack signatures, or other suspicious activity. This is highly dependent upon the existing policies of the network and firewall administration [4].

There are two basic categories of intrusion detection techniques: anomaly detection and misuse detection. *Anomaly detection* uses models of the intended behavior of users and applications, interpreting deviations from this "normal" behavior as a problem [40, 42]. A basic assumption of anomaly detection is that attacks differ from normal behavior. The main advantage of anomaly detection systems is that they can detect previously unknown attacks. In actual systems, however, the advantage of detecting previously unknown attacks is paid for in terms of high false-positive rates. Anomaly detection systems are also difficult to train in highly dynamic environments.

*Misuse (Signature-based) detection systems* essentially define what is wrong. They contain attack descriptions (or "signatures") and match them against the audit data stream, looking for evidence of known attacks [43, 44]. The main advantage of misuse detection systems is that they focus analysis on the audit data and typically produce few false positives. The main limitation of misuse detection systems is that they can detect only known attacks for which they have a defined signature. As new attacks are discovered, developers must model and add them to the signature database.

Signature-based detection is the simplest detection method because it just compares the current unit of activity, such as a packet or a log entry, to a list of signatures using string comparison operations. Signature-based detection technologies have little understanding of many network or application protocols and cannot track and understand the state of complex communications. They also lack the ability to remember previous requests when processing the current request. This limitation prevents signature-based detection methods from detecting attacks that comprise multiple events if none of the events contains a clear indication of an attack [45].

Snort [34] is a network intrusion detection tool that is plugged-in PrISM to monitor TCP/IP networks and detects a wide variety of suspicious network traffic as well as outright attacks. Snort examines network traffic against a set of rules and alerts administrators with enough data to make informed decisions on the proper course of action in the face of suspicious activity [46]. Snort features rules based logging to perform content pattern matching and detect a variety of attacks and probes, such as buffer overflows [47], stealth port scans, CGI attacks, SMB probes, and much more. The syslog alerts are sent as security/authorization messages that are easily monitored with tools such as swatch [48]. Fig. 3 shows the graphical overview of the alerts generated in PrISM by Snort.

## 3.4  Prevention Module

An IPS or Intrusion Prevention System can be an important component for protecting systems on a network. An IPS is based upon an IDS or Intrusion Detection System with the added component of taking some action, often in real time, to prevent an intrusion once detected by the IDS [45].

Traditionally, networks and systems have been protected by perimeter defense methods such as routers with access control lists (ACLs) and firewalls. Network and system intrusion attempts and attacks have been detected by network-based intrusion detection systems (NIDS) and host-based intrusion detection systems (HIDS). These types of devices continue to provide a strong and robust defense in depth strategy; however, advances in technology have made it possible for these technologies to work together. Intrusion prevention systems (IPS) combine the features of a firewall and IDS not only to detect attacks but, more importantly, to prevent them [49].

Preventive module in PrISM enables the system to automatically respond to the security events, thereby reducing burden on the administrator's shoulder. It greatly lessens incident-handling time. The block diagram of the preventive module is shown in Fig. 4.

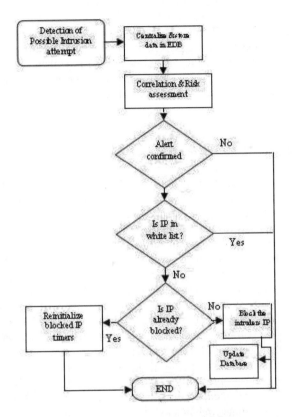

**Fig. 4.** Block Diagram of IPS Module

Our intrusion prevention module takes the approach of changing the configuration of security control devices. The system blocks any attack based on the alarm verified through correlation process. The detection module monitors network traffic for any anomalous activity and send alerts to correlation engine. The correlation engine verifies the alert and generates alarm on verification. As an alarm is generated, the prevention module comes into action. It first checks whether the IP from which the attack is generated is in White IP list. If so, then it ignores the alarm. If not in the list, then it checks whether the IP is already blocked, if so then it increases the blocked time of the attacking IP. If the attacker's IP is neither in the white list nor in the list of already blocked systems, it reconfigures the access control list of the specified security control devices to block the attacker's IP for specific amount of time and makes an entry in the database. The details of alarms generated and the attacks blocked can be seen through the Web interface. The IPS module is easily customizable and controllable through the web interface. The attacks are dealt according the responses specified in the policy. The blocking duration can also be modified through the policy. The White IP list, Security control devices interface, blocking policy is customizable through the web interface. The IPS module periodically checks for the blocked IPS whose block time is expired. As the blocked duration is expired the corresponding control devices are reconfigured to unblock the specified IP.

The module also resets TCP connection between attacker and targeted system. The server sends TCP connection reset packets to both attacker and targeted system. The sequence number is determined from the event log. A skilled intruder can change receiver's window so fast that a system may have a very hard time determining which sequence number to be used for packet reset. Therefore, TCP connection reset may not work at occasions.

## 4  Conclusions

PrISM is a fully functional Information Security Management System. It includes a wide range of services and functions including intrusion detection and prevention, integrity checks, incident management, and managerial reporting. The development of PrISM was based on the reverse engineering of Open Source Security Information Management System (OSSIM). A number of new features have been added to basic functionality, i.e. intrusion prevention capability, customized report generation facility and context sensitive help. The interface is redesigned to accommodate support for new functionality, make it more aesthetically appealing, flexible and user friendly.

For future work, it is planned to enhance the correlation engine. Improvement in correlation engine design will help in reducing the generation of false alarms. Diminution of false alarms will help the security administrator to concentrate on real event. Therefore, this enhancement will result in an overall improvement of detection and prevention process. It is also planned to incorporate new open source plug-ins for the functionality enhancement. In addition, an important activity to be initiated is the writing of new sets of rules called directives, so that PrISM will be able to protect information assets from emerging attacks.

## References

1. Guttman, J.D., Herzog, A.L.: Rigorous automated network security management. Int. J. Inf. Secur. 4, 29–48 (2005)
2. Landwehr, C.E.: Computer security. IJIS 1, 3–13 (2001)
3. Krause, M., Harold, F.T.: Handbook of Information Security Management. CRC Press LLC (2006)
4. Technical White Paper, Event Horizon™: Lanifex Intrusion Detection Solution., ver. 1.5, CSO Lanifex GmbH (2003)
5. Smirnov, A., Chiueh, T.: DIRA: Automatic Detection, Identification, and Repair of Control-Hijacking Attacks. In: Proc. of 12th Annual Network and Distributed System Security Symposium, San Diego, California (2005)

6. Anwar, M., Zafar, M.F., Ahmed, Z.: A Proposed Preventive Information Security System. In: Proceedings of International Multitopic Conference (INMIC 2006), Islamabad, Pakistan (2006)

7. Guo, F., Yu, Y., Chiueh, T.: Automated and Safe Vulnerability Assessment. In: Proceedings of 21st Annual Computer Security Applications Conference (ACSAC 2005), Tucson, USA (2005)

8. Evans, D., Guttag, J., Horning, J., Tan, Y.M.: LCLint: A tool for using specifications to check code. In: Proceedings of the ACM SIGOFT Symposium on the Foundations of Software Engineering, vol. 19(5), pp. 87–96 (1994)

9. Johnson, S.C.: Lint, a C program checker. In: AT&T Bell Laboratories. Murray Hill, NJ, USA (1978)

10. Nazario, J.: Project Pedantic – source code analysis tool(s) (2002),
    http://pedantic.sourceforge.net

11. Secure software solutions. Rough auditing tool for security, RATS 2.1,
    http://www.securesw.com/rats

12. Viega, J., Bloch, J.T., Kohno, T., McGraw, G.: ITS4: A static vulnerability scanner for C and C++ code. In: Proceeding of the 16th Annual Computer Security Applications Conference (ACSAC 2000), p. 257 (2000)

13. Wheeler, D.: Flawfinder, http://www.dwheeler.com/flawfinder

14. Vendicator. StackShield, G.C.C.: Compiler patch,
    http://www.angelfire.com/sk/stackshield

15. Chiueh, T.C., Hsu, F.H.: RAD: A compile-time solution to buffer overflow attacks. In: Proc. of 21st Intl. Conf. on Distributed Computing Systems (ICDCS 2001), pp. 4–9 (2001)

16. Cowan, C., Barringer, M., Beattie, S., Kroah-Hartman, G., Frantzen, M., Lokier, J.: Format Guard: Automatic protection from printf format string vulnerabilities. In: Proceedings of 10th USENIX Security Symposium, Washington, D.C., USA (2001)

17. Cowan, C., Pu, C., Maier, D., Walpole, J., Bakke, P., Beattie, S., Grier, A., Wagle, P., Zhang, Q., Hinton, H.: Stack-Guard: Automatic detection and prevention of buffer over flow attacks. In: Proceedings of the 7th USENIX Security Symposium, San Antonio, TX, USA (1998)

18. Etoh, H.: GCC extensions for protecting applications from stack-smashing attacks (2000),
    http://www.trl.ibm.com/projects/security/ssp

19. Frantzen, M., Shuey, M.: StackGhost: Hardware facilitated stack protection. In: Proceedings of the 10th USENIX Security Symposium, Washington, D.C., USA (2001)

20. Team, P.: Non-executable pages design and implementation,
    http://pax.grsecurity.net/~docs/noexec.txt

21. Openwall project, http://www.openwall.com

22. Hastings, R., Joyce, B.: Purify: Fast detection of memory leaks and access errors. In: Proceedings of the Winter USENIX Conference San Francisco, USA, pp. 125–138 (1992)

23. Hangal, S., Lam, M.S.: Tracking down software bugs using automatic anomaly detection. In: Proceedings of 24th Int. Conf. Software Engineering, pp. 291–301 (2002)

24. Prvulovic, M., Torrellas, J.: ReEnact: Using thread-level speculation to debug software; An application to data races in multithreaded codes. In: Proceedings of the 30th Annual International Symposium on Computer Architecture, pp. 110–121 (2003)

25. Zhou, P., Qin, F., Liu, W., Zhou, Y., Torrellas, J.: iWatcher: Efficient architectural support for software debugging. In: Proceedings of the 31st Annual International Symposium on Computer Architecture (2004)

26. Kim, H.A., Karp, B.: Autograph: Toward automated, distributed worm signature detection. In: Proceedings of 13th USENIX Security Symposium, San Diego, CA, USA (2004)

27. Toth, T., Kruegel, C.: Accurate buffer overflow detection via abstract payload execution. In: Wespi, A., Vigna, G., Deri, L. (eds.) RAID 2002. LNCS, vol. 2516, pp. 274–291. Springer, Heidelberg (2002)

28. Min, S.L., Choi, J.D.: An efficient cache-based access anomaly detection scheme. In: Proceedings of the Fourth International Conference on Architectural Support for Programming Languages and Operating Systems, Santa Clara, CA, USA, pp. 235–244 (1991)

29. Open Source Security Management, http://www.ossim.net

30. LBL Network Research Group: Arpwatch, http://www.securityfocus.com/tools/142

31. Zalewski, M.: P0f: a versatile passive OS fingerprinting tool, http://lcamtuf.coredump.cx/p0f.shtml

32. http://www.lsli.com/pad.whitepaper.html

33. Tenable Network Security, The Network Vulnerability Scanner, http://www.nessus.org

34. Sourcefire, Inc., Open Source Snort, http://www.snort.org

35. Benson, S.: Tcptrack, A sniffer to displays information about TCP connections on a network interface, http://www.rhythm.cx/~steve/devel/tcptrack/

36. Hoagland, J., Staniford, S.: SPADE (Statistical Packet Anomaly Detection Engine) Snort preprocessor plugin, http://www.securityfocus.com/tools/1767

37. ntop. A network traffic probe to show network usage, http://www.ntop.org

38. Nagios Enterprises, L.L.C.: Nagios, Open source host, service and network monitoring program, http://www.nagios.org

39. Paul, J.B.: Intrusion Detection – Evolution beyond Anomalous Behavior and Pattern Matching. Security Essentials Version 1.4 (2002)

40. Denning, D.E.: An Intrusion Detection Model. IEEE Trans. Software Eng. 13(2), 222–232 (1987)

41. Wang, T., Suckow, W., Brown, D.: A Survey of Intrusion Detection Systems. In: CSE221 course notes, Department of Computer Science, University of California, San Diego, CA, USA (2001)

42. Ghosh, A.K., Wanken, J., Charron, F.: Detecting Anomalous and Unknown Intrusions Against Programs. In: Proc. Annual Computer Security Application Conference (ACSAC 1998), pp. 259–267. IEEE CS Press, Los Alamitos (1998)

43. Ilgun, K., Kemmerer, R.A., Porras, P.A.: State Transition Analysis: A Rule-Based Intrusion Detection System. IEEE Trans. Software Eng. 21(3), 181–199 (1995)

44. Lindqvist, U., Porras, P.A.: Detecting Computer and Network Misuse with the Production-Based Expert System Toolset (P-BEST). In: 1999 IEEE Symp. Security and Privacy, pp. 146–161 (1999)

45. Scarfone, K., Mell, P.: Guide to Intrusion Detection and Prevention Systems (IDPS). Special Publication 800-94, National Institute of Standards and Technology, Gaithersburg, MD, US (2007)

46. Roesch, M.: Snort – Lightweight Intrusion Detection for Networks, http://www.snort.org/docs/lisapaper.txt

47. Aleph One: Smashing the Stack for Fun and Profit, Phrack, vol. 7(49), (1996), http://www.phrack.com

48. Hansen, S.E., Atkins, E.T.: Centralized System Monitoring with Swatch. In: USENIX Seventh Conference on Systems Administration, Monterey, California, USA, pp. 145–152 (1993)

49. Angela, O., Eric, C.: Intrusion Prevention and Active Response: Implementing an Open Source Defense. Sys. Admin. Magazine 14(3) (2005)

# A Pretty Safe Strategy for Analyzing Discrete Security Assessment Framework in Wireless Sensor Networks

Adnan Ashraf Arain[1], Bhawani Shankar Chowdhry[1,2], Manzoor Hashmani[1], Marvie Mussadiq[3], and Abdul Qadeer Khan Rajput[1]

[1] Research Faculty, IICT, Mehran University of Engineering and Technology, Pakistan
[2] Research fellow, School of Electronics and Computer Science, University of Southampton, Southampton, U.K.
[3] R&D Member, Xevious Consultants, Karachi, Pakistan
{adnanlooking@ieee.org, bsc06v@ecs.soton.ac.uk, mhashmani@yahoo.com, marvielooking@gmail.com

**Abstract.** The current research and development in the wireless sensor networks (WSNs) make them an ideal infrastructure for variety of applications. Various commercial projects are in need of an increased and on-demand security of WSNs due to varying security requirements. To fulfill these requirements, the development of application-specific security frameworks of WSNs has been an open area for research. In order to determine the security strength of various security frameworks, we formulate a cooperative strategy to perform assessment of these frameworks through (D-SAFE) Discrete Security Assessment FramEwork[1].

**Keywords:** Wireless sensor network, Security assessment framework, discrete security assessment.

## 1 Introduction

WSNs typically consist of one or more sink/user node which themselves expecting data from various sensor nodes. In general, the sink node is considered having more resources (power, memory, transmission range). All in-range user nodes pass the data to the sink node as shown in the Fig.1.

Security is considered as a major issue for wireless sensor networks (WSNs) in industries and academia alike. Due to the topology-less infrastructures of WSNs, the requirements of security varies from one application to another. The arrangement of sensors, use of different protocols, different cryptographic schemes, key distribution and management schemes, and processing power can change the level of security of a WSN. Each object in WSN has its role in building the security strength of the network. Therefore, knowing the role of each security-critical object that accelerates or decelerates the security strength is a must. Before the D-SAFE (Discrete Security Assessment FramEwork) it was impossible to even guess the security strength of a network [1] [2].

---

[1] D-SAFE research work is supported by the Ministry of Science & Tech. (Pak) in MUET (Pakistan).

D.M.A. Hussain et al. (Eds.): IMTIC 2008, CCIS 20, pp. 445–448, 2008.

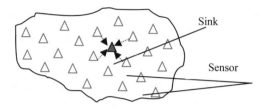

**Fig. 1.** Wireless sensor networks

Also, depending upon the type of application, the requirements of security level of WSNs appear different. The cross-border movement, monitoring inhabitants during peace/ war time, environmental monitoring, or monitoring forest fire are a few examples.

## 2  Designing Assessment Strategy for D-SAFE

Since discrete SAFE or D-SAFE is first assessment strategy of its kind, therefore it is very important to design a standard test strategy for its implementation. The D-SAFE is implemented by allocating numeric values to the critical security requirements of a WSN. Based on these assigned numerical values, D-SAFE quantifies the security strength of a WSN [1].

To design assessment strategy for D-SAFE, the possible formations of WSNs are taken into study. We primarily, categorize the network formations four types. Fig. 2 (a, b, c, d) is a comprehensive display of these categories in WSNs.

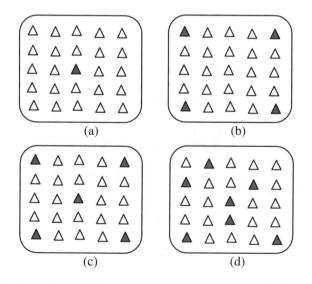

**Fig. 2.** Network formation in WSNs a) sink-centric, b) edged-sink, c) mixed-up, and the d) random-sink

These network formations are designed to avoid the effects of sensing the static or dynamic targets in real WSNs. These network formations can accommodate clustering schemes as more than one sink/user nodes are present in them.

The next stage of this strategy is to simulate the different formations of WSNs using variety of security frameworks/ approaches. These framework recommendations and security schemes are being widely used by the research communities [3][4] [5][6] [7]. Few of these have been produced in mature products [8] [9].

The results should be compared for a true analysis of the security strength of these frameworks/approaches, when WSNs implement them. At that time, the D-SAFE design should provide not only the security strength in terms of discrete numbers but also it should contain the malfunctioning security objects or process of the WSN.

## 3  Future Work

Our future direction of work will look at real-time deployment of optimal security frameworks in WSNs by obtaining discrete security values using D-SAFE. At the same time, we are aiming for greater flexibility in D-SAFE to provide the discrete value of *security requirement* of the deployment of WSNs. Similarly, these discrete values (*strength of the WSN security framework* and *security requirements of the WSN application*) shall correlate each other in designing optimized security solutions for WSNs.

## References

1. Ashraf, A., Hashmani, M., Chowdhry, B.S., Mussadiq, M., Gee, Q., Rajput, A.Q.K.: Design and Analysis of the Security Assessment Framework for Achieving Discrete Security Values in Wireless Sensor Networks. In: Proceedings IEEE Canadian Conference on Electrical and Computer Engineering, CCECE 2008, Niagara Falls, Canada, May 4-7 (2008) ISBN: 978-1-4244-1643-1, ISSN: 0840-7789
2. Arain, A.A., Hashmani, M., Chowdhry, B.S.: On Implementing Real-time Detection Techniques in Future Network Access Control. In: Proceedings of the International Conference on Information and Communication Technologies: From theory to applications, April 7-11, 2008, Damascus, Syria (2008) ISBN: 978-1-4244-1752-0. Library of Congress: 2007907105
3. Gu, L., Jia, D., Vicaire, P., Yan, T., Luo, L., Tirumala, A., Cao, Q., He, T., Stankovic, J.A., Abdelzaher, T., Krogh, B.H.: Lightweight Detection and Classification for Wireless Sensor Networks in Realistic Environments. In: Sensys. 2005, San Diego, California, USA, November 2-4, 2005, ACM, New York (2005)
4. Felemban, E., Lee, C.G., Ekici, E., Boder, R., Vural, S.: Probablistic QoS Guarantee in Reliability and Timeliness Domains in Wireless Sensor Networks, 0-7803-8968-9/05 (IEEE 2005)
5. Mallanda, C., Suri, A., Kunchakarra, V., Iyengar, S.S., Kannan, R., Durresi, A.: Simulating Wireless Sensor Networks with OMNeT++, LSU simulator version 1, 01/24/2005, NFS ITR under IIS: 0312632 (2005)

6. National Center for Biotechnology Information, `http://www.ncbi.nlm.nih.gov`
7. `http://castalia.npc.nicta.com.au/`
8. `http://wsnsrijan.googlepages.com/`
9. Mica2 and MicaMotes network with Labview support, `http://www.xbow.com`
10. `http://www.digi.com/` (802.15.4 based products)

# Renewable Energy for Managing Energy Crisis in Pakistan

Khanji Harijan[1], Muhammad Aslam Uqaili[2], and Mujeebuddin Memon[1]

[1] Department of Mechanical Engineering, Mehran University of Engineering & Technology,
Jamshoro-76062, Pakistan
[2] Department of Electrical Engineering, Mehran University of Engineering & Technology,
Jamshoro-76062, Pakistan
khanji1970@yahoo.com

**Abstract.** Only 55% of the Pakistan's population has access to electricity and per capita supply is about 520 kWh. At present, the people are facing severe load shedding/blackout problems due to shortage of about 3 GW power supply. Gas and oil have 65% share in conventional electricity generation. Indigenous reserves of oil and gas are limited and the country heavily depends on imported oil. The oil import bill is a serious strain on the country's economy. Though there is huge coal potential in the country but has not been utilized due to various reasons. This shows that Pakistan must develop renewables to manage the energy crises. This paper analyses the prospects of renewables for managing the energy crises in the country. The study concludes that there is substantial potential of renewables in the country for managing the present energy crises as well as meeting the future energy needs.

**Keywords:** Energy crisis; Electricity demand; Renewable energy; Environment friendly; Pakistan.

## 1 Introduction

Energy is an essential ingredient of socio-economic development and economic growth. Without sufficient energy in usable forms and at affordable prices, there is a little prospect of developments of improving the economy of a country and the living conditions of the people. Only 55% and 18% of the households in Pakistan have access to electricity and gas respectively. Per capita commercial energy/electricity supply in the country is one of the lowest in the world. About 68% of the country's population lives in rural areas and most of them have no access to commercial energy and use biomass such as firewood, agricultural wastes and animal dung [1][2].

At present, the people are facing severe (8-12 hours per day) load shedding/blackout problems due to shortage of more than 3 GW power supply. Gas demand grows beyond the transmission capacity in the winter and large users mainly industries, power plants and cement are curtailed during winter months to ensure supplies to domestic, commercial and small industries or fertiliser. The energy crisis in the country has forced thousands of industries to shut down operations, affecting industrial production and the livelihoods of thousands of families [3][4].

D.M.A. Hussain et al. (Eds.): IMTIC 2008, CCIS 20, pp. 449–455, 2008.

Oil and gas supply the bulk (79%) of the Pakistan's energy needs. The indigenous reserves of oil and gas are limited. The indigenous production of oil has not coped with the requirements and the country remained heavily dependent on its import. More than 80% of total oil requirements were imported from abroad by spending about US\$ 11 billion per annum [5]. The production and combustion of fossil fuels also degrades the environment. The oil import bill will further increase in the future because of rocketing oil price in the international market.

If Pakistan chooses to rely on oil and gas to meet its energy demand, it would be a constant burden on the country's foreign exchange reserves, and due to continuously increasing price of oil, our export surplus would become progressively more uncompetitive, goods for local consumption would become costlier, some industries could face closure/bankruptcy and the country could face economic stress on a wide scale. It is therefore imperative that Pakistan must develop indigenous environment friendly renewable energy sources to manage the looming energy crisis.

## 2   Energy Supply and Demand in Pakistan

The primary commercial energy supply in the country was 60.4 MTOE in 2006-07, with 86.4% fossil fuels share. Commercial energy was dominated by gas followed by oil, hydropower, coal and nuclear energy as shown in Figure 1. The major consumers of primary commercial energy in Pakistan are power, transport, industrial and domestic sectors. The power generation sector utilises about 38% of oil, 33% of gas and 2% of coal consumption in the country. Figure 2 shows the electricity generation by source in Pakistan in 2006-07. Fossil fuels, hydropower and nuclear energy have 65%, 32% and 2.5% shares respectively in the total electricity generation. The transport sector accounts for about 49% of oil and 5% of gas consumption in the country. The industrial sector utilises about 9.5% of oil, 29.5% of gas and 94% of coal consumption in the country. The residential sector consumes 16.5% of the total gas consumption in Pakistan [5].

The indigenous recoverable reserves of oil and gas are 47.4 MTOE and 605.4 MTOE respectively. At the current rate of supply, these reserves will be exhausted after 2 and 25 years respectively. Though there is huge coal potential (185 billion tones) in the country but has not been utilised to its full potential due to poor quality, financial constraints, location disadvantage, and lack of experience in modern clean coal utilization technologies. Expansion in nuclear power generation behind the present capacity is uncertain due to high capital cost, and safety and security concerns [5, 1]. The present energy demand and supply scenario clearly shows that the fossil fuels and nuclear power could not manage the energy crisis in Pakistan and there is a need to develop indigenous, cheap and environment friendly renewable energy sources such as hydropower, solar energy, wind energy and biomass energy.

With the economic development and with efforts to provide enhanced access to commercial energy, the energy demand in the country is expected to grow rapidly. It has been projected that the commercial energy demand in the country increases at an average growth rate of about 7.8% per annum and would reach at 361 MTOE by the year 2030. Figure 3 shows the energy mix plan projections of Pakistan [6].

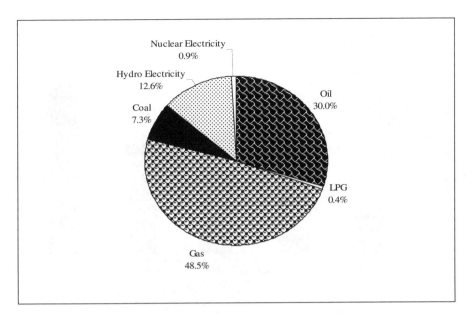

**Fig. 1.** Primary Commercial Energy Supplies in Pakistan: 2006-07

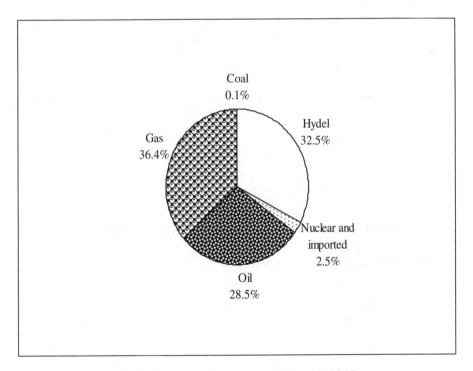

**Fig. 2.** Electricity Generation in Pakistan: 2006-07

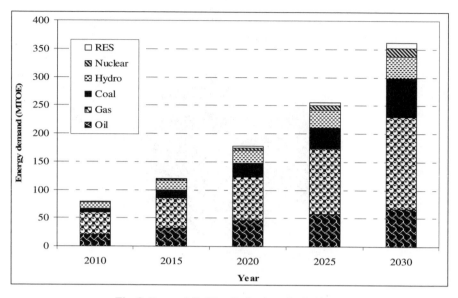

**Fig. 3.** Energy Mix Plan Projections for Pakistan

It is projected that the energy demand-indigenous supply gap would increase from 27% in 2005 to about 45% in 2025 and 57% in 2030. The shares of renewable energy sources (excluding hydropower) in commercial energy supply in 2030 would be only 2.5% as shown in Figure 3. The government of Pakistan has planned to bridge the energy demand-supply gap by imported oil and gas. It is planned that the imported oil and gas would meet about 97% and 88% of their requirements in 2030 [6]. The development of options for importing gas has been constrained by the sensitive regional security environment, special technical issues, and complexities associated with commercial and operating arrangements typical of large projects requiring inter-country agreements. This means that the country will face severe energy crises in the future too. Therefore there is a need to increase the share of indigenous environment friendly renewable energy sources in the total energy supply mix to manage the worsening energy crisis in the country.

## 3  Potential of Renewable Energy Sources in Pakistan

### 3.1  Hydropower

Pakistan is blessed with tremendous hydropower potential. The identified theoretical hydropower potential is estimated to be about 41.5 GW. Only 16% of the total theoretical hydropower potential has been exploited so far. The Northern part of the country is also rich with small hydropower resources. Other than 12 big hydropower plants, there are a large number of sites in the high terrain where natural and manageable waterfalls are abundantly available. It is estimated that the total potential of small hydropower in the northern areas of Pakistan alone is above 500 MW. The recoverable potential in micro-hydropower up to 100 kW is roughly estimated to be about 300 MW on perennial water falls in northern Pakistan. Besides, there is an immense potential for exploiting water falls in the canal network particularly in Punjab and Sindh, where low head high discharge exists on many canals [7][8].

### 3.2  Solar Energy

Pakistan receives 16-21 MJ/m$^2$ per day of solar radiation as an annual mean value, with 19 MJ/m$^2$ per day over most areas of the country. The annual mean values of sunshine duration lie between 8 and 10 hours per day all over the country, except for north parts [9]. Among all renewable energy sources, the solar energy is the most abundant and widely spread in the country. Pakistan receives approximately $15.525 \times 10^{14}$ kWh of solar energy every year, i.e. about 1715 times the current primary energy consumption in the country [8]. Solar energy can be used for power generation, water heating, cooking, drying of agricultural products, water pumping etc.

The technical potential for solar photovoltaic (PV) electricity generation in Pakistan is estimated at 3.5 PWh per year which is about 41 times the current conventional electricity generation in the country. Solar PV power potential in terms of installed capacity has been estimated as 1600 GW which is about 80 times the current installed capacity of conventional power generation in the country [8].

### 3.3  Wind Energy

Pakistan has considerable potential for wind power generation in the southern and costal areas of Sindh and Balochistan provinces. Its about 1050 km coastline has steady winds with average speeds of 5-7 m/s throughout the year. Most locations in the coastal area of Sindh and Balochistan have theoretical wind power potential of around 2000 – 3000 full load hours (FLH) and 1000 – 1400 FLH respectively. Wind power potential in terms of installed capacity has been estimated as 122.6 GW. The technical potential of GC wind power in the coastal areas of Pakistan has been estimated as 212 TWh per year, which is about 2.5 times the current total conventional power generation in the country [8].

### 3.4  Biomass Energy

There is substantial potential of biomass in the form of fuel wood, crop residues and other waste (animal, human and municipal waste) in Pakistan. Currently about 1500 m$^3$ of forest firewood and 50000 tonnes of solid waste per day are generated in the country. The estimated potential of crop residue is 225000 tonnes (with 16% bagasse and 1.7% rice husk share) per day and animal dung is more than 1 million tonnes per day. An estimated 8.8 to 17.2 billion m$^3$/year of biogas could be produced from livestock residue, equivalent to 55 to 106 TWh/year. The estimated potential of electricity generation from bagasse for the year 2005 is 5,700 GWh, which is about 6.6% of the total conventional electricity generation in Pakistan [1, 8].

## 4  Prospects of Renewables for Managing Energy Crisis in Pakistan

Renewable energy sources can be utilised for many applications but their important uses for Pakistan are presented in Table 1. Hydropower, wind energy, biomass

**Table 1.** Applications of Renewable Energy Sources in Pakistan

| Source | Application | Sector |
|---|---|---|
| Hydropower | Electricity generation | Power |
| Wind energy | Electricity generation | Power |
| | Water pumping | Domestic & Agriculture |
| Solar energy | Electricity generation | Power |
| | Water heating | Industrial, Domestic & Commercial |
| | Water pumping | Domestic & Agriculture |
| | Cooking | Domestic & Commercial |
| Biomass energy | Electricity generation | Power |
| | Cooking | Domestic & Commercial |
| | Water heating | Industrial, Domestic & Commercial |
| | Fuel | Transport and Agriculture |

(bagasse) energy and solar energy can be developed to generate electricity in the country. Electricity requirements of the rural households can be met through decentralised renewable energy systems such as mini/micro hydropower, wind home systems and solar PV home systems. All these technologies except solar PV are mature and economically viable in the country. Solar PV is expected to be competitive to other power generation options by the year 2015. These renewable energy sources have the potential to meet all the current and future electricity requirements of Pakistan.

Industrial and power sectors are dependent on steam generation through oil, coal or gas-fired boilers. Solar energy and biomass (wood and agricultural waste, MSW etc) can be used for pre-heating the water to reduce the energy resources gap. Biogas and solar energy can be used for cooking and heating purposes in the domestic and commercial sectors. In the transport sector, ethanol, bio-diesel, and renewable electricity can be used. In the long term, hydrogen produced through electricity generated from renewable energy sources via electrolysis process can be used to substitute the oil and gas. Solar and wind energy can also be used for water pumping in the domestic and agriculture sectors. Renewable energy sources such as hydropower, wind energy, solar energy and biomass, if exploited to their full potential, could manage the deepening energy crisis in the country.

## 5   Conclusion

Pakistan is facing electricity and gas shortfalls. Oil and gas supply the bulk of the country's energy needs. The indigenous reserves of oil and gas are limited and the country is heavily dependent on the import of oil. On the other hand, there is abundant potential of hydropower, wind energy, solar energy and biomass energy in the country. Hydropower, wind power, bagasse based cogeneration and solar PV have the potential to meet all the present and future electricity requirements of the country. In the transport sector, the oil can be substituted with ethanol, bio-diesel and hydrogen.

Solar energy, biogas and other biomass can contribute significantly to the energy requirements of industrial, domestic, commercial, agriculture and other sectors. These renewable energy sources should be developed for managing the present energy crisis and meeting the future energy demand in the country.

**Acknowledgements.** The authors acknowledge Department of Mechanical Engineering, Mehran University of Engineering and Technology, Jamshoro, Şindh, Pakistan, for providing the laboratory facilities to carry out this research work. The lead author also greatly acknowledges the Mehran University of Engineering and Technology Jamshoro Pakistan for permitting with study leave to carry out this research work. Research leading to this article has been financially supported by the Higher Education Commission (http://www.hec.gov.pk) of the Government of Pakistan.

# References

1. Uqaili, M.A., Harijan, K., Memon, M.D.: Prospects of Renewable Energy for Meeting Growing Electricity Demand in Pakistan. In: AIP Conf. Proc. Renewable Energy for Sustainable Development in the Asia Pacific Region, vol. 941, pp. 53–61 (2007)
2. Government of Pakistan.: Economic Survey of Pakistan 2006-2007. Economic Advisor's Wing, Finance Division, Government of Pakistan, Islamabad, Pakistan (2007)
3. Kiani, K.: Energy Crisis: Serious and Worsening,
   http://www.dawn.com/2008/01/07/ebr2.htm
4. Asif, M.: For Sustainable Energy, http://www.dawn.com/2007/09/26/op.htm
5. Government of Pakistan.: Pakistan Energy Yearbook 2007. Hydrocarbon Development Institute of Pakistan, Islamabad, Pakistan (2008)
6. Government of Pakistan.: Medium Term Development Framework 2005-2010. Planning Commission, Government of Pakistan, Islamabad, Pakistan (2005)
7. Uqaili, M.A., Mirani, M., Harijan, K.: Hydel Power Generation in Pakistan: Past Trends, Current Status and Future Projections. Mehran University Research Journal of Engineering and Technology 23(3), 207–216 (2004)
8. Harijan, K.: Modelling and Analysis of the Potential Demand for Renewable Sources of Energy in Pakistan. PhD Thesis, Mehran University of Engineering and Technology, Jamshoro, Pakistan (2008)
9. Raja, I.A.: Solar Energy Resources of Pakistan. Oxford Brookes University Press (1996)

# Author Index

Printing: Mercedes-Druck, Berlin
Binding: Stein+Lehmann, Berlin